Biodegradable Composites for Packaging Applications

"Biodegradable Composites for Packaging Applications" describes design, processing, and manufacturing of advanced biodegradable composites in packaging industry applications. It covers fundamentals of biodegradable polymers followed by introduction to biodegradable materials for food packaging industry and its processing mechanisms. Pertinent applications are explained across different chapters including intelligent packaging, applied technologies, degradation problems and its impact on environment and associated challenges.

Features

Covers biodegradable composites and targeted applications in packaging for industrial applications.

Includes exhaustive processing and characterizations of biodegradable composites.

Discusses innovative commodities packaging applications.

Reviews advanced integrated design and fabrication problems for conductive and sensors applications.

Explores various properties and functionalities through extensive theoretical and experimental modeling.

This volume is aimed at researchers and graduate students in sustainable materials, composite technology, biodegradable plastics, and food technology and engineering.

Mathematical Engineering, Manufacturing, and Management Sciences

Series Editor:
Mangey Ram,
Professor, Assistant Dean (International Affairs), Department of Mathematics, Graphic Era University, Dehradun, India

The aim of this new book series is to publish the research studies and articles that bring up the latest development and research applied to mathematics and its applications in the manufacturing and management science areas. Mathematical tools and techniques are the strength of engineering sciences. They form the common foundation of all novel disciplines as engineering evolves and develops. This series will include a comprehensive range of applied mathematics and its application in engineering areas such as optimization techniques, mathematical modelling and simulation, stochastic processes and systems engineering, safety-critical system performance, system safety, system security, high-assurance software architecture and design, mathematical modelling in environmental safety sciences, finite element methods, differential equations, and reliability engineering.

Swarm Intelligence: Foundation, Principles, and
Engineering Applications
Abhishek Sharma, Abhinav Sharma, Jitendra Kumar Pandey, and Mangey Ram

Advances in Sustainable Machining and Manufacturing Processes
Edited by Kishor Kumar Gajrani, Arbind Prasad, and Ashwani Kumar

Advanced Materials for Biomechanical Applications
Edited by Ashwani Kumar, Mangey Ram, and Yogesh Kumar Singla

Biodegradable Composites for Packaging Applications
Edited by Arbind Prasad, Ashwani Kumar, and Kishor Kumar Gajrani

Computing and Stimulation for Engineers
*Edited by Ziya Uddin, Mukesh Kumar Awasthi, Rishi Asthana,
and Mangey Ram*

For more information about this series, please visit: https://www.routledge.com/Mathematical-Engineering-Manufacturing-and-Management-Sciences/book-series/CRCMEMMS

Biodegradable Composites for Packaging Applications

Edited by
Arbind Prasad
Ashwani Kumar
&
Kishor Kumar Gajrani

CRC Press
Taylor & Francis Group
Boca Raton London New York

CRC Press is an imprint of the
Taylor & Francis Group, an **informa** business

First edition published 2022
by CRC Press
6000 Broken Sound Parkway NW, Suite 300, Boca Raton, FL 33487-2742

and by CRC Press
4 Park Square, Milton Park, Abingdon, Oxon, OX14 4RN

CRC Press is an imprint of Taylor & Francis Group, LLC

ISBN: 9781032131511 (hbk)
ISBN: 9781032131528 (pbk)
ISBN: 9781003227908 (ebk)

DOI: 10.1201/9781003227908

Typeset in Times
by codeMantra

This book is dedicated to all Mechanical, Production, Manufacturing, and Materials Engineers.

Contents

Preface

Biodegradable Composites for Packaging Applications elaborates on the production, processing, and optimization of various cost-effective biodegradable polymers and composites derived from natural and other sources, which are used in the packaging industry. Biodegradable composites are the central theme of this book. Effectively, the 16 chapters presented in this book are quite interesting and up-to-date with the scope of future research work. This has the strong potential to be valuable to researchers in engineering and material-related disciplines. The background information and the literature review provided in each chapter are an in-depth analysis of the topic covered in the chapter.

In origin, biodegradable composites can come from nature or can be synthesized in the laboratory with a variety of approaches that use metals, polymers, ceramics, or other filler materials. They are often used or adapted for packaging applications. The challenges of biodegradable plastics over conventional plastics (non-biodegradable polymers) are the mechanical strength and their optimum glass transition temperature, which make conventional plastics more familiar in packaging applications. Having high-quality and viable content for industry, it will work as a reference book for advanced polymers, which has literature review, solution, methodology, experimental procedure, results validation, and future scope. The lucid presentation of the book *Biodegradable Composites for Packaging Applications* provides a foundational link to more specialized research work in the domain of biodegradable composites for packaging applications.

Chapter 1 deals with the introduction to biodegradable polymers, market size, factors affecting their processing parameters, challenges that arise during the processing, and fruitful applications in the domestic and commercial applications. **Chapter 2** highlights food packaging. Active food packaging is a fascinating concept used in the food packaging sector for enhancing the shelf life and nutritional status of packed foodstuff. However, high cost and low thermal and mechanical properties of these biopolymers compared to that of conventional petroleum-based plastics restrict their use on a large scale. In this chapter, the authors explore the state of the art of bio-based polymers as food packaging material and highlight the functional properties of various bio-derived polymers used in the food packaging sector. **Chapter 3** introduces the reader to different processing methods adopted for biodegradable polymers. Here, the processing of biodegradable polymeric composites is classified as solution processing, melt processing, and updated processing and is discussed extensively.

Chapter 4 comprises a brief introduction on the history of the development of biodegradable composite in different applications and their recent technical advancement. It also deals with the different additives used for biocomposite fabrication. The developments in different biodegradable polymeric composites such as poly(lactic acid), polycaprolactone (PCL), polyhydroxyalkanoate (PHA), cellulose, chitosan, and starch are discussed, and their prospects are highlighted in this chapter. **Chapters 5 and 6** deal with a comprehensive review of the state-of-the-art biodegradable polymers used in food packaging applications and different nature-based biocomposites and their toxicity profile for packaging food commodities.

Chapter 7 aims at discussing the classification of biocomposites and their significance in sensor applications. It also explains the sensing mechanism of biodegradable nanocomposites and deals with matrix-based biodegradable composites for sensing environmental toxins and for sensing certain chemicals in clinical research. **Chapter 8** highlights the importance of innovative packaging. The dynamic theme in food packaging is that the packaging is an essential part of the manufacture, conservation, storage, diffusion, and planning of food sources. The properties of the food elements are simply conceivable to possess suitable packaging and persistent interaction. It is possible to have the product quality and energy essential for their commercialization and consumption.

Chapter 9 talks about edible and biodegradable materials utilized to coat or wrap food products. Edible coatings not only prevent food from mechanical, chemical, and microbial damage, but also improve their life and act as carriers of bioactive and nutraceutical compounds to enhance the functional and nutritional value of the product. They are generally classified based on their anatomical materials, i.e. proteins, lipids, polysaccharides, or composites, which act as primary raw materials for their production along with the addition of other substances, namely plasticizers and additives such as active substances, antioxidants, antimicrobials. **Chapter 10** highlights the biodegradable polymer-based smart and intelligent packaging.

Chapter 11 discusses latex manufacturing and global rubber standards for food contact. Rubber latex is not considered a food contact material due to its limitations. This chapter describes in depth the rubber components used in food contact applications. Because natural rubber latex is biodegradable, anti-degradants are necessary to protect the polymer and material properties. More migrants would reduce the need of various rubbers. **Chapter 12** deals with degradation. Degradation has a significant impact on the stability and durability of polymer materials, which can have serious repercussions for product safety and dependability. In particular, degradation results in the formation of components with reduced molecular masses. End groups that suggest a specific degradation process are frequently used to identify these organisms. The importance of comprehending these ideas in terms of deterioration will be emphasized. **Chapter 13** deals with the rheological study of biodegradable biocomposites. Rheological investigation is much more important in this case as the melt strength of the composites determines the properties such as castability, deformation, and load-bearing properties.

Chapter 14 deals with active biodegradable composite packaging materials. It assures safe handling and distribution of processed food products from farm to fork. Therefore, biopolymers (such as polycaprolactone, polylactide, carbohydrates, and proteins) and bioactive compounds (viz. natural extracts and polyphenol components) are used as precursors in the formulation and development of active biodegradable food packaging materials to improve the nutritional value and extend the shelf life of perishable food items. **Chapter 15** presents some of the most current scientific findings on these emerging pollutants, emphasizing their potentiality as well as the dangers they bring. Moreover, microplastics and nanoplastics will be discussed in terms of their health and environmental consequences on biota. Further, alternatives of conventional plastic for packaging, such as biodegradable packaging materials derived from natural sources and their future in the upcoming packaging technology, will be summarized. **Chapter 16** puts emphasis on the functions and mechanisms of the innovative food packaging technologies, their future aspects, and their contribution to enhancing food quality and safety.

Having a high-quality content, it will work as a reference book for understanding biodegradable polymers. The lucid presentation of the book *Biodegradable Composites for Packaging Applications* provides a foundational link to more specialized research work in materials engineering.

Editors
Dr. Arbind Prasad
Dr. Ashwani Kumar
Dr. Kishor Kumar Gajrani

Acknowledgements

We express our heartfelt gratitude to **CRC Press (Taylor & Francis Group)** and the editorial team for their guidance and support during the completion of this book. We are grateful to all chapter authors and reviewers for their suggestions and illuminating views on each book chapter presented in the book *Biodegradable Composites for Packaging Applications.*

Aim and Scope

The continuous development in technology also leads to the generation of many problems such as electronic waste management and plastic pollution, and the situations are very much alarming in such a way that most of the researchers and scientists are now shifting their focus from conventional research into some fruitful research orientations. The inventions of latest materials, manufacturing technologies, product design and durability, optimization of the processing parameters, and after all, methodology have shifted from traditional approach to advanced level of processes viable for industries. *Biodegradable Composites for Packaging Applications* provides in-depth knowledge to readers for easier, fast, efficient, and reliable way or method of understanding the construction of polymer, movement of polymeric chain, cross-linking behaviours, orientations of their isomers, and factors related to molecular weight, with the help of advanced techniques.

Biodegradable Composites for Packaging Applications provides a lucid way to readers for understanding advanced polymer materials for food packaging and commodities packaging industry. The book consists of 16 chapters dedicated to biodegradable plastics for packaging industry. Modern era recommends materials that are biodegradable and have vast advantages in daily life. The present book recommends advantages of progress in engineering, physical, and chemical sciences. One of the best examples of such combination is biodegradable materials used in day-to-day packaging. Biodegradable materials must be compatible with the packaging material to avoid plastic pollution, and problems with these biodegradable plastics must be resolved before a product can be used in our daily life. Production and synthesis of biodegradable composites require various technologies and methods. These methods produce best-suited materials. The aim of the proposed title is to inventory the latest achievements in the development and production of modern biodegradable composites that are used in commodities packaging and food packaging applications to increase their shelf life.

The content of the book focuses on biodegradable composites for packaging industry applications. The authors covered all the aspects starting from introduction, characterization, sustainability of the perspective application, and future scope for the biodegradable composites in packaging applications. The book presents 16 chapters with proper line-up, which are quite interesting and with the potential for further research and also able to explain them all in depth and in a clear way to all the stakeholders.

Editors
Dr. Arbind Prasad
Dr. Ashwani Kumar
Dr. Kishor Kumar Gajrani

Editors

Dr. Arbind Prasad has completed his Ph.D. in Mechanical Engineering from Indian Institute of Technology Guwahati, Assam, India. He has filed four patents out of his research work. He has numerous international journal papers, book chapters, and reputed conference papers to his credit. Dr. Arbind Prasad has obtained various prestigious awards such as Sponsored Research Industrial Consultancy (SRIC) Award from IIT Kanpur, Best Oral Presentation from American Chemical Society, and Best Paper Awards from IIT Guwahati during Research Conclave. He has been invited to deliver talks at various organizations of repute. He has coordinated various faculty development programmes, short-term courses, symposiums, and national seminars and has completed research projects sponsored under various government schemes in India. He is currently working as an Assistant Professor and Head (Mechanical Engineering) in the Department of Science and Technology, Government of Bihar, Posted at Katihar Engineering College, Katihar, Bihar, India. His main areas of interest include manufacturing, machining, polymer composites, biomaterials, materials processing, and orthopaedic biomedical applications. He is also a lifetime member of SPSI, Pune; MRSI; SBAOI; APA; and ISTE.

Dr. Ashwani Kumar received his Ph.D. (Mechanical Engineering) in the area of Mechanical Vibration and Design. He has been working as a Senior Lecturer, Mechanical Engineering (Gazetted Officer Class II) at Technical Education Department, Uttar Pradesh (Government of Uttar Pradesh), India, since December 2013. He worked as an Assistant Professor in the Department of Mechanical Engineering, Graphic Era University, Dehradun, India, from July 2010 to November 2013. He has more than 11 years of research and academic experience in mechanical and materials engineering. He is Series Editor of book series *Advances in Manufacturing, Design and Computational Intelligence Techniques* published by CRC Press (Taylor & Francis) USA. He is Associate Editor *for International*

Journal of Mathematical, Engineering and Management Sciences (IJMEMS) Indexed in ESCI/Scopus and DOAJ. He is an editorial board member of 4 international journals and acts as a review board member of 20 prestigious (Indexed in SCI/SCIE/Scopus) international journals with high impact factor, i.e. *Applied Acoustics, Measurement, JESTEC, AJSE, SV-JME, and LAJSS.* In addition, he has published 85 research articles in journals, book chapters, and conferences. He has authored/co-authored and edited 13 books of mechanical and materials engineering. He is associated with International Conferences as Invited Speaker/ Advisory Board/Review Board Member. He has delivered many invited talks in webinar, FDP, and workshops. He has been awarded Best Teacher for excellence in academic and research. He has successfully guided 12 B. Tech., M. Tech., and Ph.D. theses. In administration, he is working as coordinator for AICTE, E.O.A., Nodal officer for PMKVY-TI Scheme (Government of India), and internal coordinator for CDTP scheme (Government of Uttar Pradesh). He is currently involved in the research area of machine learning, advanced materials, machining and manufacturing techniques, biodegradable composites, heavy vehicle dynamics, and Coriolis mass flow sensor.

Dr. Kishor Kumar Gajrani is an Assistant Professor in the Department of Mechanical Engineering at the Indian Institute of Information Technology, Design and Manufacturing, Kancheepuram, Chennai, India. He has obtained M. Tech. and Ph.D. degrees from the Department of Mechanical Engineering at the Indian Institute of Technology Guwahati. Thereafter, he worked as a postdoctoral researcher at the Indian Institute of Technology Bombay. He has published 27 articles and book chapters in international journals and publishers of repute and has attended numerous conferences. His research interests are broadly related to the advancement of sustainable machining processes, advanced materials, tribology, coatings, green lubricants and coolants, and food packaging.

Contributors

V. Andal
Department of Chemistry
KCG College of Technology
Chennai, India

Sayan Kumar Bhattacharjee
Department of Chemical Engineering
Indian Institute of Technology
 Guwahati
Guwahati, India

Shasanka Sekhar Borkotoky
Department of Chemical Engineering
Assam Engineering College
Guwahati, India

Gourhari Chakraborty
Department of Chemical Engineering
National Institute of Technology
 Andhra Pradesh
Tadepalligudem, India

Sujeet Kumar Chaurasia
Center for Nanoscience and
 Technology
V.B.S. Purvanchal University
Jaunpur, India

Rajni Chopra
Department of Food Science and
 Technology
National Institute of Food Technology
 Entrepreneurship and Management
Sonipat, India

Daisy Das
Department of Fuel, Minerals and
 Metallurgical Engineering
Indian Institute of Technology (Indian
 School of Mines), Dhanbad
Dhanbad, India

Aishwarya Dhiman
Department of Food Science and
 Technology
National Institute of Food Technology
 Entrepreneurship and Management
Sonipat, India

Kishor Kumar Gajrani
Department of Mechanical
 Engineering
Indian Institute of Information
 Technology, Design and
 Manufacturing, Kancheepuram
Chennai, India
and
Centre of Smart Manufacturing
Indian Institute of Information
 Technology, Design and
 Manufacturing, Kancheepuram
Chennai, India

Meenakshi Garg
Department of Food Science and
 Technology
National Institute of Food Technology
 Entrepreneurship and Management
Sonipat, India
and
Bhaskaracharya College of Applied
 Sciences
University of Delhi
Delhi, India

V. Gayathri
Phytochemistry & Phytopharmacology
 Division
KSCSTE-Jawaharlal Nehru Tropical
 Botanic Garden and Research
 Institute Palode
Thiruvananthapuram, India

Vishwanath Jadhav
Department of R & D (NPD)
Deep Plast Industries
Gandhinagar, India

Purnima Justa
Department of Chemistry & Chemical
   Science
Central University of Himachal Pradesh
Dharamshala, India

Karthik Kannan
School of Advanced Materials Science
   and Engineering
Kumoh National Institute of
   Technology
Gumi-si, Republic of Korea

Vimal Katiyar
Department of Chemical Engineering
Indian Institute of Technology
   Guwahati
Guwahati, India

Z. Edward Kennedy
Department of Chemistry
KCG College of Technology
Chennai, India

Francis Luther King
Department of Mechanical
   Engineering
Swarnandhra College of Engineering
   and Technology
Narsapur, India

Adesh Kumar
Department of Chemistry
University of Delhi
Delhi, India

Ashwani Kumar
Department of Technical Education
   Uttar Pradesh
Kanpur, India

Hemant Kumar
Department of Chemistry
Ramjas College, University of Delhi
Delhi, India
and
Department of Chemistry
Bhaskaracharya College of Applied
   Sciences, University of Delhi
Delhi, India

Manish Kumar
Department of Biomedical Engineering
Mody University
Sikar, India

Pramod Kumar
Department of Chemistry & Chemical
   Science
Central University of Himachal
   Pradesh
Dharamshala, India

Vivek Pandey
Department of Mechanical
   Engineering
Mody University
Sikar, India

Balaram Pani
Department of Chemistry
Bhaskaracharya College of Applied
   Sciences, University of Delhi
Delhi, India

Rahul Patwa
Stokes Laboratories
Bernal Institute, University of
   Limerick
Limerick, Ireland

Atanu Kumar Paul
Department of Chemical Engineering
Indian Institute of Technology
   Guwahati
Guwahati, India

Priyanka Prajapati
National Institute of Food Technology
    Entrepreneurship and Management
Sonipat, India
and
Bhaskaracharya College of Applied
    Sciences
University of Delhi
Delhi, India

P. Shakti Prakash
Department of Biomedical Engineering
Mody University
Sikar, India

Arbind Prasad
Mechanical Engineering Department
Katihar Engineering College (Under
    the Department of Science &
    Technology, Government of Bihar)
Katihar, India

B. Sabulal
Phytochemistry & Phytopharmacology
    Division
KSCSTE-Jawaharlal Nehru Tropical
    Botanic Garden and Research
    Institute Palode
Thiruvananthapuram, India

Neha Singh
National Institute of Food Technology
    Entrepreneurship and Management
Sonipat, India

Sonika
Department of Physics
Rajiv Gandhi University
Itanagar, India

**Theivasanthi
Thirugnanasambandan**
International Research Centre
Kalasalingam Academy of Research
    and Education (Deemed University)
Krishnankoil, India
ttheivasanthi@gmail.com
https://orcid.org/0000-0002-2280-9316

Sushil Kumar Verma
Department of R & D (NPD)
Deep Plast Industries
Gandhinagar, India

1 Introduction to Biodegradable Polymers

Arbind Prasad
Katihar Engineering College (Under Department
of Science & Technology, Government of Bihar)

Gourhari Chakraborty
National Institute of Technology Andhra Pradesh

Ashwani Kumar
Department of Technical Education Uttar Pradesh Kanpur

Kishor Kumar Gajrani
Indian Institute of Information Technology,
Design and Manufacturing, Kancheepuram

CONTENTS

1.1 INTRODUCTION

The term "polymer" was derived from the Greek words polus, meaning "many, much", and meros, meaning "parts", which means that multiple repeating units form its structure. The term was coined by Jons Jacob Berzelius in 1833. The subunit of a polymer is known as a monomer. Monomers can be linked together in various ways to give linear, branched, and cross-linked polymers. Hermann Staudinger was granted the Nobel Prize for science in 1953 for his spearheading research on macromolecules [1]. His work gave the premise to understanding and planning utilizations of polymers in daily existence as commodities and daily-use

DOI: 10.1201/9781003227908-1

1

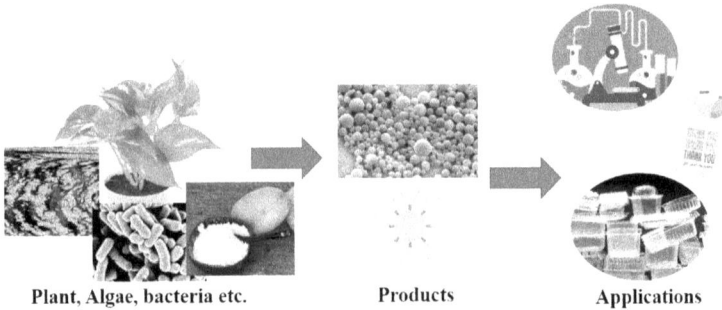

Plant, Algae, bacteria etc.   Products   Applications

FIGURE 1.1 Biodegradable materials available in nature for various applications.

materials. The present-day life is impossible without polymers. Notwithstanding every one of the advantages, plastic today is by and large seriously examined as a material liable for unsafe consequences for the environment and plastic contamination, particularly plastic particles under 5 mm in size (likewise called microplastics), remain in the environment. The 21st century is flourishing with enormous monetary development and is yet confronting environmental harms. Plastic pollution of late has been featured as a worldwide emergency at each stage, directly from its production to removal and incineration [2]. There are plenty of natural materials available in our nature. Through processing, we get various useful materials in the form of micro- and nanomaterials as products, and there onward, these products are used for fabrication in the area of packaging, biomedicine, agriculture, and household appliances as shown in Figure 1.1 [3–5].

The worldwide bioplastics creation limits are hard to assess and are typically founded on estimate because of the persistently arising scope of bio-based and biodegradable polymers and rising interests on putting resources into bioplastics area. In this time of urbanization, the depletion of petrol-based resources is at a higher recurrence. They are the most broadly utilized assets finding applications in pretty much every day of life. However, these assets are at the edge of a steady loss, at a lot higher rate than anybody could might suspect, helping toward ecological dangers [6]. The design of biodegradable plastics makes them easily degradable by normal microorganisms, giving a final result that is less destructive to the environment. In this way, biodegradable plastics are seen to be eco-friendly because of their natural advantages, which are difficult to deny compared to customary plastics. These materials are extricated from bounteously accessible natural sources. Bio-based polymers are produced from three sources, which are as follows:

1. From agriculture sources through chemical treatments, such as polysaccharides and lipids (starch, cellulose, and alginates).
2. From microorganisms (by fermentation), such as polyhydroxyalkanoates (PHAs) and polyhydroxybutyrate (PHBs).
3. Conventional synthesis, such as polylactides, PBS, PE, PTT, PPP, etc.

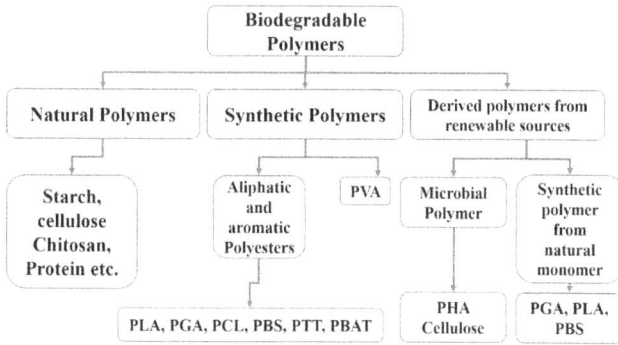

FIGURE 1.2 Classification of biodegradable polymers.

Bioplastics are mainly obtained from renewable feedstock, biomass, and waste streams. Bioplastics are naturally biodegradable, and in some sense, they are designed to be biodegradable. But they may also be non-biodegradable. The feedstock hydrolysis and microbial synthesis produce naturally biodegradable bioplastics. Sometimes, bioplastics are made biodegradable by microbial fermentation and chemical polymerization. Green packaging dependent on biodegradable composite materials has at present acquired incredible consideration in many disciplines in light of the exceptional properties when compared to conventional petrochemical-based plastics [7–9]. In addition, they are 100% biodegradable and can be completely decomposed into carbon dioxide, water, and humus. These elements can be used in numerous places, for example nano-food packaging, bio-layers, drug conveyance, and treating the soil. In this way, the essential purpose of packaging material is to retain the food quality and extend its shelf life. Biodegradable polymers can be further categorized as shown in Figure 1.2. Polymers from sustainable sources have drawn attention over the last 20 years, especially because of two significant reasons: first and foremost, the issues identified with our natural environment, and second, the restricted accessibility of petroleum fuel stock. There are many wellsprings of biodegradable plastics, from synthetic to natural polymers. Natural polymers are accessible in huge amounts from inexhaustible sources, while synthetic polymers are synthesized from non-sustainable petrol resources. Biodegradation of polymeric biomaterials includes cleavage of hydrolytically or enzymatically delicate bonds in the polymer leading to polymer erosion. An immense number of biodegradable polymers have been manufactured recently, and a few microorganisms and catalysts equipped for degrading them have been identified [10].

Petroleum-based polymers have been used as one of the principal materials in all fields, bit by bit uprooting conventional materials. The development of plastic is basically inferable from its fantastic characteristics and processing prospects. Notwithstanding, the packaging business is confronting significant difficulties such as decreasing accessibility of petrochemical feedstock, rise in the cost, and the persistence of these materials in the environment past their functional life.

Along these lines, packaging industries need to find new arrangements in delivering sustainable and harmless items to the ecosystem, for example fiber-reinforced recyclable and biodegradable packaging materials.

The rising interest in biodegradable plastic materials depends on a few variables. There is a growing natural concern, both from users who are progressively able to expend large sums on green products and from lawmakers who drive polymer ventures to look for more eco-friendly materials. There is likewise a mission to utilize natural raw materials that don't exhaust our limited fossil resources. Tragically, the accomplishment of biodegradable polymers has been restricted by many factors. Most importantly, their properties are mediocre compared to their non-degradable parts. Additionally, they are very delicate in how they are stored, processed, and utilized. Beyond question, the formation of biodegradable polymer parts requires not simply seeing them as a plastic with appropriate properties, but also significant work to tune the processing conditions to limit any deficiency of these properties.

The cutting-edge business world is being driven by the need to reduce its environmental impression, for example sustainability, while additionally controlling expenses. Luckily, it is possible to accomplish these two objectives with the expanding prominence of biodegradable polymeric packaging technologies. Petroleum-derived plastics are not promptly biodegradable, and on account of their protection from microbial debasement, they gather in the environment. Furthermore, recently, oil costs have increased extraordinarily. These realities further lead to the creation of interest in biodegradable polymers. This chapter further discusses the market trends, differences between biodegradable and non-biodegradable polymers, classification, influencing factors, latest advancements, and applications in subsequent sections.

1.2   MARKET OF BIODEGRADABLE POLYMERS

The market size is increasing continuously as the research progresses. From the report of Mordor Intelligence, in 2019, worldwide creation capacities of bioplastics reached up to around 2.11 million tons, with practically half of the volume bound to the packaging market. Biodegradability is a significant part of food packaging for perishables. Flexible packaging arrangements such as films and plate are especially reasonable for new produce, for example foods and vegetables, as they empower the time span of usability. The major players in the field of biodegradable polymers are NatureWorks LLC, Novamont SpA, Total Corbion PLA BV, Rodenburg Biopolymers, BASF SE, Biotech, etc. It has also been estimated that the biodegradable polymer market will grow at a CAGR of more than 20% over the next 5 years (https://www.mordorintelligence.com/industry-reports/europe-biodegradable-plastic-packaging-market). Asia Pacific region has the highest CAGR in 2016–2026. Nova Institute has recently reported that the worldwide bioplastics creation limit is developing at an extensive speed and that it will increase from around 2.11 million tons in 2018 to 2.62 million tons in 2023. Presently, poly lactic acid (PLA) covers around 10%–15% of the all-out plastics market.

Limited fossil fuel resources, rising petrochemical costs, reinforcement of PLA with different functional materials, environmental safety awareness, and government motivations will be key components driving the biodegradable polymers market. Reinforcement of PLA with different materials sets out immense opportunities for future. As natural fibers are of low cost and degradable in nature, its reinforcement with PLA leads to a material that is economical and ecological. The worldwide PLA market is expected to reach a market size of 5.2 billion USD, developing at a CAGR of 21.6% during the period 2013–2020. By 2020, the share of the overall industry of PLA is projected to increase from 25% to 30% [11]. Conventional polymers are generally used for packaging materials because of their simple processability, low cost, and low thickness, while synthetic conventional polymers cause natural contamination because of their non-degradable nature. Consequently, biodegradable polymers are ideal replacements for these synthetic non-biodegradable polymers in packaging applications.

1.3 NON-DEGRADABLE AND BIODEGRADABLE POLYMERS

The environmental pollution by non-degradable plastic waste attracts attention to the development of biodegradable polymers. To achieve the sustainability in packaging and domestic and commercial applications, biodegradable polymers are continuously expanding their availability as an alternative. Biodegradable polymers are widely available in natural sources (starch, corn, potatoes, collagen, dextran, gelatin, etc.), and they can be synthesized in the laboratory (polyanhydrides, polyphosphazenes, polyaminoacids, polyorthoesters, polycaprolactone, etc.) [12,13]. These materials deteriorate in the presence of CO_2, water, and microorganisms, and after some period of time, they completely degrade.

Packaging is an essential component of reaction to address the key challenges of sustainable food consumption on the worldwide scenario, which is unmistakably about limiting the ecological footprints of packed foods. An innovative supportable packaging is expected to decrease food waste and loss by preserving food quality, just as food safety issues by preventing food-borne diseases and food chemical contamination. In addition, it should address the long-term issue of environmentally unsafe plastic waste accumulation just as the saving of oil and food material resources [14,15]. Figure 1.3 shows the advantages of biodegradable polymers. Figure 1.4 shows the disadvantages of biodegradable plastics.

1.4 NEED OF BIODEGRADABLE PLASTICS

The use of biodegradable plastics is very much increasing in almost all sectors. The market is in huge demand as mentioned in previous sections. The differences are quite familiar with non-degradable polymers. One of the main advantages of biodegradable plastics is that the recyclability is easy. They consume less energy during their processing and synthesizing. The wastes produced are also reduced. By the use of biodegradable plastics produced from natural resources, the consumption of petroleum will also decrease, and finally, they will be compostable

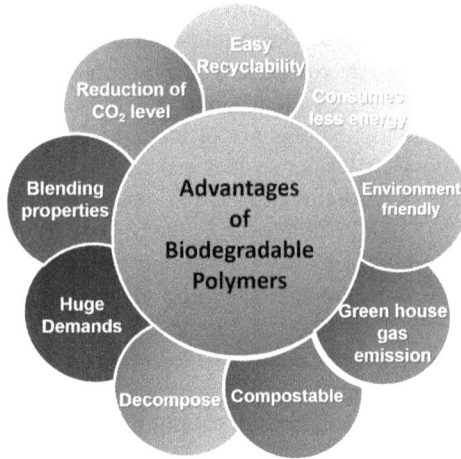

FIGURE 1.3   Advantages of biodegradable polymers.

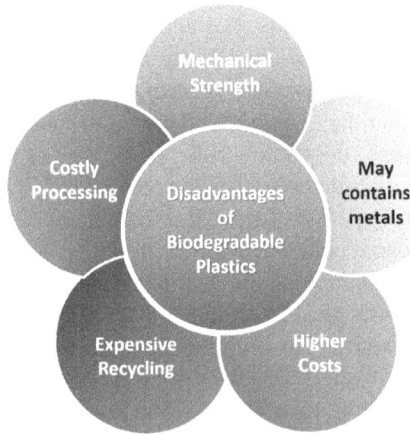

FIGURE 1.4   Disadvantages of biodegradable polymers.

in nature. There are also many benefits of biodegradable plastics, such as reduction in the emission of greenhouse gases such as CO_2 [1]. It was observed that natural biodegradable polymers degrade 10–20 times faster than conventional non-degradable polymers. It was also mentioned that the processing of biodegradable polymers creates non-toxic gases, while in case of traditional plastics, if they are burnt, toxic gases are released into the environment. The discarded non-biodegradable polymers create soil pollution, and they require very long periods to decompose. But in the case of biodegradable polymers, they take less time and also make the soil more fertile. Figure 1.5 shows the factors that make the

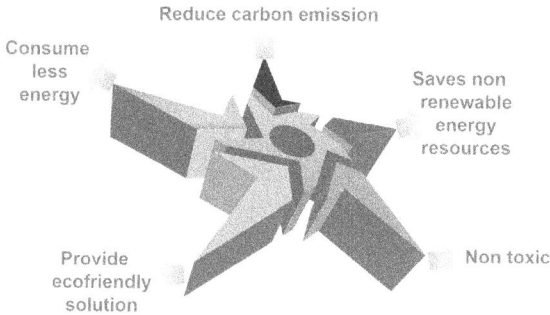

FIGURE 1.5 Factors which led to the requirement of biodegradable polymers.

development of biodegradable polymers a necessity. The processing time also consumes less energy in case of biodegradable polymers, which leads to reduction in carbon emissions. When these biodegradable polymers are used to fabricate daily-use objects, after their intended functions, they do not harm the ecological balance and ultimately provide eco-friendly solutions. There is always a demand for this kind of materials in many applications such as automobile, cutlery, packaging, biomedicine, and commodities.

1.5 FACTORS AFFECTING PROCESSING PARAMETERS AND OPTIMIZATION

Plastics are a critical material in the current life. They are adaptable, are light, and can be produced at a moderately low price. Currently, something like 1% of plastics and plastic items on the worldwide market are considered bio-based, compostable, or potentially biodegradable. Most plastics are being produced using petroleum derivatives in an interaction that increases greenhouse gas emissions [16]. To be sure, plastics contaminate starting from production to disposal during their life-cycle. The major factors associated with biodegradable polymers are mentioned in Figure 1.6. The cost-effective production of biodegradable polymers is still in the research phase. Researchers are trying hard to produce biodegradable polymers in a cost-effective manner. The processability of biodegradable polymers requires lots of expertise to perform the polymerization process. The non-toxic nature of the biodegradable polymers make them useful in various applications. The strength, especially mechanical strength, is still not high; thus, composite fabrications are widely needed to make biopolymers perform their intended functions without fail.

The reuse of plastics is low, and plastics spill over into the entire environment. Biodegradable and compostable materials can be degraded by microorganisms into water, carbon dioxide, mineral salts, and new biomass within a specified period of time [17]. Whether a biodegradable or compostable plastic biodegrades and how rapidly that occurs unequivocally relies upon the conditions it has been presented to during removal. These include temperature, time period, the presence of microorganisms, nutrients, oxygen, and dampness [18].

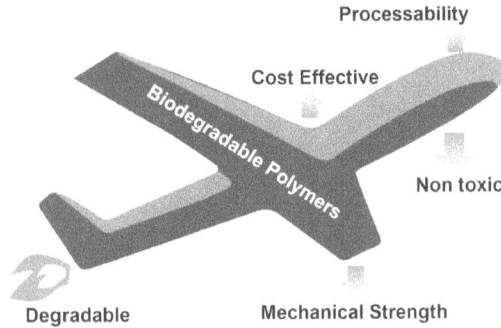

FIGURE 1.6 Factors affecting the processing of biodegradable polymers.

1.6 LATEST ADVANCEMENTS

Researchers are doing further studies on the cost-effectiveness and wide applicability of biodegradable composites [19,20]. With these advancements, many researchers have used the biodegradable coating on fruits and other eatables in order to preserve the food for longer periods. Various researchers have used biodegradable polymeric composites as cutlery for functions. The advancements in biomedicine starting from implants to sutures have been made possible by biodegradable polymers. The government agencies worldwide are creating awareness of using biopolymers in place of non-degradable composites. The negative part of plastics on the environment is that they make challenges for societies such as landfill and plastic contamination. Plastic materials for the most part require hundreds of years to break down normally in the environment. The world has seen different natural issues related to plastics; Progress has been made toward the manufacturing and utilization of biodegradable plastics. To limit environmental contamination, these types of plastics is without a doubt a superior decision, yet it accompanies its own disadvantages.

A biodegradable carboxymethyl cellulose (CMC)-based material for sustainable packaging application was reported. A value-added product of biodegradable material for sustainable packaging was developed. Waste-derived CMC mainly reduces the cost involved in the development of the film; at present, commercially available CMC is costly. During this process, CMC was mainly extracted from agricultural wastes such as sugarcane bagasse and the blends were prepared using CMC (waste derived), gelatin, agar, and varied concentrations of glycerol. The resultant product was characterized and observed to be suitable for packaging applications [21]. Nitrogen pollution impact and remediation through low-cost starch-based biodegradable polymers was reported. It was mentioned that high urea inputs raise the level of reactive nitrogen in the soil, air, and water. Unused reactive nitrogen acts as a pollutant and harms the natural resources. The use of controlled-release fertilizers for slowing down the nutrients' leaching has recently come into practice among farmers worldwide. The starches, modified with urea and borate, showed good stability and mechanical strength over time [22]. Biodegradable polymers for biomedical additive manufacturing were reported. The source of extraction, chemical modification, or synthesis route, and their physicochemical and processing properties in relationship to additive manufacturing were studied. Finally, it was discussed that biodegradable polymers can also play a vital role in the function of the materials as well as drug carriers [23]. Biodegradable polymers were used in conductive sensing applications, and it was observed that more enhanced properties were achieved, which could be more useful in the fabrication of conducting polymers and sensors [24].

Thus, advances in biodegradable or compostable plastics are very much on the stage. In future, most of the biodegradable polymers will be utilized in electronic applications, biomedical applications, agriculture, and mostly in packaging applications without compromising their mechanical strength.

1.7 APPLICATIONS OF BIODEGRADABLE POLYMERS

The future of biodegradable plastics shows great potential for fabrication as packaging materials for various products. Worldwide, government agencies are very much attentive and are showing their interest to apply these biodegradable polymers in place of conventional plastics in various applications as mentioned in Figure 1.7. As mentioned in previous sections, the use of biodegradable polymers in place of non-degradable polymers has many advantages. Plastics are being used all over the world in the form of drinking cups, disposable trays, glasses used in the making of cutleries, packaging and wrapping items, bottles, food containers, clothes, wearables, vehicle parts, electronics, pens, and water bottles. Their wide usage is because of their versatility in 3D modeling, durability, and ease of work within various manufacturing and production processes. Various studies on

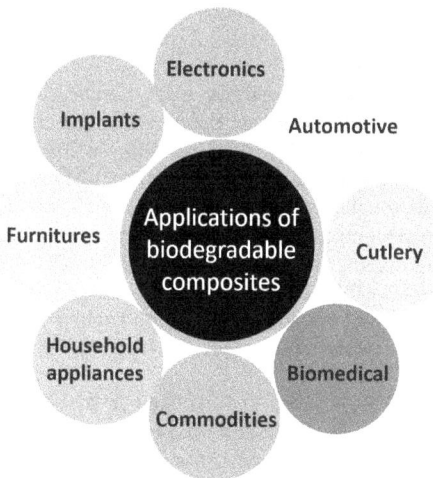

FIGURE 1.7   Applications of biodegradable composites.

the applications of biopolymers in the field of biomedicine have been performed, such as wound healing, internal fixation implants, and dental applications. These biodegradable polymers also serve as a drug carrier. Due to their bioabsorbable nature, they give relief to patients by avoiding repeated surgeries necessary for the removal of traditional implants. Electronic boards and some circuits are currently made with biodegradable composites. Various household appliances such as furniture, washing machine cabinets, water cooler cabinets, etc., are fabricated using biodegradable composite materials. The makeup structure of biodegradable plastics makes them easily degradable by natural microorganisms, giving an end product that is less harmful to the environment. As such, biodegradable plastics are perceived to be eco-friendlier due to their environmental benefits, which are hard to deny compared to ordinary plastics.

1.8   SUMMARY

Industries are looking for technologies for fast production and increased cost-effectiveness of these biopolymers in order to make them widely applicable as in the case of non-degradable plastics. Plenty of natural resources are available, which can be used as a filler or additive material to these degradable polymers in order to attain mechanical strength and enhance overall properties. The research progress in this particular field in recent years gives a clear indication that the consumers as well as the budding researchers have developed a great curiosity to develop commodities based on these raw materials. With the variety of available natural filler and biodegradable polymers, the time when almost all the daily-use products will be produced from biodegradable plastics is very near.

REFERENCES

1. Tharanathan RN (2003) Biodegradable films and composite coatings: Past, present and future. *Trends Food Sci Technol* 14:71–78. https://doi.org/10.1016/S0924-2244(02)00280-7
2. Moustafa H, Youssef AM, Darwish NA, Abou-Kandil AI (2019) Eco-friendly polymer composites for green packaging: Future vision and challenges. *Compos Part B Eng* 172:16–25. https://doi.org/10.1016/j.compositesb.2019.05.048
3. Gupta A, Prasad A, Mulchandani N, et al (2017) Multifunctional nanohydroxy-apatite-promoted toughened high-molecular-weight stereocomplex poly(lactic acid)-based bionanocomposite for both 3D-printed orthopedic implants and high-temperature engineering applications. ACS Omega 2:4039–4052. https://doi.org/10.1021/acsomega.7b00915
4. Prasad A (2021) State of art review on bioabsorbable polymeric scaffolds for bone tissue engineering. *Mater Today Proc* 44:1391–1400. https://doi.org/10.1016/j.matpr.2020.11.622
5. Prasad A, Bhasney S, Katiyar V, Ravi Sankar M (2017) Biowastes processed hydroxyapatite filled poly (lactic acid) bio-composite for open reduction internal fixation of small bones. *Mater Today Proc* 4:10153–10157. https://doi.org/10.1016/j.matpr.2017.06.339

6. Bhasney SM, Mondal K, Kumar A, Katiyar V (2020) Effect of microcrystalline cellulose [MCC] fibres on the morphological and crystalline behaviour of high density polyethylene [HDPE]/polylactic acid [PLA] blends. *Compos Sci Technol* 187:107941. https://doi.org/10.1016/j.compscitech.2019.107941

7. Savioli Lopes M, Jardini AL, Maciel Filho R (2012) Poly (lactic acid) production for tissue engineering applications. *Procedia Eng* 42:1402–1413. https://doi.org/10.1016/j.proeng.2012.07.534

8. Lim L-T, Auras R, Rubino M (2008) Processing technologies for poly(lactic acid). *Prog Polym Sci* 33:820–852. https://doi.org/10.1016/j.progpolymsci.2008.05.004

9. Devi RR, Dhar P, Kalamdhad A, Katiyar V (2015) Fabrication of cellulose nanocrystals from agricultural compost. *Compost Sci Util* 23:104–116. https://doi.org/10.1080/1065657X.2014.972595

10. Agarwal S (2020) Biodegradable polymers: Present opportunities and challenges in providing a microplastic-free environment. *Macromol Chem Phys* 221. https://doi.org/10.1002/macp.202000017

11. Briassoulis D, Pikasi A, Hiskakis M (2019) End-of-waste life: Inventory of alternative end-of-use recirculation routes of bio-based plastics in the European Union context. *Crit Rev Environ Sci Technol* 49:1835–1892. https://doi.org/10.1080/10643389.2019.1591867

12. Sousa RA, Reis RL, Cunha AM, Bevis MJ (2003) Processing and properties of bone-analogue biodegradable and bioinert polymeric composites. *Compos Sci Technol* 63:389–402. https://doi.org/10.1016/S0266-3538(02)00213-0

13. Sedlarik V, Saha N, Sedlarikova J, Saha P (2008) Biodegradation of blown films based on poly(lactic acid) under natural conditions. *Macromol Symp* 272:100–103. https://doi.org/10.1002/masy.200851214

14. Ghosh T, Bhasney SM, Katiyar V (2020) Blown films fabrication of poly lactic acid based biocomposites: Thermomechanical and migration studies. *Mater Today Commun* 22:100737. https://doi.org/10.1016/j.mtcomm.2019.100737

15. Aranaz I, Mengíbar M, Harris R, et al (2009) Functional characterization of chitin and chitosan. *Curr Chem Biol* 3:203–230. https://doi.org/10.2174/187231309788166415

16. Muniyasamy S, Mohanrasu K, Gada A (2019) Biobased biodegradable polymers for ecological applications: A move towards manufacturing sustainable biodegradable plastic products. In *Integrating Green Chemistry and Sustainable Engineering* (pp. 215–253). Wiley. https://doi.org/10.1002/9781119509868.ch8

17. Dhar P, Tarafder D, Kumar A, Katiyar V (2016) Thermally recyclable polylactic acid/cellulose nanocrystal films through reactive extrusion process. *Polymer* 87:268–282. https://doi.org/10.1016/j.polymer.2016.02.004

18. Hopewell J, Dvorak R, Kosior E (2009) Plastics recycling: Challenges and opportunities. *Philos Trans R Soc Lond B Biol Sci* 364:2115–2126. https://doi.org/10.1098/rstb.2008.0311

19. Valapa RB, Pugazhenthi G, Katiyar V (2015) Fabrication and characterization of sucrose palmitate reinforced poly(lactic acid) bionanocomposite films. *J Appl Polym Sci* 132. https://doi.org/10.1002/app.41320

20. Pugazhenthi G, Katiyar V (2016) Hydrolytic degradation behaviour of sucrose palmitate reinforced poly (lactic acid) nanocomposites. *Int J Biol Macromol* 89:70–80. https://doi.org/10.1016/j.ijbiomac.2016.04.040

21. Yaradoddi JS, Banapurmath NR, Ganachari SV, et al (2020) Biodegradable carboxymethyl cellulose based material for sustainable packaging application. *Sci Rep* 10:1–13. https://doi.org/10.1038/s41598-020-78912-z

22. Ibrahim KA, Naz MY, Shukrullah S, et al (2020) Nitrogen pollution impact and remediation through low cost starch based biodegradable polymers. *Sci Rep* 10:1–10. https://doi.org/10.1038/s41598-020-62793-3
23. Puppi D, Chiellini F (2020) Biodegradable polymers for biomedical additive manufacturing. *Appl Mater Today* 20. https://doi.org/10.1016/j.apmt.2020.100700
24. Chakraborty G, Gupta A, Pugazhenthi G, Katiyar V (2018) Facile dispersion of exfoliated graphene/PLA nanocomposites via in situ polycondensation with a melt extrusion process and its rheological studies. 46476:1–11. https://doi.org/10.1002/app.46476

2 Bio-Based Materials for Food Packaging Applications

Purnima Justa
Central University of Himachal Pradesh

Hemant Kumar
University of Delhi

Sujeet Kumar Chaurasia
V.B.S. Purvanchal University

Adesh Kumar and Balaram Pani
University of Delhi

Pramod Kumar
Central University of Himachal Pradesh

CONTENTS

DOI: 10.1201/9781003227908-2

2.1 INTRODUCTION

Packaging science and technology ensures that the items are presented to the buyer with safety and dispense the nutritional value in the case of edibles. Combination of metals, glass, paper or pulp, plastic, or composites with other materials find a wide variety of applications in packaging industry (Kumar et al. 2020). Plastic owing to its affordable cost, light weight, and exceptional protection provided to packaged materials is extensively used for packaging of food material, cosmetics, and pharmaceutical products (Reichert et al. 2020a). The global production of plastics has tremendously increased from the worldwide production of 1.5 million metric tons in the year 1950 to 368 million metric tons in 2019, representing its broad-spectrum applications in various fields (Richardson 2021). The main raw materials used in the plastic industry are fossil fuels; however, their non-recyclable and non-biodegradable nature limits their utilization. Polyvinylchloride (PVC), polyethylene terephthalate (PET), polypropylene (PP), polyamide (PA), polystyrene (PS), and ethylene vinyl alcohol (EVOH) are extensively used in the plastic business due to their small production cost and magnificent mechanical and barrier properties. Howbeit, the disposal of synthetic plastics results in the accumulation of greenhouse gases such as carbon dioxide and methane in the environment, which has serious environmental issues (Qamar et al. 2020). The longer time plastic takes to degrade (>100 years) and its disposal into sanitary landfills or oceans once they reach their end life not only affect marine creatures, but could indirectly have serious impacts on the food chain. Indeed, microplastic contamination of the oceans of the world is a hot topic these days (Nesic et al. 2020). This necessitates the need for shifting towards an alternative raw material for plastic production, which could progressively replace the use of conventional plastic.

Bioplastic shows resemblance with conventional plastic in terms of their properties and hence can be used as a solution to the menace caused by the plastic. According to the European bioplastic organization, a plastic material is described as bioplastic if it is bio-based, is biodegradable, or has both attributes. Bioplastics as the name implies are bio-based polymers produced from renewable sources and are biodegradable (Mangaraj et al. 2019). Biodegradable polymers are environmentally benign since they break down into natural products, i.e. CO_2 and N_2, water, biomass, and inorganic salts by the action of naturally occurring microorganisms such as bacteria and fungi (Kumar et al. 2020). However, renewable resources-based plastics may not necessarily be biodegradable or vice

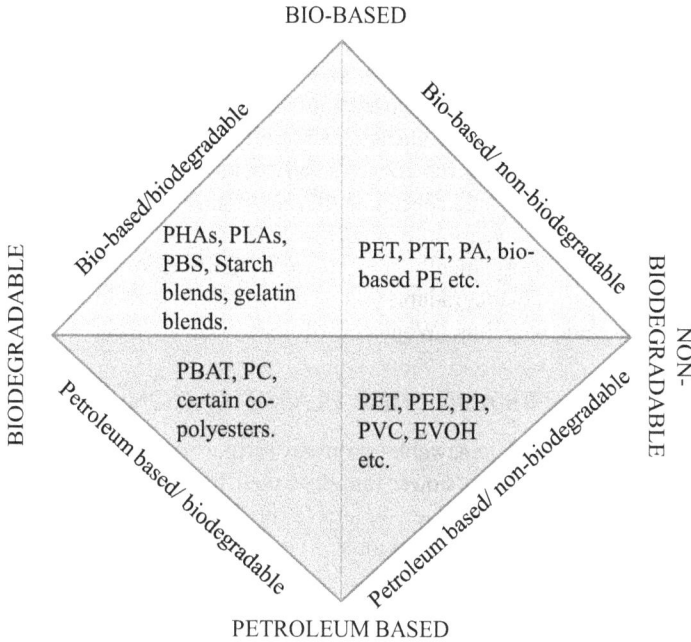

FIGURE 2.1 Different categories of plastics based on their origin and degradability.

versa (Figure 2.1 shows the origin and degradation relation) as biodegradation has a relation with the chemical structure of the compound rather than its origin (Qamar et al. 2020). Bio-based materials are a group of polymers extracted from natural material (biomass), microorganisms, or chemically synthesized from the natural monomer. The growing concern about the health and the environment acts as a driving force to produce natural and biodegradable packaging materials to protect both the environment as well as the quality of food. To match up with the conventional plastic materials for food packaging purpose, the bio-based materials have to compete with them in terms of cost and versatility, for which the bio-based plastic needs to excel in technical expectations.

2.2 NEED FOR BIODEGRADABLE PACKAGING MATERIALS

1. Growing consumer awareness towards a healthy lifestyle has intimidated packaging industries to move towards bio-based packaging materials as conventional packaging material may contaminate the food by undergoing various chemical changes over a period of time (Herbes et al. 2018).
2. The key responsibility of the food industry is to guarantee food safety and provide long-term alternatives to enhance the shelf life of food and trim down waste. The bio-based sensors along with bio-based polymers help in overcoming this problem. They help in monitoring the

physiochemical changes in these bio-based packaging materials that improve the quality of food. Bio-based smart food packaging provides a single platform for food safety and reduces the dependence on non-renewable resources for food packaging applications.

3. Management of waste products is a huge challenge faced by packaging markets. Dumping the conventional packaging waste is hazardous to both marine and land species as they are non-biodegradable and may take thousands of years to decompose. Bioplastics can be reused, and mechanically or organically recycled, and energy recovery is also possible. The use of biodegradable and compostable bio-based products promotes organic recycling (Wojnowska-Baryła et al. 2020).

2.3   BIOPOLYMERS AND THEIR CLASSIFICATIONS

Bio-based polymers have renewable resources as their raw material. Renewable resources are consumed at a slower rate than their replenishment (Niaounakisn 2015). Sugars and polysaccharides, vegetable oils, lignin, pine resin derivatives, and proteins contribute to the preparation of bio-based polymers. Easy accessibility, low toxicity, and effective cost make vegetable oil the widely used bio-based feedstock in the polymer industry (Garrison et al. 2016). Bio-based polymers can be broadly classified into three main categories (Figure 2.2 shows the classification of biodegradable polymers based on their origin). The three main classes are as follows: (i) The first class comprises naturally derived biomass polymers, which are directly extracted from biomass; however, they may undergo slight chemical modifications, e.g. cellulose, cellulose acetate, chitin, starches, and modified starch; (ii) the second class consists of bioengineered polymers, i.e. biosynthesized by plants and microorganisms, for example polyhydroxyalkanoates (PHAs) and polyglutamic acid; (iii) synthetic polymers form the third class, e.g. bio-polyolefins and bio-PET (Nakajima et al. 2017). Biopolymers have proved to be excellent candidates for encapsulating antimicrobial agents, thus providing huge benefits in food packaging applications (Muñoz-Bonilla et al. 2019).

2.3.1   Biomass-Derived Biopolymers

Polysaccharides (starch, cellulose, gums, chitosan, and chitin), proteins, and lipids (cross-linked triglycerides) are the polymers that are directly extracted from the biomass (Weber et al. 2002). Cellulose, hemicelluloses, starch, and their derivatives have been elaborated as edible film and coating materials (De Azeredo et al. 2014). The very first biopolymer made from cellulose is cellophane, which has widely been exploited for food wrapping applications.

2.3.1.1   Polysaccharide-Derived Polymers
2.3.1.1.1   *Starch-Based Polymers*
Because of its vast accessibility from plant sources, renewability, film-forming ability, non-toxicity, and low cost, starch is the most promising biodegradable

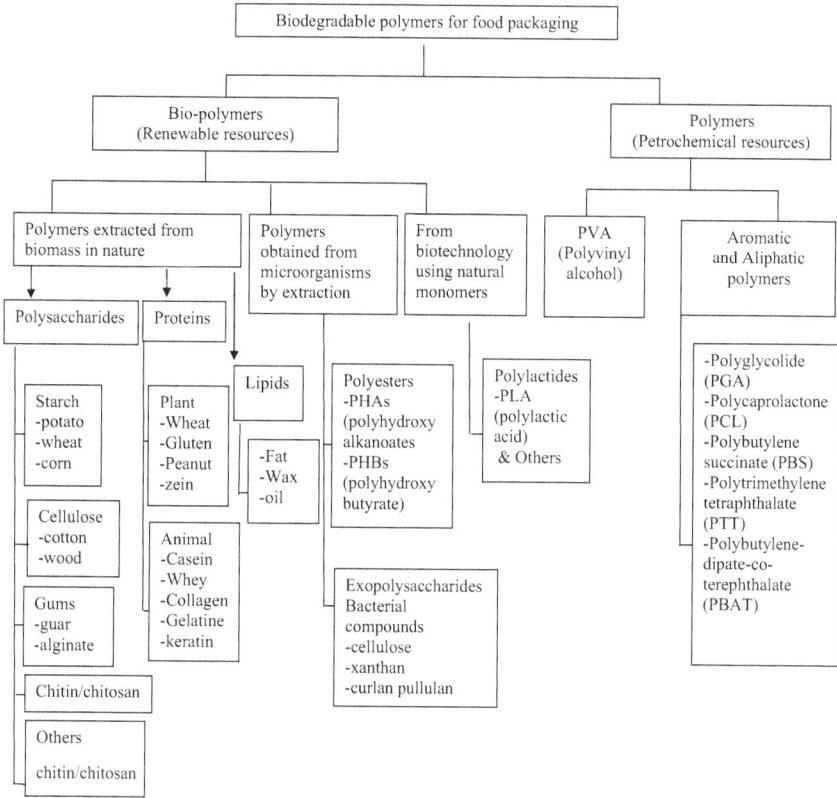

FIGURE 2.2   Different categories of biodegradable polymers considering their origin.

and biocompatible substance among other natural polymers that have garnered more attention (Jha 2020). Starch is the major source of energy in plants, which is stored as carbohydrates. The key constituents of starch include two homopolymers of D-glucose. Regulation of banning disposable plastics by many countries has spurred the applications of starch foam, which could be an alternative for the traditional plastic type of foam (Jiang et al. 2020). Desirable packaging material should have a reasonable elongation at break and high TS and elastic modulus to handle stress and strain during processing and shipping. Plant-derived additives such as green tea extract and citric acid incorporation into starch were found to provide all these features (Lauer et al. 2020). The synergistic effect in terms of improving antibacterial and mechanical properties was deduced by combining Ag/ZnO/Cu nanoparticles in the formulation of starch-based films (Peighambardoust et al. 2019). Starch blend with polyvinyl alcohol (PVA) is capable of evaluating pH changes and suppressing awful microbial growth in food (Liu et al. 2017). Thermoplastic corn starch/chitosan oligomer blends also possess antimicrobial properties (Castillo et al. 2017).

2.3.1.1.2 Cellulose

Cellulose is a linear polymer having anhydrous glucose as a repeating unit linked by β-1,4-linkages. It is the most profuse polymer on earth and has been explored for a wide range of applications, including biomedical implants and composite materials. Cellulose films exhibit good tensile strength (Sirviö et al. 2013), flexibility, and transparency (Hermawan et al. 2019). Cellulose-based nanomaterials have widely been exploited for food packaging applications as they possess tremendous mechanical strength and biocompatibility (Amalraj et al. 2018, Othman and Procedia 2014, de Oliveira et al. 2019, Youssef and El-Sayed 2018). Incorporation of nanocellulose improves the mechanical and barrier properties as well as characterization rates of PLA (Yu et al. 2017). Silver nanoparticles along with surface-modified nanocellulose further contribute to the antibacterial effect (Fortunati et al. 2012). However, using nanocellulose as a filler in PLA has limitations such as non-uniform nanocellulose dispersion inside the polymer, which can be resolved by using a twin-screw extrusion technique (Ariffin et al. 2018). Bacterial cellulose has gained a lot of interest recently in the food industry because of its antimicrobial properties, thus increasing food shelf life (Cazón and Vázquez 2020). However, technologies for processing BC for food purposes still need to be explored (Azeredo et al. 2019). Cellulose acetate in collaboration with nanocomposite films acts as a host for antioxidant/antimicrobial agents, thus forming effective food packaging material (Marrez et al. 2019, Rodríguez et al. 2014, Dairi et al. 2019).

2.3.1.1.3 Chitosan

The exoskeleton of crustacean shell upon acid treatment gives chitin and, on further interaction with NaOH or enzyme, forms chitosan, which after cellulose is the second most copious polysaccharide in nature. It is a deacetylated derivative of chitin and is a linear polysaccharide consisting of (1,4)-linked-2-amino-deoxy-β-D-glucan (Tripathi et al. 2008). Biocompatibility, antimicrobial activity, non-toxicity, biodegradability, and chelating potential are certain attributes possessed by chitosan, which makes it suitable in almost every field such as medicine, food, and agriculture. Various methods by which chitosan films can be fabricated are coating, layer-by-layer assembly, casting, etc., which lead to chitosan-based films with numerous functionalities (Wang et al. 2018). The ability of chitosan to bind transition metal ions helps in the controlled release of metallic nanoparticles, thus providing prolonged antimicrobial activity. It shows antimicrobial activity against bacteria, fungi, and yeast. Chitosan develops a layer around food material that blocks gas transport and minimizes bacterial growth (Priyadarshi et al. 2020). Grafting of natural extracts, e.g. propolis (Siripatrawan and Vitchayakitti 2016) and rosemary essential oil (Abdollahi et al. 2012), onto chitosan produces film with improved physical properties, antioxidant property, and reduced water vapour and oxygen permeability. Chitosan has been used in the packaging of dairy products, fruits, cereals, meat, juices, eggs, etc. As dairy products have a short shelf life, they can be improved by embedding chitosan films with fruit-derived essential oils (Priyadarshi et al. 2018).

2.3.1.1.4 Gums

Gums are naturally occurring polysaccharides having gel-forming and emulsion-stabilizing ability, leading to their widespread industrial application. The term "hydrocolloids" refers to water-soluble gums tremendously used as coating agents, texture modifiers, stabilizers, emulsifiers, etc. (Salehi 2020). Blending of gums with other polysaccharide films, e.g. starch, improves the mechanical and barrier properties of these films (Sapper et al. 2019, Kim et al. 2015). Gums are usually classified into three groups: plant-, microbial-, and seaweed-based gums. Among these microbial gums, xanthan, pullulan, gellan, and curdlan are considered suitable for medicine, pharmacy, and food applications by Food and Drug Administration (Alizadeh-Sani et al. 2019, da Rosa Zavareze et al. 2019). Plant-based gums because of their brittle nature and poor mechanical strength are less significant in food packaging applications. However, it has been shown that the use of esterifying agent could be useful in overcoming these limitations (Venkateshaiah et al. 2021).

2.3.1.2 Proteins

Amino acids are the basic components of proteins. The 20 amino acids combine in various manners to give different polymer units with discernible physiochemical properties. Variation in functional groups makes it possible to evolve them chemically or physically, and enzymatically, thus obtaining material offering good prospects for the fabrication of packaging materials (Chen et al. 2019). Good film-forming ability, high nutritional value, and relative abundance are a few additional benefits owing to protein-based packaging materials. However, protein-based films have poor water barrier properties. This issue could be resolved by the application of plasticizers and reinforcement with nanocomposites (Zubair et al. 2020).

2.3.1.2.1 Gluten

The wheat starch industry has wheat gluten (WG) as its by-product having both food and non-food applications. WG has unique charge transfer and gas exchange properties. High permeability to CO_2 and O_2 improves the shelf life of food, thus making it a suitable material for food packaging applications (Bibi et al. 2017). The factors which strongly influence the mechanical and moisture barrier properties of gluten films are ethanol, pH, and gluten content. Moreover, grafting of plasticizers such as sorbitol or glycerol can mitigate the brittleness of these films (Liu et al. 2018). Forming nanocomposites with cellulose nanocrystals and metal oxide nanoparticles (TiO_2) enhances the antimicrobial properties (El-Wakil et al. 2015). Studies show that incorporating ZnO into gluten films also introduces antimicrobial properties within them (Rezaei et al. 2020). The introduction of essential oils into gluten films creates an antimicrobial environment, thus preventing the growth of microorganisms, a huge problem in food packaging (Gómez-Heincke et al. 2016). Hybrid WG/chlorophyll films enhance the antioxidant property of gluten by 5%–20% (Chavoshizadeh et al. 2020).

2.3.1.2.2   Corn Zein

Zein is extracted from corn relatively in pure form. Being biodegradable, it acts as a raw material for film and coating. However, it is hydrophobic although its potential to undergo physical changes with moisture acts as a major limitation to its application (Lawton 2002). To enhance the barrier properties of PP films, corn zein nanocomposite coating could replace conventional synthetic polymer, reducing both the oxygen permeability and moisture sensitivity (Ozcalik and Tihminlioglu 2013). Corn zein along with plasticizers such as polyethylene glycol and glycerol also reduces water vapour and oxygen sensitivity of PP film (Tihminlioglu et al. 2010, Doğan Atik et al. 2008).

2.3.1.2.3   Soy Protein

Soy protein (SP) has emulsifying and texturizing properties, making it a major contributor to the food packaging industry. Soy protein isolate (SPI) is a by-product of the soybean industry. SPI-based films and coatings are soft, transparent, and flexible. The presence of polar amino groups makes it hydrophilic, contributing to its brittleness in a wet state, poor barrier properties towards moisture, and degraded mechanical properties. Blending with synthetic polymers, cross-linking, and use of plasticizers are some alternatives used to overcome these problems (González and Igarzabal 2013). The melding of SPI with collagen could lead to film with unique properties, which on further addition of plasticizers becomes more flexible and extensible. The mechanical, physical, optical, and barrier properties are a function of blend composition (Ahmad et al. 2016).

2.3.1.2.4   Casein

Casein is a milk-derived protein obtained by precipitating milk at a slightly acidic pH (Liang et al. 2020). The film-forming ability of casein is due to its potential to form electrostatic, hydrophobic, and hydrogen bonds and its random coil nature (Saez-Orviz et al. 2017). As protein-based film possesses poor mechanical and water barrier properties, some modifications are required to make it a suitable candidate for food packaging applications. Plant-derived phenolic compounds, e.g. tannic acid cross-linked with casein, yield films with improved physicochemical properties (Picchio et al. 2018). Cross-linking of casein with collagen fibre-based films resulted in films with better structure, stability, and packaging characteristics (Wu et al. 2019).

2.3.1.2.5   Whey Proteins

Whey proteins (WP) are globular proteins particularly rich in β-lactoglobulin, which undergoes denaturation to form films and coatings. They have a hydrophilic character, but possess low oxygen permeability, which makes their desirable application in the packaging of oxygen-sensitive products. An intensive cross-linked network contributes to the great mechanical properties of these protein-based films. However, this is often accompanied by less flexibility as elongation is lower because of cross-networking (Schmid and Müller 2019). Formation of plant protein (zein)-embedded whey protein isolate bio-nanocomposites offers several advantages such as enhanced tensile strength and reduced moisture sensitivity (Oymaci and Altinkaya 2016). The applicability of WP as additives in the agro-food industry and as edible films on salmon, fruits, peanuts, or cereals has been reported (Schmid et al. 2012). Incorporation of WP films with polyacrylic acid/lysozyme results in controlled release of lysozyme, thus producing long-lasting antimicrobial films (Ozer et al. 2016). At room temperature, PVA/WPI blends are best suited for packaging applications due to decreased water sorption (Lara et al. 2020). It has been found that PVA blending with whey matrix also enhances the flexibility of the film (Lara et al. 2019). The blending of essential oils increases the antimicrobial and antioxidant capacity of WP films (Ribeiro-Santos et al. 2017).

2.3.2 SYNTHETIC BIODEGRADABLE POLYMERS

2.3.2.1 Polylactic Acid (PLA)

Bacterial fermentation of carbohydrates produces lactic acid (2-hydroxypropionic acid), the basic component of PLA. High molecular weight polymers of PLA are widely used in the food packaging industry. The fish meal waste, kitchen waste, and paper sludge contribute to the lactic acid production (Jamshidian et al. 2010). Biocompatibility, biodegradability, good transparency, good mechanical properties, and cost-effectiveness ensure its existence in the food packaging sector. On the other hand, the high oxygen permeation and brittleness of PLA films restrict its use in food packaging. To match the level of conventional plastics, the PLA-based films and coatings need modifications to get desired chemical, mechanical, and biological characteristics. Blending with polysaccharides (Muller et al. 2017) and proteins (González and Igarzabal 2013) complements each other in giving better-quality films with improved characteristics. Curcumin (a bioactive natural compound)-incorporated PLA films exhibit excellent antioxidant and antibacterial property and good UV barrier property (Roy and Rhim 2020). Nanoclays such as montmorillonite and halloysite nanotubes (HLN) are used to prepare bio-nanocomposites for food packaging applications. PLA/HLN nanoconjugates were found to enhance the shelf life of cherry tomatoes (Risyon et al. 2020).

2.3.2.2 Polyglycolic Acid (PGA)

Glycolic acid is the basic monomer of PGA, a biodegradable polymer derived from both renewable and non-renewable sources. The commercial production of PGA is hindered by its hydrophilic nature, fast degradation, brittleness, and insolubility in most organic solvents (Samantaray et al. 2020). It has been shown that the degradation and mechanical properties of PLA can be enhanced by mixing PGA (Jem et al. 2020). Bio-based polymers have low thermal and mechanical resistance and exhibit poor barrier properties. The formation of bio-nanocomposites provides a novel approach to forming films with enhanced physiological properties. Reinforcing PGA with cellulose nanofillers shows a substantial improvement in oxygen barrier properties at relatively different humidity conditions (Vartiainen et al. 2016).

2.3.2.3   Polyvinyl Alcohol (PVA)

PVA is a polar polymer exhibiting odourless and non-toxic characteristics. It is highly resistant to oxygen and aroma. It is widely used as a water-soluble paper adhesive coating (Jha et al. 2019). PVA finds applications in biomaterial due to its effortless preparation, resistance to chemicals, biodegradability, and great mechanical properties. PVA blends with polysaccharides, and proteins have been found to boost the structural, barrier, and antibacterial properties of these films, thus obtaining effective food packaging materials (Tripathi et al. 2010, Narasagoudr, Hegde, Vanjeri, et al. 2020). PVA/chitosan blends possess low thermal stability and non-satisfactory mechanical properties. The modification with certain tree extract (boswellic acid) and flavonol (rutin) has solved the problem by enhancing these properties (Narasagoudr, Hegde, Chougale, Masti, and Dixit 2020, Narasagoudr, Hegde, Chougale, Masti, Vootla, et al. 2020). Ag/PVA nanocomposites used as surface coatings for fruits such as strawberry and lemon improved the shelf life of products (Kowsalya et al. 2019). Microbial growth in packed food is one of the major threats faced by the food packaging industry.

2.3.2.4   Polycaprolactone (PCL)

PCL is a synthetic biodegradable aliphatic polyester made from crude oil. It is hydrophobic and semi-crystalline in nature. It is highly resistant to water, chlorine, oil, and solvent. Low melting point, good solubility, and excellent blending properties are certain attributes that have gained interest in its potential in packaging applications. Biodegradability of PCL is slow, so it mainly finds a wide range of applications in drug delivery in the long term (Mohamed and Yusoh 2016). It is used as a coating material on certain gelatin-based films and exhibits antimicrobial properties (Figueroa-Lopez et al. 2018). Studies revealed that nanostructure carriers encapsulating bioactive compounds loaded onto PCL/chitosan nanofibres improve its thermal stability and also induces antimicrobial and antibacterial activities in the films (Zou et al. 2020). Bacteriophages have extensively been exploited in the food industry for their antibacterial properties. Researchers have shown the use of phage T4 in association with modified PCL films as functional antibacterial food packaging material against *E. coli* (Choi et al. 2021). Coating of PCL on acetylated hemicelluloses film could be used as active packaging for fatty foods (Mugwagwa et al. 2020).

2.3.3   MICROBIALLY DERIVED POLYMERS

2.3.3.1   Bacterial Cellulose

Some bacteria (e.g. acetic acid bacteria) can also produce cellulose by undergoing oxidative fermentation in both synthetic and non-synthetic media. This could be used as an alternative source of cellulose without causing much harm to the environment. The bacterial cellulose (BC) exhibits an intense degree of polymerization, high water retaining capacity, good mechanical strength, and high crystallinity. High purity, *in situ* secretion of colour and flavour, distinct shape-forming ability, and nanoscale range are certain pros of BC compared to that of other dietary fibres. Thus, it finds application as a raw material for food and is used as a thickening, stabilizing, gelling, and suspending agent. It can also make food ingredients that are minimal in calories and cholesterol (Shi et al. 2014). BC nanofibrils incorporating protein zein nanoparticles forming multifunctional nanocomposites with improved tensile strength, biocompatibility, and thermal properties were also obtained, having sustainable food applications (Li, Gao, et al. 2020). Blends of BC with other synthetic or natural biodegradable polymers help in handling certain limitations, i.e. poor mechanical strength, antimicrobial, and barrier properties of these polymers, thus uplifting their food packaging applications (Haghighi et al. 2021, Fabra et al. 2016, Albuquerque et al. 2021).

2.3.3.2 Polyhydroxyalkanoates (PHAs)

Most bacterial cells store energy in the form of polyester polyhydroxyalkanoates (PHAs). The commercial development of PHA took place after the 1970s oil crisis, which stimulated researchers to look for some other substitute based on bio-derived products. Like other bio-based materials, they are renewable, are biodegradable, and most importantly, have good water barrier properties, which make them superior to other natural polymers. However, the cons associated with PHAs are their high cost, brittleness, and poor thermal stability. Properties of PHAs can be tailored by blending with other biopolymers without substantially affecting their biodegradability (Plackett and Siró 2011). Among various family members of PHAs, those that have been exploited to a large extent are polyhydroxybutyrate (PHB) and poly(hydroxybutyrate-co-hydroxyvalerate) (PHBV), which is a copolymer (Pardo-Ibáñez et al. 2014). The reinforcement of polymer matrix with nanocomposites, i.e. fibres, clay, or particles, having one dimension in nanometre range (10^{-9} nm) helps in modifying the degradation rate, crystallinity, barrier potential, mechanical properties, thermal properties, and the morphology of the resulting material. "Nanopaper" is a term used to indicate cellulose nanofibrils and lignocellulose nanofibrils when they form a gel-like transparent material with the removal of water from gel, i.e. dewatering. These nanopapers exhibit ultimate mechanical strength and very good barrier properties in dry conditions. However, one serious flaw associated with them is their hydrophilicity, which limits their application in food packaging. It has been investigated that these nanopapers when coated on both sides by electrospun PHB and PHBV to form multilayered film shows an improvement in water contact angle, morphology, mechanical, and barrier properties (Cherpinski et al. 2018).

2.4 DIFFERENT TYPES OF BIODEGRADABLE PACKAGING MATERIALS

Different types of biodegradable packaging materials used for the preservation, transport, and storage of food are films, bags, boxes with lids, gels, and trays.

2.4.1 Biodegradable Gel

The biodegradable gels have antimicrobial activity and avert food material against microbial contamination (Ivonkovic et al. 2017). The use of hydrogels in food packaging decreases water activity, slows down microbial growth, and reduces softening of crispy products (Batista et al. 2019). Keratin-based hydrogel mixed with citric acid is suitable as an environmentally compatible food packaging material (McLellan et al. 2019).

2.4.2 Biodegradable Bags and Pouches

Dextrose corn polyester is the main raw material of biodegradable bags. These bags find great applicability in packaging due to their stiffness, flexibility, strength, and moisture and temperature resistivity. Bio-based pouches are made up of pure organic fruits and vegetable pulps. Mater-Bi bioplastic bags are already in use in grocery stores of Italy for fruits and vegetables (Reichert et al. 2020b).

2.4.3 Biodegradable Films and Trays

In an attempt to replace the conventional polyethylene (PE), biodegradable films came into existence. Unlike PE films, they are biocompatible and are environmentally friendly. Biodegradable films are fabricated using polysaccharides, proteins, lipids, and essential oils. Blending of these materials with each other results in the enhancement of properties of these films. Various additives can be mixed with these films to make robust their mechanical and barrier properties. Reinforcement of these films with essential oils induces antimicrobial/antioxidant activity within the film (Atarés et al. 2016). Biodegradable foam trays made from cassava starch via extrusion have been studied for food packaging (Brant et al. 2018). PLA trays have been assessed for food packaging (Ingrao et al. 2015). It has been investigated that starch/PVA blend biodegradable trays reinforced with clove and oregano essential oils show high antimicrobial activity, lower water sorption, and more flexibility (Debiagi et al. 2014).

2.5 NANOTECHNOLOGY AND PACKAGING SCIENCE

Nanomedicine is the most explored branch of nanoscience, which is becoming more popular with each passing day. The use of nanotechnology in the agro-food industry is undoubtedly one of the most promising advantages of nanotechnology in the food sector. The main focus of researchers, inventors, and organizations is to look for new techniques and protocols that would avail the direct benefit of nanotechnology in food products. The performance, processing, and cost are three main challenges associated with biodegradable polymers, and these are interrelated. The development of nanocomposites could provide new opportunities to tackle these problems. They have the potential to improve the mechanical, barrier, thermal, and physicochemical properties of these polymers (Figure 2.3

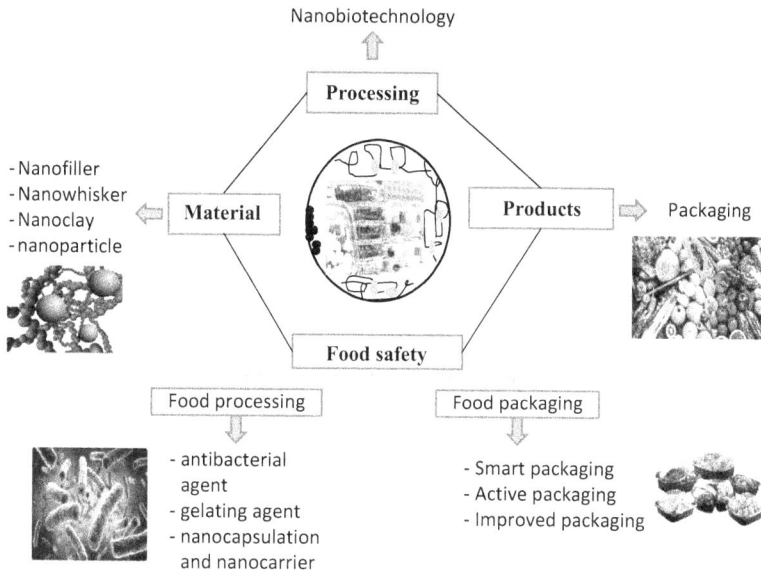

FIGURE 2.3 Nanotechnology and food packaging.

shows the role of nanotechnology in food packaging applications). The use of nanotechnology in food packaging is divided into three parts: (i) improved packaging, i.e. enhancement of barrier and physical properties; (ii) active packaging, i.e. antimicrobial/antioxidant character development; and (iii) smart/intelligent packaging, i.e. introduction of freshness- and microbial growth-detecting indicators (Kuswandi 2016). Nanofibres, nanoplates, and nanoparticles are also used in nano-package foods. Nanosensors are of great importance in food packaging, which could track the deterioration of food products, pallets, and containers during the entire process of food supply. Silica nanoparticles have been used to check OP, moisture maintenance to keep food fresh, as mould growth regulator inside the refrigerator to prevent spoilage of food (Adeyeye and Science 2019). Nanoclays could replace charged polymers in developing multifunction thin films by layer-by-layer assembly technique as it provides a great oxygen barrier (Lindström et al. 2020). The agricultural sector, meat processing industry, dairy products, bakery industry, etc., are various sectors availing the benefits of edible nano-coatings that provide flavour, antioxidants, enzymes, colours, etc., to the products. Nano-laminate coatings made from adsorbing bio-based polyelectrolytes (polysaccharides and proteins) could improve layer properties and act as a suitable host for active compounds (antimicrobials, flavours, odours, etc.), thus enhancing the shelf life of packaged food items. Various metal and metal oxide (TiO_2, ZnO, Ag, etc.) nanoparticles because of their high antimicrobial activity act as promising candidates for active food packaging. They are often combined with antimicrobial agents to provide antimicrobial active packaging.

It has been investigated that the fabrication of PVA-based nanocomposites using Ag nanoparticles grafted with montmorillonite clay and ginger extract improves film properties (Primožič, Knez, and Leitgeb 2021). Pea starch/PVA blend incorporating cellulose nanocrystals (CNC) has been found to show improvement in mechanical and barrier properties along with CNC enhancing the stability of the film. CNC/silver nanoparticles blend introduced into PLA matrix show antibacterial effect against *S. aureus* and *E. coli*, thus providing active food packaging. The crystallization, mechanical, thermal, and barrier behaviour of the PHBV matrix has been considerably improved by carbon nanotubes, which possess high elastic modulus and great tensile strength (Fortunati et al. 2018). Although nanotechnology is seamlessly conquering the food packaging sector, the safety and ethical issues associated with the development and use of these nanocomposites are of immense importance. The synergistic work of scientists and companies producing these nanocomposites could help in the fruitful use of these nanoparticles in food packaging to provide healthier, tastier, and safer food supplies to the world.

2.6 CONCLUSIONS

Bio-based biodegradable polymers provide novel innovations replacing the conventional petroleum-based material in food packaging applications. It serves as a sustainable approach towards a green economy as it minimizes the environmental mutilation caused by the production and processing of plastics. Government regulations on the use of petroleum-based plastics and consumers' agitation towards health and safety have been fostering packaging industries to adopt bio-derived materials for packaging applications. However, the physical, mechanical, barrier, and rheological properties of bio-based plastics are not comparable with that of fuel-derived plastic. Upgrading these in the form of composites/blends helps in matching the level of commercial plastic conquering market for ages. Nanotechnology comes up as an innovative technique, affecting the future of green packaging extensively by improving its performance. This chapter mainly covers the functional properties and state-of-the-art and emerging trends of bio-based polymers such polysaccharides, proteins, and aliphatic and aromatic polyesters (PGA and PCL) in the food packaging sector. It also emphasizes the loopholes associated with these natural bio-derived materials and methods to resolve their drawbacks. The challenges that need to be mentioned include the management of raw materials, cost production, improvement in processing techniques, and performance of bio-based materials. Bioplastic waste management is also one major issue as certain bioplastics are designed in a way to undergo degradation in specifically managed conditions, thus creating environmental vandalization on its disposal in a non-favourable environment. Synergistic efforts of microbiologists, biomass researchers, synthetic chemists, and process engineers are required for the future development of sustainable packaging material for the betterment of both humanity and Mother Nature.

ACKNOWLEDGEMENT

The authors would like to sincerely acknowledge the start-up funding (SERB-DST) provided by the Central University of Himachal Pradesh, India.

REFERENCES

Abdollahi, M., Razaee, M., and Farzi, G. 2012. Improvement of active chitosan film properties with rosemary essential oil for food packaging. *International Journal of Food Science and Technology* 47: 847–53.

Adeyeye, S.A.O. 2019. Food packaging and nanotechnology: Safeguarding Consumer Health and Safety. *Nutrition, and Food Science* 49, no. 6: 1164–79.

Ahmad, M., Nirmal, N.P., Danish, M., et al. 2016. Characterisation of composite films fabricated from collagen/chitosan and collagen/soy protein isolate for food packaging applications. *RSC Advances* 6: 85.

Albuquerque, R.M., Meira, H.M., Silva, I.D., et al. 2021. Production of a bacterial cellulose/poly (3-hydroxybutyrate) blend activated with clove essential oil for food packaging. *Polymers and Polymer Composites* 29, no. 4: 259–70.

Alizadeh-Sani, M., Ehsani, A., Moghaddaskia, E., and Khezerlou, A. 2019. Microbial gums: Introducing a novel functional component of edible coatings and packaging. *Applied Microbially Biotechnology* 103: 6853–66.

Amalraj, A., Gopi, S., Thomas, S., and Haponiuk, J.K. 2018.Cellulose nanomaterials in biomedical, food, and nutraceutical applications. A review. *Paper presented at the Macromolecular Symposia* 380, no. 1: 1800115.

Ariffin, H., Norrahim, M.N.F., and Yasim- Anuar T.A.T. 2018. Oil palm biomass cellulose-fabricated polylactic acid composites for packaging applications. In *Bionanocomposites for Packaging Applications*, 95–105. Springer Cham.

Atarés, L. and Chiralt, A. 2016. Essential oils as additives in biodegradable films and coatings for active food packaging. *Trends in Food Science and Technology* 48: 51–62.

Azeredo, H.M., Baraud, H., Farinas, C.S., et al. 2019. Bacterial cellulose as a raw material for food and food packaging applications. *Frontiers in Sustainable Food Systems* 3: 7.

Batista, R.A., Espitia, P.J.P., Quintans, J., et al. 2019. Hydrogel as an alternative structure for food packaging systems. *Carbohydrate Polymers* 205: 106–16.

Bibi, F., Guillaune, G., Gontard, N., and Sorli, B. 2017. Wheat gluten, a bio-polymer to monitor carbon dioxide in food packaging: Electric and dielectric characterization. *Sensors Actuators B: Chemical* 250: 76–84.

Brant, A.J.C., Naime, N., Lugão, A.B., and Ponce, P. 2018, Influence of ionizing radiation on biodegradable foam trays for food packaging obtained from irradiated cassava starch. *Brazilian Archives of Biology Technology* 61: 1–16.

Castillo, L. A., Farenzena, S., Pintos, E., et al. 2017. Active films based on thermoplastic corn starch and chitosan oligomer for food packaging applications. *Food Packaging and Shelf Life* 14: 128–36.

Cazón, P. and Vázquez, M. 2020. Bacterial cellulose as a biodegradable food packaging material: A review. *Food Hydrocolloids* 113: 106530.

Chavoshizadeh, S., Chang, Y., Kim, S.Y., and Han, J. 2020. Sesame oil oxidation control by active and smart packaging system using wheat gluten/chlorophyll film to increase shelf life and detecting expiration date. *European Journal of Lipid Science and Technology* 122: 1900385.

Chen, H., Wang, J., Cheng, Y., et al. 2019. Application of protein-based films and coatings for food packaging: A review. *Polymers* 11: 2039.

Cherpinski, A., Torres-Giner, S., Vartiainen, J., et al. 2018. Improving the water resistance of nanocellulose-based films with polyhydroxyalkanoates processed by the electrospinning coating technique. *Cellulose* 25: 1291–307.

Choi, I., Chang, Y., Kim, S.Y., and Han, J. 2021. Polycaprolactone film functionalized with bacteriophage T4 promotes antibacterial activity of food packaging toward Escherichia coli. *Food Chemistry* 346: 128883.

da Rosa Zavareze, E., Kringel, D.H., and Dias, A.R.G. 2019. Loong-Tak Lim & Michael Rogers (Eds.). Nano-scale polysaccharide materials in food and agricultural applications. In *Advances in Food and Nutrition Research*, 85–128, Elsevier. https://doi.org/10.1016/bs.afnr.2019.02.013

Dairi, N., Ferfera-Harrar, H., Ramos, M., and Garrigós, M.C. 2019. Cellulose acetate/AgNPs-organoclay and/or thymol nano-biocomposite films with combined antimicrobial/antioxidant properties for active food packaging use. *International Journal of Biological Macromolecules* 121: 508–23.

De Azeredo, H.M.C., Rosa, M.F., Souza Filhoet., et al. 2014. The use of biomass for packaging films and coatings. *Advances in Biorefineries*: 819–74.

de Oliveira, J.P., Bruni, G.P., El Halal, S.L.M., et al. 2019. Cellulose nanocrystals from rice and oat husks and their application in aerogels for food packaging. *International Journal of Biological Macromolecules* 124: 175–84.

Debiagi, F., Kobayashi, R.K.T., Nakazato, G., et al. 2014.Biodegradable active packaging based on cassava bagasse, polyvinyl alcohol and essential oils. *Industrial Crops and Products* 52: 664–70.

Doğan Atik, İ., Özen, B., and Tihminlioğlu, F. 2008. Water vapour barrier performance of corn-zein coated polypropylene (PP) packaging films. *Journal of Thermal Analysis and Calorimetry* 94, no. 3: 687–93.

El-Wakil, N.A., Abou-Zeid, R.E., and Dufresne, A. 2015. Development of wheat gluten/nanocellulose/titanium dioxide nanocomposites for active food packaging. *Carbohydrate Polymers* 124: 337–46.

Fabra, M.J., López-Rubio, A., Ambrosio-Martín, J., and Lagaron, J.M. 2016. Improving the barrier properties of thermoplastic corn starch-based films containing bacterial cellulose nanowhiskers by means of phaelectrospun coatings of interest in food packaging. *Food Hydrocolloids* 61: 261–68.

Figueroa-Lopez, K.J., Castro-Mayorga, J.L., Mahecha, M.M., et al. 2018. Antibacterial and barrier properties of gelatin coated by electrospun polycaprolactone ultrathin fibers containing black pepper oleoresin of interest in active food biopackaging applications. *Nanomaterials* 8: 199.

Fortunati, E., Armentano, I., Zhou, Qi., et al. 2012. Multifunctional bionanocomposite films of poly (lactic acid), cellulose nanocrystals and silver nanoparticles. *Carbohydrate Polymers* 87: 1596–05.

Fortunati, E., Luzi, F., Yang, W., et al. 2018. Miguel Ângelo Cerqueira, Jose Maria Lagaron, Lorenzo Miguel Pastrana Castro & António Augusto Vicente (Eds.). Bio-based nanocomposites in food packaging. In *Nanomaterials for Food Packaging*, 71–110. Elseiver. https://doi.org/10.1016/C2016-0-01251-2

Garrison, T.F., Murawski, A., and Quirino, R.L. 2016. Bio-based polymers with potential for biodegradability. *Polymers* 8: 262.

Gómez-Heincke, D., Martinez, I., Partal, P., et al. 2016. Development of antimicrobial active packaging materials based on gluten proteins. *Journal of the Science of Food Agriculture* 96: 3432–38.

González, A. and lgarzabal, C.I.K. 2013. Soy protein–poly (lactic acid) bilayer films as biodegradable material for active food packaging. *Food Hydrocolloids* 33: 289–96.

Haghighi, H., Gullo, M., La China, S., et al. 2021.Characterization of bio-nanocomposite films based on gelatin/polyvinyl alcohol blend reinforced with bacterial cellulose nanowhiskers for food packaging applications. *Food Hydrocolloids* 113: 106454.

Herbes, C., Beuthner, C., and Ramme, I. 2018. Consumer attitudes towards biobased packaging–a cross-cultural comparative study. *Journal of Cleaner Production* 194: 203–18.

Hermawan, D., Lai, T.Z., Jafarzadeh, S., et al. 2019. Development of seaweed-based bamboo microcrystalline cellulose films intended for sustainable food packaging applications. *Bio Resources* 14: 3389–410.

Ingrao, C., Tricase, C., Cholewa-Wójcik, A., et al. 2015. Polylactic acid trays for fresh-food packaging: A carbon footprint assessment. *Science of the Total Environment* 537: 385–98.

Ivonkovic, A., Zelijko, K., Talic, S., and Lasic, M. 2017. Biodegradable packaging in the food industry. *Journal of Food Safety and Food Quality* 68: 26–38.

Jamshidian, M., Tehrang, E.A, Imran, M., et al. 2010. Poly-lactic acid: Production, applications, nanocomposites, and release studies. *Comprehensive Reviews in Food Science and Food Safety* 9: 552–71.

Jem, K.J. and Tan, B. 2020. The development and challenges of poly (lactic acid) and poly (glycolic acid). *Advanced Industrial and Engineering Polymer Research* 3: 60–70.

Jha, A. and Kumar, A. 2019. Biobased technologies for the efficient extraction of biopolymers from waste biomass. *Bioprocess and Biosystems Engineering* 42: 1893–901.

Jha, P. 2020. Effect of plasticizer and antimicrobial agents on functional properties of bionanocomposite films based on corn starch-chitosan for food packaging applications. *International Journal of Biological Macromolecules* 160: 571–82.

Jiang, T., Duan, Q., Zhu, J., Liu, H., and Yu, L. 2020. Starch-based biodegradable materials: Challenges and opportunities. *Advanced Industrial Engineering and Polymer Research* 3: 8–18.

Kim, S.R.B., Choi, Y.G., Kim, J.Y., and Lim, S.T. 2015. Improvement of water solubility and humidity stability of tapioca starch film by incorporating various gums. *Food Science and Technology* 64: 475–82.

Kowsalya, E., Mosa Christas, K., Balashanmugam, P., and Rani, J.C. 2019. Biocompatible silver nanoparticles/poly (vinyl alcohol) electrospun nanofibers for potential antimicrobial food packaging applications. *Food Packaging and Shelf Life* 21: 100379.

Kumar, S. 2020. Biodegradable and recyclable packaging materials: A step towards a greener future. 328–337.

Kuswandi, B. 2016. Nanotechnology in food packaging. Shivendu Ranjan, Nandita Dasgupta & Eric Lichdouse (Eds.). In *Nanoscience in Food and Agriculture* 1, 151–83. Springer. https://doi.org/10.1007/978-3-319-39303-2_6

Lara, B.R.B. 2019. Morphological, mechanical and physical properties of new whey protein isolate/polyvinyl alcohol blends for food flexible packaging. *Food Packaging and Shelf Life* 19: 16–23.

Lara, B.R.D., Dias, M.V., Junior, M.G., et al. 2020. Water sorption thermodynamic behavior of whey protein isolate/polyvinyl alcohol blends for food packaging. *Food Hydrocolloids* 103: 105710.

Lauer, M. and Smith, R.C. 2020. Recent advances in starch-based films toward food packaging applications: Physicochemical, mechanical, and functional properties. *Comprehensive Reviews in Food Science and Safety* 19: 3031–83.

Lawton, J.W. 2002. Zein: A history of processing and use. *Cereal Chemistry* 79: 1–18.

Li, Q., Gao, R., Wang, L., et al. 2020. Nanocomposites of bacterial cellulose nanofibrils and zein nanoparticles for food packaging. *ACS Applied Nanomaterials* 3: 2899–910.

Liang, L. and Lou, Y. 2020. Casein and pectin: Structures, interactions, and applications. *Trends in Food Science and Technology* 97: 391–403.

Lindström, T. 2020. Evolution of biobased and nanotechnology packaging–A review. *Nordic Pulpand Paper Research Journal* 35: 491–515.

Liu, B., Xu, H., Zhao, H., et al. 2017. Preparation and characterization of intelligent starch/PVA films for simultaneous colorimetric indication and antimicrobial activity for food packaging applications. *Carbohydrate Polymers* 157: 842–49.

Liu, R., Cong, X., Song, Y., Wu, T., and Zhang, M. 2018. Edible gum–phenolic–lipid incorporated gluten films for food packaging. *Journal of Food Science* 83: 1622–30.

Marrez, D., Abdelhamid, A.E., and Darwessh, O.M. 2019. Eco-friendly cellulose acetate green synthesized silver nano-composite as antibacterial packaging system for food safety. *Food Packaging and Shelf Life* 20: 100302.

Mangaraj, S., Yadav, A., Bal, LM., Dash, SK., et al. 2019. Application of biodegradable polymers in food packaging industry: A comprehensive review. *Journal of Packaging Technology and Research* 3: 77–96.

McLellan, J., Thornhill, S.G., Shelton, S., and Kumar, M. 2019. Swati Sharma & Ashok Kumar (Eds.). Keratin-based biofilms, hydrogels, and biofibers. In *Keratin as a Protein Biopolymer*, 187–200. Springer. Cham. https://doi.org/10.1007/978-3-030-02901-2_7

Mohamed, R.M. and Yusoh, K. 2016. A review on the recent research of polycaprolactone (PCL). *Advanced Material Research* 1134: 249–55.

Mugwagwa, L.R. and Chimphango, A.F. 2020. Enhancing the functional properties of acetylated hemicellulose films for active food packaging using acetylated nanocellulose reinforcement and polycaprolactone coating. *Food Packaging and Shelf Life* 24: 100481.

Muller, J., González-Martinez, C., and Chiralt, A. 2017. Combination of poly (lactic) acid and starch for biodegradable food packaging. *Materials* 10: 952.

Muñoz-Bonilla, A., Sonseca, A., Arrieta, M.P., et al 2019. Bio-based polymers with antimicrobial properties towards sustainable development. *Materials* 12: 641.

Nakajima, H., Dijkstra, P., and Loos, K. 2017. The recent developments in biobased polymers toward general and engineering applications: Polymers that are upgraded from biodegradable polymers, analogous to petroleum-derived polymers, and newly developed. *Polymers* 9: 523.

Narasagoudr, S.S., Hedge, V.G., Chougale, R.B., et al. 2020. Influence of boswellic acid on multifunctional properties of chitosan/poly (vinyl alcohol) films for active food packaging. *International Journal of Biological Macromolecules* 154: 48–61.

Narasagoudr, S.S., Hedge, V.G., Vootla, S., et al. 2020. Physico-chemical and functional properties of rutin induced chitosan/poly (vinyl alcohol) bioactive films for food packaging applications. *Food Hydrocolloids* 109: 106096.

Narasagoudr, S.S., Hedge, V.G., Vanjeri, V.N., et al. 2020. Ethyl vanillin incorporated chitosan/poly (vinyl alcohol) active films for food packaging applications. *Carbohydrate Polymers* 236: 116049.

Nesic, A., Castillo, C., Castano, P., et al. 2020. Charis M.Galanakis (Ed.). Bio-based packaging materials. In *Biobased Products and Industries*, 279–309. Elsevier. https://doi.org/10.1016/B978-0-12-818493-6.00008-7

Niaounakisn, M. 2015. *Biopolymers: Applications and Trends*. United Kingdom, U.K. William Andrew.

Othman, S.H. 2014. Bio-nanocomposite materials for food packaging applications: Types of biopolymer and nano-sized filler. *Agriculture, and Agricultural Science Procedia* 2: 296–303.

Oymaci, P. and Altinkaya, S.A. 2016. Improvement of barrier and mechanical properties of whey protein isolate based food packaging films by incorporation of zein nanoparticles as a novel bionanocomposite. *Food Hydrocolloids* 54: 1–9.

Ozcalik, O. and Tihminlioglu, F. 2013. Barrier properties of corn zein nanocomposite coated polypropylene films for food packaging applications. *Journal of Food Engineering* 114: 505–13.

Ozer, B.B.P., Uz, M., Oymaci, P., and Altinkaya, S.A. 2016. Development of a novel strategy for controlled release of lysozyme from whey protein isolate based active food packaging films. *Food Hydrocolloids* 61: 877–86.

Pardo-Ibáñez, P., Lopez-Rubio, A., Martinez-Sanz, M., et al. 2014. Keratin–polyhydroxy-alkanoate melt-compounded composites with improved barrier properties of interest in food packaging applications. *Journal of Applied Polymer Science* 131:39947.

Peighambardoust, S.H, Pournasir, N., and Paldel, P.M. 2019. Properties of active starch-based films incorporating a combination of Ag, Zno and Cuo nanoparticles for potential use in food packaging applications. *Food Packaging and Shelf Life* 22: 100420.

Picchio, M.L., Linck, Y.G., Monti, G.A., et al. 2018. Casein films crosslinked by tannic acid for food packaging applications. *Food Hydrocolloids* 84: 424–34.

Plackett, D. and Siró, L. 2011. Jośe-María Lagrón (Ed.). Polyhydroxyalkanoates (PHAs) for food packaging. In *Multifunctional and Nanoreinforced Polymers for Food Packaging*, 498–526. Elsevier. https://doi.org/10.1533/9780857092786.4.498

Primožič, M., Knez, Ź., and Litgeb, M. 2021. (Bio) Nanotechnology in food science— Food Packaging. *Nanomaterials* 11: 292.

Priyadarshi, R., Kumar, B., Deeba, F., et al. 2018. Chitosan films incorporated with apricot (prunus armeniaca) kernel essential oil as active food packaging material. *Food Hydrocolloids* 85: 158–66.

Priyadarshi, R. and Rhim, J.W. 2020. Chitosan-based biodegradable functional films for food packaging applications. *Innovative Food Science and Emerging Technologies* 62: 102346.

Qamar, S.A., Bilal, H., Asgher, M., et al. 2020. Bio-based active food packaging materials: Sustainable alternative to conventional petrochemical-based packaging materials. *Food Research International* 137: 109625.

Reichert, C.L., Bugnicourt, E., Coltelli, M.B., et al. 2020. Bio-based packaging: Materials, modifications, industrial applications and sustainability. *Polymers* 12: 1558.

Rezaei, M., Pirsa, S., and Chavoshizadeh, S. 2020. Photocatalytic/antimicrobial active film based on wheat gluten/ZnO nanoparticle. *Journal of Inorganic OrganometallicPolymers and Materials* 30: 2654–65.

Ribeiro-Santos, R., Sanches-Silva, A., Motta, J.F.G., et al. 2017. Combined use of essential oils applied to protein base active food packaging: Study in vitro and in a food simulant. *European Polymer Journal* 93: 75–86.

Richardson, A.C. 2021. Enforcing climate accountability through sanctions. *New Annales* 3:16.

Risyon, N.P., Othman, S.H., Basha, R.K., and Talib, R.A. 2020. Characterization of poly-lactic acid/halloysite nanotubes bionanocomposite films for food packaging. *Food Packaging and Shelf Life* 23: 100450.

Rodríguez, F.J., Torres, A., Peñaloza, Á., et al. 2014. Development of an antimicrobial material based on a nanocomposite cellulose acetate film for active food packaging. *Food Additives and Contaminants* 31: 342–53.

Roy, S. and Rhim, J.W. 2020. Preparation of bioactive functional poly (lactic acid)/ curcumin composite film for food packaging application. *International Journal of Biological Macromolecules* 162: 1780–89.

Saez-Orviz, S., Laca, A., Rendueles, M., and Diaz, M. 2017. Approaches for casein film uses in food stuff packaging. *Afinidad* 74: 26–29.

Salehi, F. 2020. Edible coating of fruits and vegetables using natural gums: A review. *International Journal of Fruit Science* 20: S570–S89.

Samantaray, P.K., Little, A., Haddleton, D.M., et al. 2020. Poly (glycolic acid)(PGA): A versatile building block expanding high performance and sustainable bioplastic applications. *Green Chemistry* 22: 4055–81.

Sapper, M., Talens, P., and Chiralt, A. 2019. Improving functional properties of cassava starch-based films by incorporating xanthan, gellan, or pullulan gums. *Journal of Polymer Sciences* 2019: 1–8.

Schmid, M., Dallmann, K., Bugnicourt, E., et al. 2012. Properties of whey-protein-coated films and laminates as novel recyclable food packaging materials with excellent barrier properties. 2012.

Schmid, M. and Müller, K. 2019. Hilton C. Deeth & Nidhi Bansal (Eds.). Whey protein-based packaging films and coatings. In *Whey Proteins*, 407–437. Elsevier. https://doi.org/10.1016/B978-0-12-812124-5.00012-6

Shi, Z., Zhnag, Y., Phillips, G.O., and Yang, G. 2014. Utilization of bacterial cellulose in food. *Food Hydrocolloids* 35: 539–45.

Siripatrawan, U. and Vitchayakitii, W. 2016. Improving functional properties of chitosan films as active food packaging by incorporating with propolis. *Food Hydrocolloids* 61: 695–702.

Sirviö, J.A., Liimatainen, H., Ninimäki, J., and Hormi, O. 2013. Sustainable packaging materials based on wood cellulose. *RSC Advances* 3: 16590–96.

Tihminlioglu, F., Atik, Í.D., and Özen, B. 2010. Water vapor and oxygen-barrier performance of corn–zein coated polypropylene films. *Journal of Food Engineering* 96: 342–47.

Tripathi, S., Mehrortra, G.K., and Dutta, P.K. 2008. Chitosan based antimicrobial films for food packaging applications. *e-Polymers* 8, no.1: 1–7.

Tripathi, S., Mehrortra, G.K., and Dutta, P.K. 2010. Preparation and physicochemical evaluation of chitosan/poly (vinyl alcohol)/pectin ternary film for food-packaging applications. *Carbohydrate Polymers* 79: 711–16.

Vartiainen, J., Shen, Y., Kaljunen, T., et al. 2016. Bio-based multilayer barrier films by extrusion, dispersion coating and atomic layer deposition. *Journal of Applied Polymer Science* 133: 42260(1–6).

Venkateshaiah, A., Havlíček, K., Timmins, R.L., et al. 2021. Alkenyl succinic anhydride modified tree-gum kondagogu: A bio-based material with potential for food packaging. *Carbohydrate Polymers* 266: 118126.

Wang, H., Qian, J., and Ding, F. 2018. Emerging chitosan-based films for food packaging applications. *Journal of Agricultural and Food Chemistry* 66: 395–413.

Weber, C.J., Haugaard, V., Festersen, R., and Bertelsen, G. 2002. Production and applications of biobased packaging materials for the food industry. *Food Additives and Contaminants* 19: 172–77.

Wojnowska-Baryła, I., Kullkowska, D., and Bernat, K. 2020. Effect of bio-based products on waste management. *Sustainability* 12: 2088.

Wu, X., Luo, Y., Liu, Q., et al. 2019. Improved structure-stability and packaging characters of crosslinked collagen fiber-based film with casein, keratin and SPI. *Journal of the Science of Food and Agriculture* 99: 4942–51.

Youssef, A.M. 2018. Bionanocomposites materials for food packaging applications: Concepts and future outlook. *Carbohydrate Polymers* 193: 19–27.

Yu, H.Y., Zhang, H., and Song, M.L. 2017.From cellulose nanospheres, nanorods to nanofibers: Various aspect ratio induced nucleation/reinforcing effects on polylactic acid for robust-barrier food packaging. *ACS Applied Materials and Interfaces* 9: 43920–38.

Zou, Y., Zang, C., Wang, P., et al. 2020. Electrospun chitosan/polycaprolactone nanofibers containing chlorogenic acid-loaded halloysite nanotube for active food packaging. *Carbohydrate Polymers* 247: 116711.

Zubair, M. and Ullah, A. 2020. Recent advances in protein derived bionanocomposites for food packaging applications. *Critical Reviews in Food Science and Nutrition* 60: 406–34.

3 Processing of Biodegradable Composites

Gourhari Chakraborty
National Institute of Technology Andhra Pradesh

Arbind Prasad
Katihar Engineering College (Under the Department of Science & Technology, Government of Bihar)

Ashwani Kumar
Department of Technical Education Uttar Pradesh Kanpur

CONTENTS

DOI: 10.1201/9781003227908-3

3.1 INTRODUCTION

Biodegradable polymers have become an essential ingredient for products in all fields of application because of their properties comparable to conventional polymers, bio-based nature, biocompatibility, and eco-friendliness [1]. In recent times, significant attention has been paid to their development of polymeric technology alongside polymer development. Biodegradable polymers have versatile applications in packaging, biomedicine, conductive materials, household goods, coating, cosmetics, etc., which reveals the requirement of different application-oriented processing methods of biodegradable polymers [2]. Biodegradable polymers have some limitations, such as low thermal stability, low melt strength, and low mechanical processability, which have been considered in the modern processing strategies for the successful production of improved biodegradable polymers [3]. Different research groups have carried out the composites and blends preparation using biodegradable polymers through different processing routes [4–6]. Different biofillers such as cellulose, starch, and chitosan were utilized to develop bio-nanocomposites for various applications [7–8]. Inorganic fillers such as carbon, metal oxides, inorganic compounds, and hydroxyapatite were used by different groups for the development of biodegradable polymer-based composites [9–10]. In most cases, either solution-based approaches or melt processing approaches were used for the respective product development. Solution processing is easy to handle but involves solvent recovery issues. On the other hand, melt processing has commercial production ability of biodegradable polymeric composites with dispersion and molecular weight deterioration challenges [11]. Alongside this, sometimes application-oriented casting of polymer solution or melt is required to generate specific products such as thin films, fibres, sheets, and moulds [12].

In view of the above, in this chapter, the details of different processing techniques are discussed, corroborating recent works carried out by other research groups in biodegradable polymer processing. It is divided into three parts: in the first part, one of the conventional polymer processing techniques, i.e., solution processing, is discussed comprehensively, including different solution processing techniques. In the second part, melt processing techniques are discussed, including various polymer melt processing techniques with their respective description. In the third part, recent updated techniques are discussed, such as in situ polymerization, master batch dilution technique, and 3D printing technique, either an improved version of a conventional technique or a combination of both conventional techniques. Finally, challenges associated with the processing of biodegradable polymeric composites are discussed and future prospects of the processing technologies are also highlighted. In the end, an overall summary of the chapter is appended.

3.2 PROCESSING TECHNIQUES

Processing of composites of biodegradable polymers is associated with the dispersion of fillers into the polymer matrix. Filler materials can have distinct morphology, surface nature, and compatibility with the polymeric matrix. In most of the laboratory-scale production, solution processing techniques are utilized to develop initial information of a particular composite system. In case of commercial production, the polymer technology that is mostly involved is the melt processing route. In some cases, other facile techniques are used to develop more useful, improved, and application-oriented polymeric products. Here, biodegradable processing techniques are broadly classified into (i) solution processing, (ii) melt processing, and (iii) updated processing techniques and are discussed in the following sections.

3.2.1 SOLUTION PROCESSING

This is the conventional polymer processing technique where a solvent system is used to solubilize polymer prior to product-oriented operations. In nanocomposite fabrication, a similar solvent is used for filler and the polymer matrix. Solution processing requires the selection of solvent that is stable within the processing temperature and is recoverable under elevated temperature or vacuum. Solution processing techniques are mostly used to understand the interaction and impact of properties in case of new matrix/filler combinations, less thermally stable polymers, and thin-film electronics fabrication. The presence of solvent adds one additional solvent recovery unit operation after product fabrication, which sometimes increases the production cost and environmental hazards. The most used solution processing techniques are described in the following subsections.

3.2.1.1 Solution Casting

This processing protocol is a fundamental and widely utilized solution processing technique. This technique consists of dissolution of the polymer matrix in a suitable solvent followed by mixing with desired filler, then casting into required shape and size and finally solvent recovery. This is used for laboratory-scale thin-film casting of biodegradable polymers for normal interaction study. In case of membrane casting, in most of the cases solution casting is used. In case of thin-film casting of conductive films made of a biodegradable polymer; solution casting is used. It is an easy-to-handle approach for fabricating composite films. It requires similar surface nature for matrix and filler for better mixing. In some cases, sonication and stirring (magnetic or mechanical) help in the better dispersion and less settling of the filler. Exfoliated graphene (GR) was incorporated into polylactic acid (PLA) matrix by taking chloroform as solvent by Valapa et al. [13].

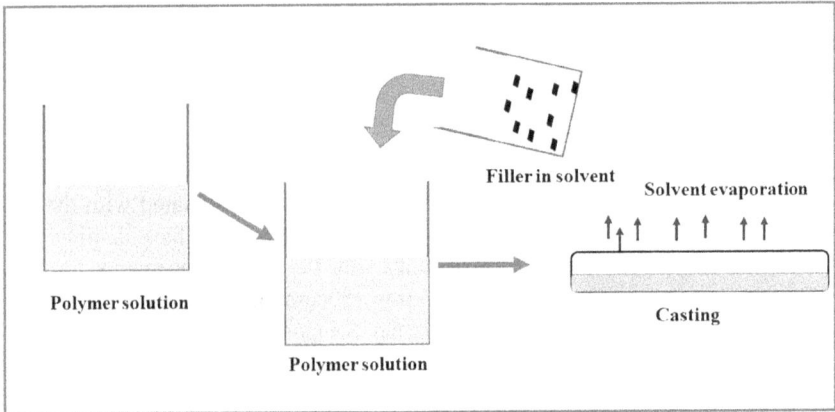

FIGURE 3.1 Solution casting of biodegradable polymeric composites.

Sonication was utilized, and different sonication times were used to optimize the dispersion. It was observed that the incorporation of GR improved the thermal stability of the PLA/GR composites. Barrier property also enhanced due to the incorporation of filler through the solution casting technique. Similarly, for the processing of cellulose nanocrystal, clays with biodegradable polymers solution casting were used. Conductive thin films from PLA by incorporating GR and magnetic CNC for sensor application were fabricated by Chakraborty et al. [14] using solution casting. The film turned out to be a potential biomaterial for sensor application (Figure 3.1).

Porous composite development through solvent casting followed by leaching of the porogen is called casting–leaching technique. This is also one application of solvent casting technique where porogens such as salt and sugar are incorporated into polymer composite before casting. Finally, using solvents in which porogen easily dissolves, fine pores are generated in the polymeric material. Fabrication of porous scaffolds for cell culture is one of the application areas of such scaffolds. Borokotoky et al. [15] fabricated a CNC-based PLA nanocomposite porous scaffold using casting–leaching technique using sucrose as a porogen and 1,4-dioxane as a casting solvent. Through morphological and density investigations, a highly porous foam-like structure was observed.

3.2.1.2 Electrospinning

This process is used to prepare polymer fibre from solution by the applying electric force. An Electrospinning setup is an assembly of (i) high-voltage DC electric supply, (ii) syringe pump, (iii) spinneret (needle), and (iv) collector. The biodegradable polymer dissolution procedure and mixing with respective filler are similar to solvent casting. After complete mixing of polymer composite, the fluid is then taken into the syringe tube for fibre spinning. In the electrospinning method, the electric force is applied between the needle and the collector electrode. Depending on the magnitude of the electric field, due to the repulsive electrostatic force, the charged jet of the polymer solution is ejected from the tip of the Taylor cone. Fibre diameter and length depend on the voltage difference, fluid viscosity, surface tension, filler morphology, filler loading, needle tip diameter and distance between the needle, and collector. Uniformity of the fibre and type of fibre collection depends on the nature of application and electrospinning setup. Different electrospinning techniques based on setup are as follows: (i) multi-jet electrospinning methods, (ii) multi-needle electrospinning methods and (iii) needleless electrospinning methods. Depending on the nature of the collection, (i) single fibre, (ii) nanofibre bag, and (iii) core–shell-type fibres are fabricated by different research groups. Electrospun fibres are mostly used in biomedical applications, sensor applications, optical applications, and catalysis [16]. Electrospinning of biodegradable polymers and composites has also been widely carried out in various dimensions. Qasim et al. [17] showed that chitosan-based electrospun fibres are widely applicable in tissue engineering and regenerative medicine field. Dobrovolskaya et al. [18] used a 0.3-mm electrode and 18 kV voltage difference to prepare a chitosan-based composite solution in water. With 20 wt.% loading of PEO, well-structured fibres were formed, which have not appeared in low loading. In case of thermoplastic biodegradable polymers such as polylactic acid, Huang and Thomas [19] fabricated electrospun fibres using different solvents such as chloroform, acetone, ethanol, and dimethyl sulphoxide (DMSO). Fibre uniformity and porosity were observed to be dependent on the solvent combination. The fibre mat was successfully applied for oil separation. Electrospun fibres are also used in active packaging applications. Altan et al. [20] fabricated PLA-based composite nanofibres loaded with zein and carvacrol through electrospinning technique.

3.2.1.3 Dry Spinning

In case of polymers in which heating is not possible due to the degradation of the polymer chain, dry spinning is carried out for polymer fibre fabrication. This technique is also based on a solvent system where polymer along with fillers is taken into solvent and, after passing through spinneret, dried using hot air. Dry spinning involves the dissolution of polymer into the solvent followed by loading into the spinneret unit. Then the fibres are drawn through the spinneret into a hot-air chamber and immediately wound onto bobbins. It is suitable for heat-sensitive applications; however, flammability and additional operation cost of solvent recovery are the associated disadvantages of this process. Clarkson and Youngblood [21] had synthesized CNC/PLA fibres using the dry spinning technique. Multiple temperature profiles were maintained during the spinning. DMF was used as a solvent, and solvent recovery was carried out at 70°C in a vacuum.

3.2.1.4 Wet Spinning

This is one of the oldest techniques for fibre drawing using solvent system. In this method, polymer is dissolved in a non-volatile solvent and then the solvent is transformed into fibre through spinneret either by simple washing or with the aid of chemical reaction. Then using coagulation, the bath solvent is removed. At last, filament yarn is either immediately wound onto bobbins or treated for obtaining desired characteristics for end-use, such as colouring, thermal treatment, or surface treatment. Through wet spinning technique, large tows can be handled. However, slow production rate, impurity, and solvent recovery are the drawbacks of this process. In case of the textile industry, primarily this process is used for fibre drawing. PLA-based fibres were fabricated through wet spinning by Giełdowska et al. [22], and hydrolytic degradation analysis was carried out for the fibres. Cellulose-based wet spun fibres were prepared by Liu et al. [23]. In this case, 5 wt.% alginate solution was used as a base, and different loadings of CNC were incorporated and, in coagulation, a bath of 5 wt.% $CaCl_2$ solutions was used. The draw ratio was maintained at 1.2, and washing was done with air and alcohol before air drying. Dry jet wet spinning is one of the variants of wet spinning, where the spinning solution passes through an air gap before being submerged into the coagulation bath. This method is used in lyocell spinning of dissolved cellulose and can lead to higher polymer orientation due to the higher stretchability of the spinning solution against the precipitated fibre.

3.2.1.5   Gel Spinning

This process involves the extrusion of polymer gel through a spinneret, and then either by dry or wet technique, this is transformed into fibres. In case of gel-spun fibres, tensile strength becomes higher. It is a medium-speed process and is suitable for liquid crystalline polymers. However, the possibility of hazard generation and the requirement of an additional solvent recovery unit are the drawbacks associated with gel spinning process. In some of the biodegradable polymer-based composite fabrication methods, this technique is used. Marquez-Bravo et al. [24] fabricated bio-nanocomposites of chitosan filled with cellulose nanofibre (CNF) through the gel spinning technique. It was observed that at 0.4 wt.% loading of CNF, Young's modulus increased up to 8 GPa.

3.2.1.6   Coating

To prevent the decay of food items such as fruits and vegetables, coating using biodegradable composites is carried out. This coating improves the shelf life of the food item and preserves it for a long time. Either spray coating or dip coating is carried out over the food items. Chitosan and chitosan-based composites are mostly used as coating materials. In modern times, active coating materials are used, which possess antimicrobial activity and thus create resistance to microbial attack [25]. Polysaccharides such as chitosan, cellulose, and pectin are widely used as edible coating materials. In some cases, essential oils are used as a coating material for food items [26,27]. In some applications, proteins are also used as a coating material [28].

3.2.1.7   Freeze-Drying

This is a low-temperature dehydration process and is also termed as lyophilization. Using freeze-drying, water or other solvents are removed from the composite.

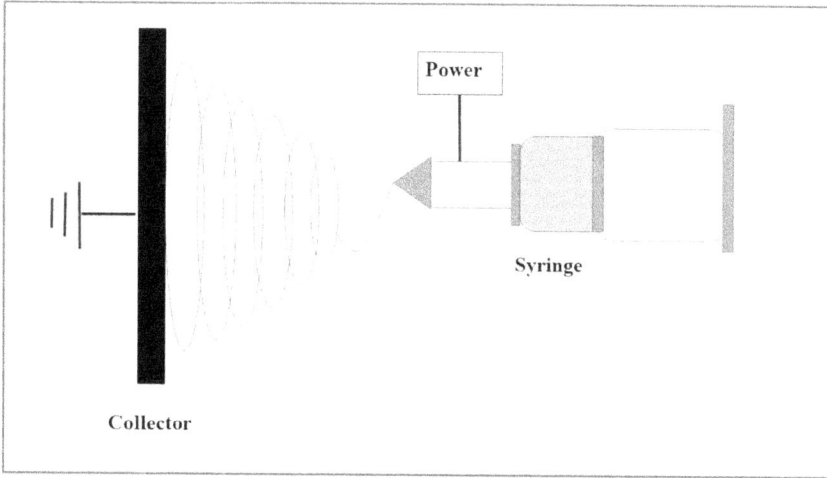

FIGURE 3.2   Electrospinning of biodegradable polymeric composites.

The lyophilization process is applicable for producing heat-sensitive materials. This procedure is used for the fabrication of scaffolds for biomedical applications. Freeze-drying process involves freezing (−80°C) the sample containing water or solvent, and then the sample is placed in a high vacuum to dry the material. In some cases, secondary heating is also done to increase the temperature above 0°C. Biodegradable polymer-based scaffolds are made through this technique in some cases [29] (Figure 3.2).

3.2.2 MELT PROCESSING

Melt processing techniques turn a liquid melt into a solid object with a definite shape and structure. Liquid melts have a wide range of physical and chemical properties, just like their solid counterparts. Some melts crystallize, while others go through a glass transition as they cool. The granule features, the material's hygroscopic behaviour (whether it absorbs water), the flow properties, the thermal properties (such as thermal stability and heat transmission), shrinkage, crystallization behaviour, and molecule orientation are all considerable issues [30,31] (Figure 3.3).

3.2.2.1 Hygroscopic Properties

When a polymer complex contains water or other low-boiling point ingredient, the heat necessary for processing can raise the temperature above the boiling point. When the pressure drops, visible bubbles form within the thermoplastic material, similar to when it emerges from an extrusion die. When processing temperatures are greater, the amount of water that may be tolerated is considered to be lower. This is because greater temperatures produce more steam from the same amount of water. Typically, commodity thermoplastics do not suffer from water-related issues to the same extent as the engineering thermoplastics.

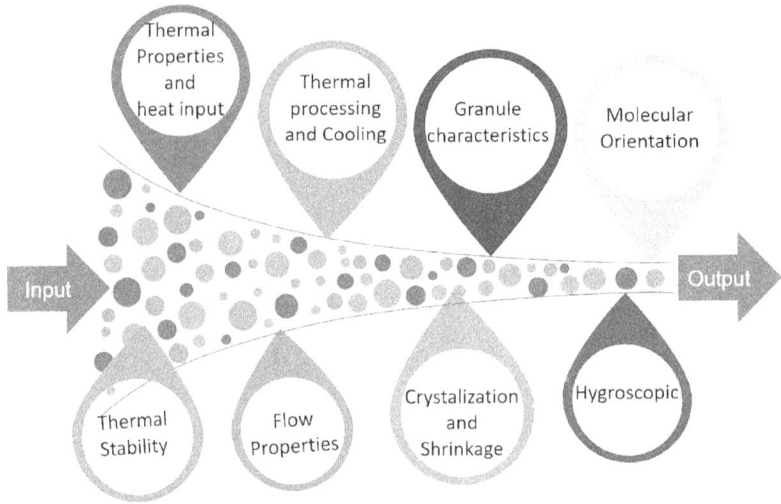

FIGURE 3.3   Factors affecting the melt extrusion process.

3.2.2.2   Granule Characteristics

The Material in granular form is widely used as the feed in processes including extrusion moulding, injection moulding and blow moulding. If the item is provided in more than one feed form, feeding complications will develop. Spherical granules (about 3 mm (0.125″) in diameter) are the most efficient feeding efficiency, while tiny powder is the least efficient. Regranulated material can be nearly as hazardous as ungranulated material since it can have a wide range of particle sizes. Although cube-cut granules are preferable, lace-cut granules are superior because they are made by cutting strands with a circular cross-section. The machines must be fed with a consistent raw material blend due to the feeding variances of various granulates [32]. This is especially true for master batch blends.

3.2.2.3   Melt Mixing and Roll Milling

Hot spin mixing, freeze-drying, and roll mixing or co-milling are the most important technologies for the production of solid dispersions. An indicator is that melt extrusion technology's life cycle management is a crucial driver. Roll mills are used to combine the composite materials. The nip gap, roll rotation speed, friction ratio between the roller and composite materials, and sustained temperature throughout the process are variables for roll mills. The majority of roll mills, in general, have a set speed and friction ratio. In general, the distance between the roller and the temperature is altered. The melt mixing-cum-rolling milling technique allows for the precise control of the raw polymer's residence duration and temperature. The polymer composites are mixed continuously during the procedure. The twin-screw mixer provides several advantages, including the ability to change process parameters because it includes a programmable mode [33]. There is no risk of polymer flow stagnation, and residence duration and melt temperature may all be controlled. Some of the disadvantages include its average pressure generation capability and the size of the polymer chunks that can be fed through the extruder being limited by the feed port opening.

3.2.2.4 Extrusion

Extrusion is a process that involves forcing molten polymer through a die to create components with a set cross-sectional area, such as tubes and rods. Extrusion is the technique of forcing softened polymer through a die with an opening to produce long products having a consistent cross-section (rods, sheets, pipes, films, and wire insulation coating). Through a hopper, the polymer material in the form of pellets is fed into an extruder. The material is then driven into a die by a feeding screw, turning it into a continuous polymer product. The polymer is softened and melted by heating sources positioned above the barrel. The material's temperature is controlled by thermocouples. Blown air or a water bath is used to cool the product as it exits the die. Polymer extrusion (unlike metal extrusion) is a continuous operation that lasts as long as feedstock pellets are available [34,35]. Extrusion is primarily used for thermoplastics, but it can also be utilized for elastomers and thermosets. Cross-linking occurs as the material is heated and melted in the extruder.

3.2.2.5 Moulding

Plastic moulding is the process of pouring liquid plastic into a specific container or mould and allowing it to harden into the desired shape. Extrusion moulding, compression moulding, blow moulding, injection moulding, and rotational moulding are the five types [36].

3.2.2.6 Thermoforming

Thermoforming is a two-step procedure for shaping flat thermoplastic sheets: first, softening the sheet with heat and forming it in the mould cavity. Because of their cross-linked structure, elastomers and thermosets cannot be created using thermoforming processes because they do not soften when heated.

Thermoplastics which may be processed by the thermoforming method are as follows:

Polypropylene (PP), polystyrene (PS), polyvinyl chloride (PVC), low-density polyethylene (LDPE), high-density polyethylene (HDPE), cellulose acetate, poly(methyl methacrylate) (PMMA) and acrylonitrile butadiene styrene (ABS).

In the food packaging sector, thermoforming is commonly used to make ice cream and margarine tubs, meat trays, microwave containers, snack tubs, sandwich packs, among other things. Thermoforming is also used to make small tools, fasteners, toys, boat hulls, blister and skin packs, and pharmaceutical and electrical items.

There are three thermoforming methods, differing in the technique used for the forming stage:
- Vacuum thermoforming
- Pressure thermoforming
- Mechanical thermoforming.

A vacuum is created in the mould cavity space, which is used to shape a preheated thermoplastic sheet. The soft sheet is forced to deform in accordance with the hollow shape by ambient pressure. The plastic cools and hardens when it comes in contact with the mould surface.

3.2.2.7 Hot Isostatic Pressing

Internal microporosity in metal castings and other materials is eliminated using a manufacturing process known as hot isostatic pressing (HIP). Metal, polymer, ceramic, and composite powders can also be densified in the solid-state using HIP. Both of these approaches produce superior material characteristics.

3.2.2.7.1 Benefits
- Removes all internal voids caused by additive manufacturing technologies in castings and metals components.
- Reduces the number of castings that are rejected during inspection.
- Improves the uniformity of the product.
- Improves the soundness and mechanical qualities of castings (fatigue life, ductility and impact strength), potentially allowing for a more streamlined design.
- Improves the vacuum-tightness and surface polish of castings.

It creates high-density materials from metal, composite, polymer, or ceramic particles without melting. Due to tiny, homogeneous grain size and isotropic structure, solid materials with exceptional characteristics are created from powders. It allows for creating solids from unusual powder mixtures that would otherwise be impossible to create using normal manufacturing methods. Powders can be used to make complex-shaped solid components. Metal injection moulded (MIM) parts' toughness, ductility, fatigue strength and uniformity are all improved. Without the use of temperature-limiting adhesives, it is possible to bind incompatible metals. HIP bonding is used to create clad components.

3.2.2.7.2 Applications & Materials

Hot section and structural gas turbine components (both dynamic and static); aerospace structural and engine parts; implantable medical devices; automotive engine components; valve bodies and other petrochemical processing equipment; critical munition pieces; tooling, die, and general engineering parts; sputter targets; and PM alloy billets and near net shapes are just a few examples of parts hot-isostatic-pressed in large quantities. In a single process cycle, HIP can create several diffusion bonds. HIP cladding is a method of coating premium materials with better qualities, such as corrosion and wear resistance, onto more cost-effective substrates so that the part can be constructed efficiently. Alloys based on nickel, cobalt, tungsten, titanium, molybdenum, aluminium, copper and iron; oxide and nitride ceramics; glasses; intermetallics; and premium plastics can all be hot-isostatic-pressed [37,38].

3.2.2.7.3   Process Details
- Automated HIP cycles that are repeatable and adapted to the demands of consumers.
- Door-to-door component traceability.
- Purity criteria for inert gas (argon).
- Customer, military and/or industry criteria must be met or exceeded.
- HIP of new materials or alloys requires technical assistance.

Tooling that is compatible with the HIP cycle and customer material specifications.

3.2.2.8   Melt Drawing
In this process, hot film extrudate is drawn to the coder rollers. The shrinkage at edges will lead to the formation of neck, leading to beading. The shrinkage at edges depends upon polymer melt temperature and the polymer itself [39].

3.2.3   Updated Processing Techniques

In this section, the processes that are different from the conventional solution processing and melt processing are discussed. These processes are termed as 'updated processing techniques'.

3.2.3.1   In Situ Polymerization
This technique is utilized for composite synthesis where the surface nature of the filler is different from the polymer, but similar to a monomer. This process is bulk polymerization of monomer in the presence of filler. Generally, this is used for condensation polymers. Depending upon the presence of functional group, polymer chain is grafted into the surface of the filler and the surface property of the filler altered. In another way where no functional group is present, non-covalent wrapping of polymer chain over the filler surface takes place. Both these ways improve the compatibility of the filler for further processing through melt or solution processing. Li et al. [40] fabricated a graphene oxide (GO)/PLLA composite using *in situ* polymerization technique, where grafting of PLLA chain with GO was carried out prior to the solution processing of composite [40]. It was also noticed that compatibility, dispersion and strength of the composite increased due to this facile procedure. This technique is also utilized for the dispersion of biofillers inside hydrophobic polymers and carbon fillers inside hydrophilic polymers.

3.2.3.2   Master Batch Dilution Technique
Master batch dilution technique is a technique by which proper utilization of the filler is possible, and this makes the process more economical. This process

consists of two steps involving preparation of master batch and dilution to required loading in the final product. Master batch preparation techniques are in situ polymerization, covalent grafting through click chemistry, Π–Π stacking, etc. Final master batch dilution is carried out either by solution processing or by melt processing. It is termed as dilution technique as the filler loading during master batch synthesis is higher compared to the loading in the final product.

3.2.3.3 3D Printing

This technique is used to fabricate three-dimensional (3D) products from polymer nanocomposite. This process has become popular in recent days because of the application-specific design and manufacture. The Utilization of biodegradable polymers such as PLA for 3D printed product manufacturing reduces the solid waste disposal. 3D printed products such as toys and biomedical products, for example, artificial hearts and kidneys, are making the market day by day. This process consists of three parts: first is the design of the product using software, second is the making of printing filament using polymer, and third is printing by melting filament. PLA-based filaments are available on the market and already in use for product manufacture. Development of 3D printing filament and technology is going on, particularly for biodegradable polymers and composites.

3.3 CHALLENGES IN BIODEGRADABLE COMPOSITES PROCESSING

The solid waste disposal problem has led the entire polymer technology towards biodegradable polymers. This increasing demand of biodegradable polymers for product manufacturing also opens up the requirement of application-specific technologies and materials. Biodegradable polymers have some inherent limitations, which can be addressed by reinforcing with different fillers and additives based on the requirement. The following issues appear during the processing of biodegradable polymers along with fillers. Control of the microbial attack is also a problem associated with these composites.

- In case of particulate materials such as silicate and clay, lack of compatibility phase separation and agglomeration appear.
- In case of bio-nanocomposite fabrication, high shear and less thermal stability of filler degradation take place.
- Biodegradable polymers degrade during melt processing, and thus, recycling of these polymers is always associated with loss in property.
- Sometimes, for better dispersion, functionalization of filler is carried out prior to covalent grafting of polymer. This reduces the effectivity of the fillers such as carbon allotropes towards conductive applications.
- Machinability of biodegradable polymeric materials is less, and due to low melt strength sometimes it becomes difficult to keep intact the same product structure after cooling.

- Casting defects are also a problem of biodegradable composite-based products, such as film shrinking, cracks, breakage of fibre, brittleness and non-uniform surface.

3.4 FUTURE PROSPECTS OF BIODEGRADABLE COMPOSITE PROCESSING

Biodegradable polymers are mostly versatile and processable through conventional techniques such as solvent processing and melt processing. Biodegradable composites also can be used for application-oriented product casting. Nowadays, problems associated with the flow property are addressed by using different kinds of chain modifiers and plasticizers. The compatibility issue is addressed by emerging techniques related to the filler type and nature of the matrix. The weight degradation problem associated with the melt processing of degradable polymers is fixed by introducing the reactive extrusion technique, which helps to keep the mechanical strength between each processing cycle. However, more real field property evaluation for the materials are required for biodegradable composites before application. More studies on the degradation and composite management after use also need to be studied. Some of the future applications where the possibility of the application of such composites lies are as follows:

- Biodegradable polymer-based membrane for fuel cell and separation.
- Thin-film conductive composite for sensor application.
- Supercapacitor based on biodegradable polymeric electrolytes.
- Biodegradable polymer-based packaging for short-term application.
- Biodegradable polymeric material for biomedical applications such as orthopaedics, tissue engineering, and suture.
- Bio-based adhesives.
- Biodegradable polymer-based compost and manure.
- 3D printing of organs using biocompatible nanocomposite made up of biodegradable polymer.

3.5 SUMMARY

This current chapter describes different processing techniques for biodegradable polymeric composites. Processing depends on the type of the polymer, filler, and application. In some cases, application-specific processing is required. Solution processing considers solvent selection, polymer-solvent interaction, and solvent-filler interaction. In some instances, sonication and stirring help in the dispersion of solution-processed composites. Melt processing considers the thermal stability and melt flowability of the polymer composite material. Because of this, all the processing techniques are classified into three parts.

- In part one, different solution processing techniques are discussed extensively, including casting, electrospinning, spinning, coating, and freeze-drying. In respective subclasses, all processes are briefly discussed.
- In part two, different melt processing techniques are described with possible drawbacks. The basic process of melt mixing, extrusion, moulding, thermoforming and melt drawing is described in this section.
- In part three, some updated processing techniques are described, corroborating biodegradable polymeric composites.
- Finally, different challenges associated with biodegradable polymeric composites are highlighted, and the possible future of such materials is also discussed in briefly.

REFERENCES

1. Reddy, M. M., Vivekanandhan, S., Misra, M., Bhatia, S. K., & Mohanty, A. K. (2013). Biobased plastics and bionanocomposites: Current status and future opportunities. *Progress in Polymer Science*, 38(10–11), 1653–1689.
2. Rhim, J. W., Park, H. M., & Ha, C. S. (2013). Bio-nanocomposites for food packaging applications. *Progress in Polymer Science*, 38(10–11), 1629–1652.
3. Ojijo, V., & Ray, S. S. (2013). Processing strategies in bionanocomposites. *Progress in Polymer Science,* 38(10–11), 1543–1589.
4. Lim, L. T., Auras, R., & Rubino, M. (2008). Processing technologies for poly (lactic acid). *Progress in Polymer Science*, 33(8), 820–852.
5. Chieng, B. W., Ibrahim, N. A., Yunus, W. M. Z. W., Hussein, M. Z., & Loo, Y. Y. (2014). Effect of graphene nanoplatelets as nanofiller in plasticized poly (lactic acid) nanocomposites. *Journal of Thermal Analysis and Calorimetry*, 118(3), 1551–1559.
6. Dhar, P., Tarafder, D., Kumar, A., & Katiyar, V. (2016). Thermally recyclable polylactic acid/cellulose nanocrystal films through reactive extrusion process. *Polymer*, 87, 268–282.
7. Qi, H., Mäder, E., & Liu, J. (2013). Unique water sensors based on carbon nanotube–cellulose composites. *Sensors and Actuators B: Chemical*, 185, 225–230.
8. Babu, R. P., O'connor, K., & Seeram, R. (2013). Current progress on bio-based polymers and their future trends. *Progress in Biomaterials*, 2(1), 1–16.
9. Liu, L., Zachariah, M. R., Stoliarov, S. I., & Li, J. (2015). Enhanced thermal decomposition kinetics of poly (lactic acid) sacrificial polymer catalyzed by metal oxide nanoparticles. *RSC Advances*, 5(123), 101745–101750.
10. Yoon, O. J., Sohn, I. Y., Kim, D. J., & Lee, N. E. (2012). Enhancement of thermo-mechanical properties of poly (D, L-lactic-co-glycolic acid) and graphene oxide composite films for scaffolds. *Macromolecular Research*, 20(8), 789–794.
11. Wang, N., Zhang, X., Ma, X., & Fang, J. (2008). Influence of carbon black on the properties of plasticized poly (lactic acid) composites. *Polymer Degradation and Stability*, 93(6), 1044–1052.
12. Mezghani, K., & Spruiell, J. E. (1998). High speed melt spinning of poly (L-lactic acid) filaments. *Journal of Polymer Science Part B: Polymer Physics*, 36(6), 1005–1012.
13. Valapa, R. B., Pugazhenthi, G., & Katiyar, V. (2015). Effect of graphene content on the properties of poly (lactic acid) nanocomposites. *RSC Advances*, 5(36), 28410–28423.

14. Chakraborty, G., Dhar, P., Katiyar, V., & Pugazhenthi, G. (2020). Applicability of Fe-CNC/GR/PLA composite as potential sensor for biomolecules. *Journal of Materials Science. Materials in Electronics*, 31(8), 5984–5999.
15. Borkotoky, S. S., Dhar, P., & Katiyar, V. (2018). Biodegradable poly (lactic acid)/ Cellulose nanocrystals (CNCs) composite microcellular foam: Effect of nanofillers on foam cellular morphology, thermal and wettability behavior. *International Journal of Biological Macromolecules*, 106, 433–446.
16. Xue, J., Wu, T., Dai, Y., & Xia, Y. (2019). Electrospinning and electrospun nanofibers: Methods, materials, and applications. *Chemical Reviews*, 119(8), 5298–5415.
17. Qasim, S. B., Zafar, M. S., Najeeb, S., Khurshid, Z., Shah, A. H., Husain, S., & Rehman, I. U. (2018). Electrospinning of chitosan-based solutions for tissue engineering and regenerative medicine. *International Journal of Molecular Sciences*, 19(2), 407.
18. Dobrovolskaya, I. P., Yudin, V. E., Popryadukhin, P. V., Ivan'kova, E. M., Shabunin, A. S., Kasatkin, I. A., & Morgantie, P. (2018). Effect of chitin nanofibrils on electrospinning of chitosan-based composite nanofibers. *Carbohydrate Polymers*, 194, 260–266.
19. Huang, C., & Thomas, N. L. (2018). Fabricating porous poly (lactic acid) fibres via electrospinning. *European Polymer Journal*, 99, 464–476.
20. Altan, A., Aytac, Z., & Uyar, T. (2018). Carvacrol loaded electrospun fibrous films from zein and poly (lactic acid) for active food packaging. *Food Hydrocolloids*, 81, 48–59.
21. Clarkson, C. M., & Youngblood, J. P. (2018). Dry-spinning of cellulose nanocrystal/ polylactic acid composite fibers. *Green Materials*, 6(1), 6–14.
22. Giełdowska, M., Puchalski, M., Szparaga, G., & Krucińska, I. (2020). Investigation of the influence of PLA molecular and supramolecular structure on the kinetics of thermal-supported hydrolytic degradation of wet spinning fibres. *Materials*, 13(9), 2111.
23. Liu, X., Lu, X., Wang, Z., Yang, X., Dai, G., Yin, J., & Huang, Y. (2021). Effect of bore fluid composition on poly (lactic-co-glycolic acid) hollow fiber membranes fabricated by dry-jet wet spinning. *Journal of Membrane Science*, 640, 119784.
24. Marquez-Bravo, S., Doench, I., Molina, P., Bentley, F. E., Tamo, A. K., Passieux, R.,... & Osorio-Madrazo, A. (2021). Functional bionanocomposite fibers of chitosan filled with cellulose nanofibers obtained by gel spinning. *Polymers*, 13(10), 1563.
25. Kumar, S., Ye, F., Dobretsov, S., & Dutta, J. (2019). Chitosan nanocomposite coatings for food, paints, and water treatment applications. *Applied Sciences*, 9(12), 2409.
26. Galus, S., Arik Kibar, E. A., Gniewosz, M., & Kraśniewska, K. (2020). Novel materials in the preparation of edible films and coatings—A review. *Coatings*, 10(7), 674.
27. Anis, A., Pal, K., & Al-Zahrani, S. M. (2021). Essential oil-containing polysaccharide-based edible films and coatings for food security applications. *Polymers*, 13(4), 575.
28. Mihalca, V., Kerezsi, A. D., Weber, A., Gruber-Traub, C., Schmucker, J., Vodnar, D. C.,... & Pop, O. L. (2021). Protein-based films and coatings for food industry applications. *Polymers*, 13(5), 769.
29. Sayed, M., Mahmoud, E. M. M., Bondioli, F., & Naga, S. M. (2019). Developing porous diopside/hydroxyapatite bio-composite scaffolds via a combination of freeze-drying and coating process. *Ceramics International*, 45(7), 9025–9031.

30. Kovarova, L., Kalendová, A., Simonik, J., Malac, J., Weiss, Z., & Gerard, J. F. (2004). Effect of melt processing conditions on mechanical properties of poly-vinylchloride/organoclay nanocomposites. *Plastics, Rubber and Composites*, 33(7), 287–294.
31. Chavarria, F., Shah, R. K., Hunter, D. L., & Paul, D. R. (2007). Effect of melt processing conditions on the morphology and properties of nylon 6 nanocomposites. *Polymer Engineering & Science*, 47(11), 1847–1864.
32. Rusanowska, P., Cydzik-Kwiatkowska, A., Świątczak, P., & Wojnowska-Baryła, I. (2019). Changes in extracellular polymeric substances (EPS) content and composition in aerobic granule size-fractions during reactor cycles at different organic loads. *Bioresource Technology*, 272, 188–193.
33. El Achaby, M., Arrakhiz, F. E., Vaudreuil, S., el Kacem Qaiss, A., Bousmina, M., & Fassi-Fehri, O. (2012). Mechanical, thermal, and rheological properties of graphene-based polypropylene nanocomposites prepared by melt mixing. *Polymer Composites*, 33(5), 733–744.
34. Tzoganakis, C. (1989). Reactive extrusion of polymers: A review. *Advances in Polymer Technology: Journal of the Polymer Processing Institute*, 9(4), 321–330.
35. Hyvärinen, M., Jabeen, R., & Kärki, T. (2020). The modelling of extrusion processes for polymers—A review. *Polymers*, 12(6), 1306.
36. Giboz, J., Copponnex, T., & Mélé, P. (2007). Microinjection molding of thermoplastic polymers: A review. *Journal of Micromechanics and Microengineering*, 17(6), R96.
37. Gul, R. M., & McGarry, F. J. (2004). Processing of ultra-high molecular weight polyethylene by hot isostatic pressing, and the effect of processing parameters on its microstructure. *Polymer Engineering & Science*, 44(10), 1848–1857.
38. van de Werken, N., Koirala, P., Ghorbani, J., Doyle, D., & Tehrani, M. (2021). Investigating the hot isostatic pressing of an additively manufactured continuous carbon fiber reinforced PEEK composite. *Additive Manufacturing*, 37, 101634.
39. Coates, P. D., & Ward, I. M. (1981). Die drawing: Solid phase drawing of polymers through a converging die. *Polymer Engineering & Science*, 21(10), 612–618.
40. Li, W., Xu, Z., Chen, L., Shan, M., Tian, X., Yang, C.,...& Qian, X. (2014). A facile method to produce graphene oxide-g-poly (L-lactic acid) as an promising reinforcement for PLLA nanocomposites. *Chemical Engineering Journal*, 237, 291–299.

4 Challenges and Perspectives of Biodegradable Composites

Shasanka Sekhar Borkotoky
Assam Engineering College

CONTENTS

4.1 INTRODUCTION

Recent days, biodegradable polymeric composites have been gaining importance due to the growing environmental concerns. From the last decade, the polymeric materials have been used in almost every aspect of daily life. However, the present market is mainly dominated by the petro-based, non-degradable plastic materials. The ultimate disposal is a major concern for the non-degradable plastics. Most of the current market-dominating polymers such as polystyrene (PS), polypropylene (PP), polyethylene terephthalate (PET) and polyethylene (PE) undergo very slow degradation and require many centuries to degrade completely in the environment. Some measures have recently been taken for the reuse and recycling of plastics for better utilization. Another important alternative way to deal with this problem is the development of bio-based and biodegradable polymers and introduce them to the current market as a replacement to the current petro-based

DOI: 10.1201/9781003227908-4

polymers [1]. The major advantages of using these polymers are their environmental friendliness and bio-compatibility. Along with the advantages, the biodegradable polymers have some limitations to compete with the current non-degradable polymers dominating the market today. These limitations are mainly the lack of mechanical and thermal properties to compete with petro-based polymers. Recent researches on biodegradable polymers have mainly focused on the improvement of these properties and on the establishment of them as a greener alternative to the conventional petro-based polymers [2]. Recent researchers have identified some promising bio-based and biodegradable polymers that can be an alternative to the conventional polymers present in the market. Biodegradable polymers such as polylactic acid (PLA), polycaprolactone (PCL) and polyhydroxyalkanoates (PHAs) have the potential to penetrate the market in a greater way in different applications. The improvement in the properties of biodegradable polymers can be achieved by introducing different nanomaterials such as clay [3] and carbon nanotubes [4]. Recently, in a greener approach, various bio-based and bio-derived nanomaterials such as cellulose-derived cellulose nanocrystals (CNCs) [5], gum arabic [6], silk nanoparticles [7] and chitosan-based nanomaterials [8] have been synthesized effectively, and improvements in different properties have been reported. Biodegradable composites can be a better replacement for the benefit of the environment. One of the challenges associated with the biodegradable composites is to retain their effective properties for the duration of the useful period before disposal into the environment for degradation. It is often noticed that the term "biodegradable" is mistakenly used by various organizations in different consumer products. A biodegradable composite must undergo biodegradation under certain environmental conditions within a specified time limit [9]. Standard protocols and test methods such as ASTM D6400, ASTM D5338 and ASTM D5929 are available for the determination of biodegradability of a composite. The degradability of plastics under controlled environment conditions of water and soil has been reported by many researchers. The biodegradability of composites also includes the degradability of the added materials. One of the most promising biodegradable polymers for composite is PLA. The mechanical properties of PLA are almost comparable with some of the conventional polymers used in the day-to-day life. PLA composites have a greater prospect in the near future, and many interesting results are published [10]. It is also observed that various new plant-based and agro-waste sources for bio-based nanoparticles are utilized for producing biodegradable composites. Biodegradable foam composites have also been of focus in recent researches. Foamed biodegradable composites have some added advantages such as lightweight and less usage of materials [11]. PLA composite foams are gaining importance due to their improved and comparable properties with different non-degradable foams. Scaffolds are generally utilized in various fields of applications, such as biomedicine, tissue culture and some commodity applications. Some of the other useful porous biodegradable composites include PCL [12], starch [13] and poly(3-hydroxybutyrate-co-3-hydroxyvalerate) (PHBV) [14]. Different approaches are taken for the fabrication of biodegradable foams and will be discussed here in this chapter. In this chapter, we are going to discuss various biodegradable composites, sources of nanomaterials, various aspects and challenges of these composites.

4.2 OVERVIEW OF BIODEGRADABLE POLYMERS

The growing industrialization and civilization also have some bad effects on the society. For a prosperous society, we need to develop basic structures and needs. In the recent past, it was observed that the growing use of plastics in the society leads to the generation of wastes and the problem of ultimate disposal. Since petroleum-based conventional polymers need centuries for their complete degradation, researches are focused on the solution to this problem. The growing technical development in human society also increases the e-wastes disposal, majority of which are manufactured by conventional polymers. To overcome the problem, scientists are focused on finding a greener way towards sustainability. Plant-based and biocompatible polymers are introduced into the market, which are biodegradable. Biodegradable polymers can also be made from petroleum-based sources [15]. However, more focus is given to the renewable sources for the production of biopolymers. Biodegradable polymers came to existence in the market in 1980s, when they were introduced in the market to compete with conventional plastics. Since then, researches are going on for the technical development along with improvement in different properties for various applications. The degradation process of a biodegradable polymer is associated with enzymatic or chemical deterioration by living organisms under specified conditions. Generally, two steps are involved in the process of degradation. First, the polymer is fragmented to lower molecular weight species with the action of hydrolysis, photodegradation or the attack of microorganisms. The second and final step involves the bio-assimilation of the fragments followed by mineralization (Figure 4.1). They mainly disintegrated to produce carbon dioxide (CO_2), methane, water and small fraction of other products. Different degradation studies such as hydrolytic, enzymatic, photodegradation, thermal degradation and biodegradation are also going on for biodegradable polymers such as PLA and PCL [16]. Biodegradable polymers can broadly be classified into two groups: (i) natural biologically derived biopolymers and (ii) synthetic biodegradable polymers. Table 4.1 indicates the classification of biopolymers.

Natural biologically derived biopolymers are abundant in nature, and these can be synthesized from renewable sources and from living organisms. On the other hand, synthetic biodegradable polymers are derived from non-renewable sources such as petroleum. However, there is no clear boundary that can separate these two classes of biopolymers as some biopolymers can be derived from both renewable and non-renewable resources. PLA is a chemically synthesized biopolymer and can be derived from agro-based sources. PLA can be synthesized from renewable sources such as corn and sugar beets by condensation polymerization of D- and L-lactic acids along with other methods such as ring opening polymerization of lactide [2]. PLA has excellent mechanical properties and a high melting point comparable to conventional plastics. It is extensively used

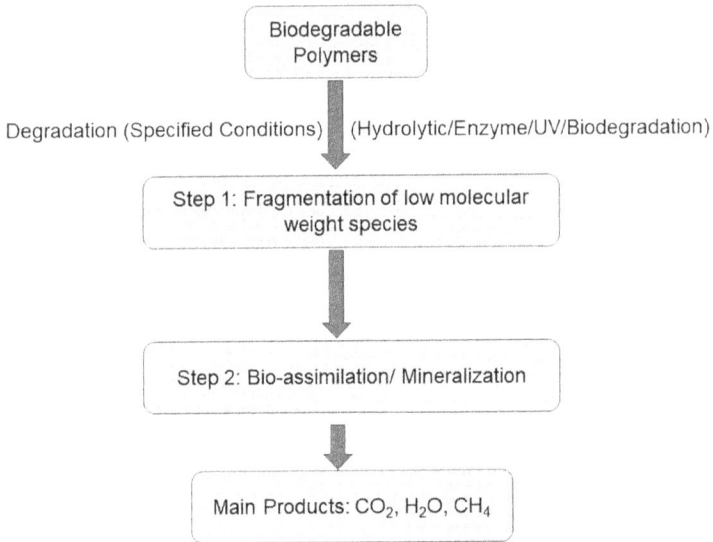

FIGURE 4.1 Schematic representation of the biodegradation process.

TABLE 4.1
Classification of Biodegradable Polymers with Examples

Biodegradable Polymers	
Natural Biopolymers	**Synthetic Biopolymers**
Polymers through fermentation route: poly(3-hydroxybutyrate) and poly(3-hydroxybutyrate-co-3-hydroxyvalerate) (PHBV)	Polycaprolactone (PCL)
	Polylactic acid (PLA)
Polymers from chemical modification of natural products: starch, chitin/chitosan and cellulose-based biopolymers	Polybutylene succinate (PBS)
	Poly(glycolic acid) (PGA)
	Poly(vinyl alcohol) (PVA)

in biomedical applications and biocomposites for different types of applications. PLA is utilized in different fields such as packaging, extrusion coating and commodity products.

Another important biodegradable polymer is polycaprolactone (PCL). PCL is basically a polyester derived from ring opening polymerization (ROP) of cyclic monomer ε-caprolactone. PCL is synthesized from petroleum-based sources. PCL scaffolds have been used in fields such as tissue engineering, drug delivery, biodegradable packaging applications and consumer electronics. Biocomposites of PCL with different biofillers have been reported in recent investigations [17]. PCL and its composites have recently been projected as eco-sustainable materials for different types of applications.

Cellulose is the most abundant biopolymer source in nature. Cellulose-derived monomers are utilized for the production of sustainable biopolymers. It is a bio-based and bio-derived polymer. The synthesis routes of cellulose-based bio-polymers can be categorized as follows [18].

a. Cellulose deconstruction monomer-derived biopolymers
b. Natural cellulose fibres and cellulose-derived biopolymers
c. Nanocellulose-derived biopolymers.

The synthesis route includes the pre-treatment to separate lignocellulose from lignin present in cellulose. Cellulose has a high potential to replace the petro-leum-based polymers with low production cost and availability. Cellulose-based biopolymers are widely used in solar cells, fuel cells, biosensors, batteries, and so on.

One of the oldest biopolymers used in various fields is starch and starch-based biodegradable polymers. Starch is the second largest abundant source of biomass in nature. Starch and its derivatives are the bio-based, bio-derived and biodegrad-able polymer obtained from greener agro-based sources [19]. Starch biopolymers can be synthesized at comparatively low costs. However, they have some limita-tions such as poor mechanical strength and solubility in water. Due to the limita-tions, starch-based researches are going on to technically enhance the properties by blending and fabricate biocomposites of starch-based biopolymers along with the introduction of chain extenders, plasticizers and biofillers in the matrix [20].

Another abundant source of biopolymer is chitin. Chitosan-/chitin-based bio-polymers are generally derived from sea-based sources [21]. Chitosan fibres are also utilized in biocomposites. Chitosan-based scaffolds are also utilized in some sophisticated applications such as tissue culture and biomedical applications. Chitin-based biopolymers have some limitations such as insolubility and intracta-bility. Researchers are focusing to improve different properties of chitosan-based biopolymers for different applications [22].

One of the important biodegradable polyesters, polybutylene succinate (PBS) can be synthesized by polycondensation reaction between succinic acid and butanediol [23]. The synthesis reactions are carried out in two steps: initially the esterification between the diacid and diol followed by polycondensation reactions at higher temperature to get PBS. PBS has some excellent properties such as ther-mal and chemical resistance and processability [24]. It is mainly used in packag-ing, films, commodity products, etc. [25].

Poly(glycolic acid) (PGA) is an interesting biopolymer which can be degraded faster compared to other biopolymers. PGA undergoes chemical hydrolysis in the absence of enzymes. PGA can be derived from different chemical routes. Generally, it can be synthesized by the polycondensation of glycolic acid. However, to achieve high molecular weight PGA, industrially ROP of glycolide is used. Due to its chemical structure, it has excellent thermal stability, very high gas barrier property and high mechanical properties. PGA is mainly used in the form of its copolymer poly(lactic-co-glycolic acid) (PLGA). PGA biocompos-ites and their blends with other biopolymers are utilized effectively for different types of required applications. PGA has the high potential to replace conventional polymers in high thermal and gas barrier applications in near future [26,27]. Poly(vinyl alcohol) (PVA) is another interesting water-soluble biopolymer generally used as a paper additive and in textile industries. PVA is synthesized from the polyvinyl acetate (PVAc) by hydrolysis [28]. Other important biopolymers include alginate [29] and agar [30]. The improvements in these biopolymers need to be carried out to make them comparable in properties with conventional polymers [31].

Recently, the focus is mainly given to the development of bio-based and bio-derived nanobiofillers, which can be used effectively in the polymer matrix due to their added advantages towards the environment. Green nanobiofillers are mainly obtained from natural sources. These materials are abundantly found in nature; thus, it will also reduce the cost of production of the nanobiofillers. The biodegradable nature of these fillers also helps them in finding their place in the current environmental scenario. Nanobiofillers are generally obtained from three sources [32]:

a. Animal-based sources
b. Natural sources
c. Plant-based sources.

Animal-based sources mainly include the extraction of biofillers from seashells and from silkworms. Chitin can be extracted from crab shells, shrimps, etc. [33]. Chitin is an important material, and chitin-based nanobiofillers are utilized in different biopolymer matrices for property tuning. Chitosan is one of the most important biopolymers, and it is also utilized as a nanobiofiller. Chitosan is obtained from chitin, and modified chitosan [34] is reported for the property tuning of PLA-based biocomposites and biocomposite foams. Surface modification of chitosan is required to make it suitable for hydrophobic polymer systems. Nanosilk particles are extracted from silkworms and are utilized in various polymers. Silk nanocrystals (SNCs) are reported to increase the mechanical properties of PLA biocomposites and biocomposite foams [35]. Silk nanofibres (SNFs) and silk nanodiscs (SNDs) are utilized also in some investigations.

Natural sources of biofillers include biomass, carbon and other naturally available clays. Biomass-based nanofillers are introduced. This has the ability to solve the environmental issues.

Plant-based sources of nanobiofillers include plant fibres, wood and starch. Nanocellulose can be extracted from the plant-based sources. CNCs are one of the most used biofillers for different biocomposites [36]. The unique properties of cellulose-based nanomaterials make them suitable for different fields of applications. Cellulose is the most abundant source in nature. Cellulose nanofibres are also utilized in different polymer biocomposites. Bagasse is also utilized as a biofiller in polymer biocomposites. Nanofibres from wood are also extracted to incorporate in a polymer matrix. Sometimes, starch is also utilized as a biofiller in polymer matrix.

Another biofiller used in biocomposites is gum arabic [37]. It can be extracted from trees such as *Acacia senegal* and *Acacia seyal*. Gum arabic has advantages such as non-toxicity and biodegradability. It is approved by the United States Food and Drug Administration (USFDA). Modified gum arabic (MG) has been reported for the hydrophobic polymer systems, and interesting results are obtained in biopolymer foam matrix [38]. Gum-based biofillers have the ability to tune different desired properties of biocomposites.

Depending on the desired properties and the market demand, sometimes various other additives apart from the above-discussed additives have to be incorporated in the polymer matrix. These additives include colourants, photodegradants, antistatic agents and antifoaming agents. Colourants are utilized in the polymer matrix at a desired level to get the desired colour in the final product. Some examples of colourants are CdS (yellow), ZnO, Congo red (red), etc. On the other hand, photodegradants such as iron dithiocarbamate are utilized to enhance the photodegradation process in some specific application-based polymers. Antistatic agents are utilized to prevent the formation of static charge in the polymers during their service life for some applications. Generally, quaternary ammonium salts are used for this purpose. Antifoaming agents such as silicon oil and organo-phosphates are used to prevent the generation of gas or cellular structure in the polymer matrix.

4.3   PROSPECTS OF BIOCOMPOSITES AND THEIR PROCESSING

In this section of this chapter, we are going to discuss different approaches and advancements in the field of biocomposites from recent investigations and their prospects and challenges in the current socio-economic scenario. We will also briefly discuss the fabrication and processing techniques of the biocomposites.

4.3.1   RECENT DEVELOPMENTS IN BIOCOMPOSITES

Bio-based and biodegradable polymeric biocomposites are the need of the hour due to their positive impacts on the environment. Biodegradable polymers such as PLA, PCL, PBS, PGA and PVA are utilized in different aspects of our daily life. Hydroxyapatite (HA)/polymer biocomposites have recently gained attention due to their promising ability as a biomedical implant. To improve the desired properties of these biopolymers so that they can replace the petroleum-based non-degradable polymers, investigations are going on by adding various biofillers and other additives. Bio-based and bio-derived polymers have renewable feedstock, which is generally available abundantly in nature. We will discuss here some promising biopolymer composites and their recent developments in technical and environmental aspects.

One of the promising biopolymers is PLA, which has comparable mechanical properties to that of conventional market-dominating ones such as polystyrene. Recent investigations in PLA biocomposites showed that they have the ability not only to replace, but also to dominate the current market due to their positive environmental impacts. Nanofillers such as CNCs, MG, chitosan, silk nanoparticles and graphene oxide are successively utilized in the PLA biocomposites. Different types of fibres such as flax fibre, wood fibre, glass fibre, kenaf fibre and cellulose nanofibre are also utilized in the biocomposites of PLA. The effects of different types of nanoclays and carbon nanotubes on the PLA biocomposite matrix are also reported in the literature. It is observed that these nanofillers mainly affect the mechanical and thermal properties of the biocomposite. Some nanobiofillers such as CNCs and chitosan also affect the degradation of the PLA as reported in the literature. Applications of PLA biocomposites range from biomedical applications to automotive applications. PLA biocomposites are also utilized for green film packaging applications. For sustainable packaging, biofillers such as MG and cellulose nanocrystals in PLA biocomposite are reported. In biomedical applications, silk-based nanobiofillers are more prominent in PLA biocomposite scaffolds for tissue culture. Some of the recent investigations are discussed below in the fields of PLA and other biopolymer composites.

Orellana et al. [39] reported the effects of surface-modified CNCs on the PLA matrix. It is observed from the investigation that CNCs improve the mechanical and optical properties of the base polymer. The surfactant is utilized to change the functionality of the CNC. However, at higher loadings of CNCs deterioration in the mechanical properties of PLA biocomposite is observed.

Improvements in the gas barrier property of the PLA biocomposite were reported by Tripathy et al. [40] by the incorporation of lactic acid-*grafted* gum arabic. They reported a reduction of ~tenfold oxygen permeability in case of the biocomposite compared with neat PLA counterparts.

Fathima et al. [41] reported the effects of nanochitosan on the PLA matrix. They utilized the PLA biocomposite films for the packaging application for Indian white prawn (Fenneropenaeus indicus). They have used cross-linking agents and plasticizers as additives to the biocomposite. Antimicrobial properties of the PLA/chitosan films have been confirmed by this research. They utilized the solvent casting technique for the fabrication of biocomposite films.

Wang et al. [42] investigated the silk fibroin–PLA biocomposites. They observed that silk fibroin can control the crystallinity and biodegradability of the biocomposite. It also provides thermostability to the biocomposite and also increases the antibacterial properties of the biocomposite.

The effects of graphene oxide (GO) and thermally reduced graphene oxide on the PLA matrix were reported by Arriagada et al. [43]. They noticed that the PLA/GO-based biocomposites have potential applications in biomedical fields.

Teymoorzadeh et al. [44] reported the flex fibre/PLA biocomposites and the effect of the biofiller on the PLA matrix properties. They observed a 142% increase in the flexural modulus of the biocomposites with 40% flex fibres compared to the neat PLA counterpart.

Huda et al. [45] investigated the wood fibre/PLA biocomposites along with wood fibre/polypropylene composites under similar conditions. They noticed that wood fibre/PLA composites have comparable strength and mechanical properties to conventional wood fibre/polypropylene composites. A significantly higher mechanical property is observed for PLA biocomposites compared to neat PLA. The flexural modulus of wood fibre/PLA biocomposites is also found comparable to the traditional wood fibre/polypropylene composites. From the investigation, it can be reassured that PLA biocomposites have the potential to be a replacement of conventional composites for different applications.

A glass fibre (GF)-reinforced PLA biocomposite was reported by Wang et al. [46]. They noticed improvements in strength, rigidness and toughness of the biocomposite. High dispersion and interactions of the glass fibres with the PLA matrix were reported. It is interesting to observe that GF also improves the foam-ability of the base polymer according to the investigation reported.

Incorporation of kenaf fibre in the PLA matrix resulted in an improvement in heat-resistant property along with the crystallization as per the investigation carried out by Serizawa and co-workers [47]. They also observed improvements in the impact strength of the biocomposite.

Latest research on the biocomposite of PLA/cellulose nanofibre reported that the oxygen barrier property of the polymer increased up to ~47.3% on increasing the loading by 0.25%, 0.5% and 1% of CNF in the PLA biocomposite [48]. An improvement in the thermal stability of the biocomposite is also observed.

The orientation and dispersion of the carbon nanotubes (CNTs) in the PLA matrix was reported by Zhou et al. [49]. Interestingly, it was reported in the investigation that the orientation of the CNT might improve the conductivity of the biocomposite.

Ramesh et al. [50] studied the effects of MMT clay on PLA biocomposites and hybrid PLA biocomposites. They mainly investigated the thermal, mechanical and barrier properties of the biocomposites. MMT also improved the decomposition temperature of the PLA biocomposites. Water-resistant properties of the biocomposites also improved due to the incorporation of MMT clay nanofillers.

Organo-modified montmorillonite clay (OMMT) was also incorporated in the biopolymer PCL, and their effects were studied by Malik and co-workers [51]. They prepared the biocomposites utilizing solution casting technique. They reported that the moisture absorption capacity of the polymer biocomposite is reduced rapidly from 34.4% to 22.3% on the incorporation of the nanoclay.

Recent studies on PCL biocomposite by Dhakal et al. [52] utilized leaf sheath date palm fibre waste biomass to incorporate in the polymer matrix, and they investigated the properties of the biocomposite. A huge increment in tensile strength and tensile modulus was observed compared to the neat PCL.

Bio-based bamboo fibres (BFs) were also utilized as a biofiller in the PBS biopolymer in a recent investigation reported by Pivsa-Art and co-workers [53]. Young's modulus and water absorption capacity of the biocomposite were increased on the incorporation of BF, while elongation at break decreased with an increase in the concentration of BF. The PBS biocomposite obtained in the investigations can be applied in packaging and furniture tools.

Abral et al. [54] reported the PVA/cassava starch (with or without ultrasonic probe treatment) biocomposites and characterized different properties of the biocomposite. They also utilized short bacterial cellulose fibres in the biocomposite. They reported increments in the thermal and moisture resistances of the biocomposite after the incorporation of fibres.

4.3.2   Recent Developments in Porous/Cellular Biocomposite Foams/Scaffolds

One of the important sectors in biocomposite research is the investigation of cellular biocomposite foams/scaffolds. The microcellular and nanocellular porous foam morphology has some unique characteristics such as lightweight, and it consumes lesser amount of materials as almost 75%–95% of the material consists of gaseous voids [55]. Though the polymeric foam development started in 1930s, the market establishment and technical development in the processing and fabrication of foams was observed from 1980s. Because of the unique properties, polymer foams have a wide range of applications from packaging to biomedical applications along with commodity applications. Biocomposite foams are utilized for some sophisticated applications where the material content and weight are of major concern, such as aerospace applications [56]. Biocomposite foams have also found wide applications in insulation industries and as a sound absorbing material, and due to their unique cushioning effects, they can also be utilized in some high-end applications. Recent advancements in biodegradable polymer foams have also been observed from various investigations. The tuning of desired properties of biocomposite foams make them more suitable for the current environmental issues caused by conventional non-degradable composite foams. Recently, nanobiofillers have been introduced in biopolymer foam matrices and improvements and tuning in different required properties are observed so that they can be a green replacement for the society. The effect of the biofillers on the foam matrix and on the morphology of cell structure is a major area of the current research. Some of the useful biopolymers utilized for biocomposite foam are PLA, PCL, starch, PHAs, and so on. PLA biocomposite cellular foams/scaffolds are very promising in different applications as suggested by some recent investigations. Some recent developments in different biocomposite foams will be discussed here in this section.

Bionanofillers such as CNCs [57], MG [38], modified chitosan [58] and silk nanocrystals (SNCs) [35] were incorporated in the PLA foam matrix by Borkotoky and co-workers. They utilized the modified casting and leaching (C/L) technique for the fabrication of the biocomposite foams. They investigated different properties of biocomposite foams such as cell size, cell density, thermal properties, wettability and mechanical properties. They reported that both CNC and SNC biofillers impart thermal stability to the biocomposite. The thermal degradation kinetic study of the developed PLA/CNC biocomposite foams was also carried out [59]. The plasticizing effects of the biofillers modified chitosan and MG greatly affect the thermal stability of the biocomposite foams. They also suggested different biomedical applications such as tissue engineering and green packaging of the fabricated biocomposite microcellular PLA-based foams.

A latest investigation on PLA/PBS open-cell foam has been reported by Li et al. [60]. They utilized supercritical CO_2 as a blowing agent for foaming. They achieved a porosity of 97.7% for the fabricated foams. They achieved an open cell content in the fabricated foam of 98.2%, which is very beneficial for the oil absorption process. They suggested the application of the fabricated foams for oil–water separation process.

The effects of microcellulose fibrils (MCFs) on the PLA foam matrix were reported by Oluwabunmi and co-workers [61]. They fabricated a biocomposite foam utilizing supercritical CO_2 as a physical blowing agent. They reported concentration-dependant effects of MCFs on the biocomposite foam. They observed that the glass transition temperature decreased with the increase in the concentration of the biofiller. An increase in the mechanical properties was also observed in the initial lower loadings of MCFs.

Kaisangsri and co-workers [62] reported the fabrication of foam trays of cassava starch blended with natural fibre and chitosan. They investigated the properties of the foam and observed that the results are comparable with conventional polystyrene foams. Water absorption index (WAI) and water solubility index (WSI) of the fabricated biocomposite foams were found to be greater than those of the conventional polymer foam as per the investigation. They concluded that the fabricated biocomposite foam tray can be an alternative to the polystyrene foam trays in near future.

Similar investigations are in progress on biopolymers such as PCL [63] and PVA. The effects of biofillers on biopolymer foam matrix are reported in many publications. The selection of biofiller for biocomposite foaming is a challenging task. Depending on the final end usage of the biocomposite foam, the biofillers are generally selected. The toxicity and cost analysis must be performed before any scale-up applications.

4.3.3 PROCESSING METHODS OF BIOCOMPOSITES AND BIOCOMPOSITE FOAMS

In this section, we will briefly discuss different techniques of processing of biocomposites and biocomposite foams.

4.3.3.1 Processing of Biocomposites

The polymer biocomposites can be processed using different techniques and depending upon the properties of the biopolymer [64]. Biocomposites are generally processed using techniques such as melt processing, solution processing, emulsion processing and rubbery stage processing. The glass transition temperature (T_g) of the base polymer is important while considering the processing.

In melt processing, the polymer must have good flowability at relatively lower temperatures of processing. The process is very sensitive to the moisture content present in the initial stages. The polymer must be dried before processing so that there will be no entrapped moisture in it. In this method, processing instruments such as extruder, Brabender Plasti-Corder and injection moulding are utilized for the processing of the biocomposite.

In solution processing, polymers are dissolved and mixed efficiently in a solvent along with the additives or biofillers and then the mixture is casted with or without the help of some sophisticated instruments such as Doctor's blade and casting knife and allowed to dry. After the evaporation of the solvent, the

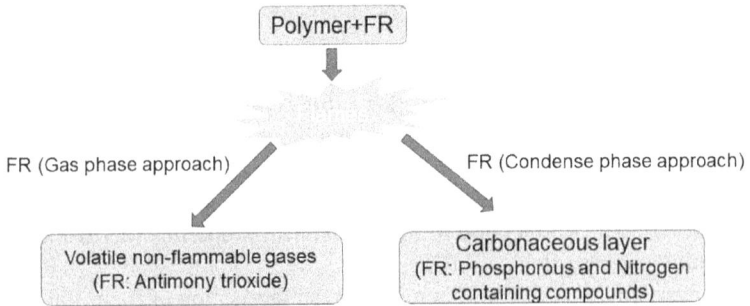

FIGURE 4.2 Schematic representation of the processing methods.

biocomposite films can be collected. This method is mainly used for finished products whose thickness is very less. Solution processing is generally utilized in paint industries, thin film fabrication, and so on.

In emulsion processing, the additives are dispersed with polymer emulsion with the addition of some dispersing agents. Different techniques such as spraying and electrodeposition have been utilized for giving the proper desirable shape to the polymer biocomposites. In the final stage, drying by physical or chemical means has to be performed.

In rubbery stage processing, the base polymer must have excellent rubbery characteristics and the T_g must be below the room temperature while the temperature of decomposition of the polymer is high. This type of processing can be achieved by using extruder (single screw/double screw), Banbury, two roll mill, etc.

The biocomposite processing techniques are generally performed by extrusion, casting, spinning, pultrusion, and using Banbury and two roll open mill mixing.

Different moulding techniques utilized for biocomposite fabrication include compression moulding, blow moulding, injection moulding, transfer moulding and coating. Sometimes after the processing, a finishing step is also required, such as painting, printing and policing, for the final end use applications to the market. Figure 4.2 shows a schematic representation of the biocomposite processing.

4.3.3.2 Processing of Cellular Biocomposite Foams

Polymer biocomposite foams and scaffolds can be fabricated by using two different approaches, i.e. (i) batch foaming process and (ii) continuous foaming process.

In batch foaming process, the fabrication of the biocomposite foam can be achieved in a batch-wise manner. Different processing parameters can be strictly maintained in this process. Generally, physical blowing agents such as CO_2 and N_2 are utilized in this process. Polymer foaming is done by pressurization at high temperature and sudden depressurization process. High-pressure vessels can be utilized in this technique.

In continuous process of foaming, the fabrication of biocomposite foams is performed in a continuous manner. Extruders are utilized in this process and inert gas is supplied during the processing and, at the die head section, the pressure is released.

Two different techniques of foaming are generally utilized for biocomposites.

a. Physical foaming
b. Chemical reactive foaming.

Physical foaming can be further categorized into

a. Casting and leaching (C/L)
b. Foaming using gases
c. Thermally induced phased separation (TIPS).

In C/L technique, porogens are used in a highly volatile polymer solvent mixture. After the casting of the polymer composite with the porogen, it is allowed to evaporate and then the porogens can be leached out of the polymer matrix using a porogen-soluble medium. Generally, salts and sugars are utilized as porogen in the polymer matrix. The ultimate porous cellular structure of biocomposites can be obtained. For foaming using gases, physical foaming agents (PBA) are used in the molten biocomposite matrix. The inert gases are sometimes used in super-critical state. PBA should be non-toxic and non-reactive to the system.

In TIPS method, two steps are followed. In the first step, polymer beads are partially foamed using PBAs, and then in the second step, these partially foamed beads are placed in a mould and the foaming process is carried out. In the chemical reactive foaming techniques, chemical blowing agents (CBAs) are introduced in the biocomposite matrix, which will generate gas in the polymer matrix, and foam structure is obtained.

4.3.4   CHALLENGES IN BIOPOLYMER PROCESSING

Biopolymer composite processing at industrial level is a challenging task. The thermal stability and the selection of processing temperature are critical in case of biocomposite processing. Another challenge in the development of biocomposites is the inconsistency of the bio-based fillers and fibres. The properties of plant-based and agro-based biofillers vary from season to season, and it also depends on various ecological factors such as the amount of rain and sunlight intensity. The properties of biofibres also depend on the part of the tree from which they are collected, soil environment, the pre-treatment of the fibres, and so on. The hydrophilicity of most of the natural bio-based and bio-derived biofillers has been one of the barriers for the incorporation of them in a hydrophobic polymeric system. Hydrophilic biofillers have to be modified to make them hydrophobic so that they can be incorporated in the polymeric matrix. Gum arabic and chitosan have been modified and incorporated in hydrophobic PLA systems successfully as reported in recent investigations. This is a challenging task for the researchers. Natural biofibres absorb water and moisture from the surroundings, and this moisture has to be eliminated before further processing of biocomposites. The compatibility of the biofillers with the polymer is also an obstacle in the path of development in biocomposite. Some mechanical, chemical and biological pre-treatments in bio-fillers are necessary. Sometimes, additives such as coupling agents and plasticizers have to be incorporated in the system to make it compatible.

The selection of proper materials and biofillers for specific application needs vast knowledge and expertise in the processing field. Sometimes, even if we get the proper combination of materials, still the processing method selection also plays a vital role in achieving the desired level of properties of the biocomposites. Improper material selection will lead to the loss of money and wastage of efforts. Thus, interdisciplinary knowledge is required in proper utilization and development of biocomposites.

4.4   CONCLUSIONS

From the above discussion, it is observed that biocomposites are utilized in different aspects of life and they have a wide area of applications in various fields. The biocomposites have the ability to compete with conventional petro-based polymer composites, and recent investigations establish biocomposites as a greener alternative to them. Bio-based fillers and additives can be incorporated to biopolymers for the better use of abundantly available natural green resources. However, some obstacles are also there in the development of biocomposites. The main challenge is the compatibility of the biofiller and the polymer. It is difficult to incorporate hydrophilic biofillers into hydrophobic systems. However, investigations are going on for the improvement in desired properties of biocomposites. The sustainable green biocomposite foams are also gaining attention from different aspects. The utilization of sustainable biofillers derived from biomass, plant and animals is widely increasing day by day. A greener approach to the fabrication of sustainable biocomposites utilizing biofillers from renewable resources has greater prospects in near future in environmental and socio-economic issues. It is the need of the hour to choose the cost-effective and greener approach for the development of technology and processing of the biocomposites in the current environmental conditions to make them more viable to the benefit of the society.

REFERENCES

1. Kosior, E., Braganca, R. M., & Fowler, P. (2006). Lightweight compostable packaging: literature review. *Waste & Resources Action Programme*, 26, 1–48.
2. Auras, R. A., Lim, L.-T., Selke, S. E. M., & Tsuji, H. (2011). *Poly (Lactic Acid): Synthesis, Structures, Properties, Processing, and Applications.* John Wiley & Sons, New York.
3. Di, Y., Iannace, S., Di Maio, E., & Nicolais, L. (2005). Poly (lactic acid)/organoclay nanocomposites: thermal, rheological properties and foam processing. *Journal of Polymer Science Part B: Polymer Physics*, 43, 689–698.
4. Yang, L., Li, S., Zhou, X., Liu, J., Li, Y., Yang, M.,… & Zhang, W. (2019). Effects of carbon nanotube on the thermal, mechanical, and electrical properties of PLA/CNT printed parts in the FDM process. *Synthetic Metals*, 253, 122–130.
5. Kamal, M. R., & Khoshkava, V. (2015). Effect of cellulose nanocrystals (CNC) on rheological and mechanical properties and crystallization behavior of PLA/CNC nanocomposites. *Carbohydrate Polymers*, 123, 105–114.

6. Onyari, J. M., Mulaa, F., Muia, J., & Shiundu, P. (2008). Biodegradability of poly (lactic acid), preparation and characterization of PLA/gum Arabic blends. *Journal of Polymers and the Environment*, 16(3), 205–212.

7. Chomachayi, M. D., Jalali-arani, A., Beltrán, F. R., de la Orden, M. U., & Urreaga, J. M. (2020). Biodegradable nanocomposites developed from PLA/PCL blends and silk fibroin nanoparticles: study on the microstructure, thermal behavior, crystallinity and performance. *Journal of Polymers and the Environment*, 28(4), 1252–1264.

8. Haaparanta, A. M., Järvinen, E., Cengiz, I. F., Ellä, V., Kokkonen, H. T., Kiviranta, I., & Kellomäki, M. (2014). Preparation and characterization of collagen/PLA, chitosan/PLA, and collagen/chitosan/PLA hybrid scaffolds for cartilage tissue engineering. *Journal of Materials Science: Materials in Medicine*, 25(4), 1129–1136.

9. Pitt, C. G., & Schindler, A. (2019). Biodegradation of polymers. In *Controlled Drug Delivery*, Ed. Stephen D. Bruck (pp. 53–80). CRC Press, Boca Raton.

10. Murariu, M., & Dubois, P. (2016). PLA composites: from production to properties. *Advanced Drug Delivery Reviews*, 107, 17–46.

11. Lee, S.-T., Park, C. B., & Ramesh, N. S. (2006). *Polymeric Foams: Science and Technology*. CRC Press, Boca Raton.

12. Rivas-Rojas, P. C., Ollier, R. P., Alvarez, V. A., & Huck-Iriart, C. (2021). Enhancing the integration of bentonite clay with polycaprolactone by intercalation with a cationic surfactant: effects on clay orientation and composite tensile properties. *Journal of Materials Science*, 56(9), 5595–5608.

13. Molavi, H., Behfar, S., Shariati, M. A., Kaviani, M., & Atarod, S. (2021). A review on biodegradable starch based film. *Journal of Microbiology, Biotechnology and Food Sciences*, 2021, 456–461.

14. Chen, W., Li, Y., Huang, Y., Dai, Y., Xi, T., Zhou, Z., & Liu, H. (2021). Quercetin modified electrospun PHBV fibrous scaffold enhances cartilage regeneration. *Journal of Materials Science: Materials in Medicine*, 32(8), 1–10.

15. Abdel-Raouf, M. E., Keshawy, M., & Hasan, A. M. (2021). Green polymers and their uses in petroleum industry, current state and future perspectives. In *Crude Oil-New Technologies and Recent Approaches*. IntechOpen, London.

16. Gu, J. D. (2021). Biodegradability of plastics: the issues, recent advances, and future perspectives. *Environmental Science and Pollution Research*, 28(2), 1278–1282.

17. Siqueira, D. D., Luna, C. B. B., Ferreira, E. S. B., Araújo, E. M., & Wellen, R. M. R. (2020). Tailored PCL/Macaíba fiber to reach sustainable biocomposites. *Journal of Materials Research and Technology*, 9(5), 9691–9708.

18. Shaghaleh, H., Xu, X., & Wang, S. (2018). Current progress in production of biopolymeric materials based on cellulose, cellulose nanofibers, and cellulose derivatives. *RSC Advances*, 8(2), 825–842.

19. Singh, T., Gangil, B., Patnaik, A., Biswas, D., & Fekete, G. (2019). Agriculture waste reinforced corn starch-based biocomposites: effect of rice husk/walnut shell on physicomechanical, biodegradable and thermal properties. *Materials Research Express*, 6(4), 045702.

20. Soykeabkaew, N., Laosat, N., Ngaokla, A., Yodsuwan, N., & Tunkasiri, T. (2012). Reinforcing potential of micro-and nano-sized fibers in the starch-based biocomposites. *Composites Science and Technology*, 72(7), 845–852.

21. Khattak, S., Wahid, F., Liu, L. P., Jia, S. R., Chu, L. Q., Xie, Y. Y.,... & Zhong, C. (2019). Applications of cellulose and chitin/chitosan derivatives and composites as antibacterial materials: current state and perspectives. *Applied Microbiology and Biotechnology*, 103(5), 1989–2006.

22. Sahoo, D., Sahoo, S., Mohanty, P., Sasmal, S., & Nayak, P. L. (2009). Chitosan: a new versatile bio-polymer for various applications. *Designed Monomers and Polymers*, 12(5), 377–404.

23. Jiang, L., & Zhang, J. (2017). 7-Biodegradable and biobased polymers. In Myer Kutz (ed.), *Plastics Design Library, Applied Plastics Engineering Handbook* (Second Edition, pp. 127–143). William Andrew Publishing, ISBN 9780323390408, https://doi.org/10.1016/B978-0-323-39040-8.00007-9.

24. Rudnik, E. (2013). 13-Compostable polymer properties and packaging applications. In Sina Ebnesajjad (ed.), *Plastics Design Library, Plastic Films in Food Packaging* (pp. 217–248). William Andrew Publishing, ISBN 9781455731121, https://doi.org/10.1016/B978-1-4557-3112-1.00013-2.

25. Fujimaki, T. (1998). Processability and properties of aliphatic polyesters, "BIONOLLE", synthesised by polycondensation reaction. *Polymer Degradation and Stability*, 59, 209.

26. Samantaray, P. K., Little, A., Haddleton, D. M., McNally, T., Tan, B., Sun, Z.,... & Wan, C. (2020). Poly (glycolic acid) (PGA): a versatile building block expanding high performance and sustainable bioplastic applications. *Green Chemistry*, 22(13), 4055–4081.

27. Magazzini, L., Grilli, S., Fenni, S. E., Donetti, A., Cavallo, D., & Monticelli, O. (2021). The blending of poly(glycolic acid) with polycaprolactone and poly(l-lactide): promising combinations. *Polymers*, 13(16), 2780, https://doi.org/10.3390/polym13162780.

28. Chiellini, E., Corti, A., D'Antone, S., & Solaro, R. (2003). Biodegradation of poly (vinyl alcohol) based materials. *Progress in Polymer Science*, 28(6), 963–1014, ISSN 0079-6700, https://doi.org/10.1016/S0079-6700(02)00149-1.

29. Mignon, A., Snoeck, D., D'Halluin, K., Balcaen, L., Vanhaecke, F., Dubruel, P.,... & De Belie, N. (2016). Alginate biopolymers: counteracting the impact of superabsorbent polymers on mortar strength. *Construction and Building Materials*, 110, 169–174, ISSN 0950-0618, https://doi.org/10.1016/j.conbuildmat.2016.02.033.

30. Smitha, S., & Sachan, A. (2016). Use of agar biopolymer to improve the shear strength behavior of sabarmati sand. *International Journal of Geotechnical Engineering*, 10(4), 387–400, doi: 10.1080/19386362.2016.1152674.

31. Karak, N. (2009). *Fundamentals of Polymers: Raw Materials to Finish Products.* PHI Learning Pvt. Ltd, New Delhi.

32. Mohan, T. P., & Kanny, K. (2020). Green nanofillers for polymeric materials. In Ahmed, S., & Ali, W. (eds.), *Green Nanomaterials. Advanced Structured Materials*, Vol. 126. Springer, Singapore, https://doi.org/10.1007/978-981-15-3560-4_5.

33. Wijesena, R., Tissera, N., & Karunanayake, L. (2012). Preparation and characterization of α-Chitin nanofibers from crab shells of Portunus pelagicus (blue swimmer crab). In *Proceedings of International Polymer Science and Technology Symposium*, Vol. 1, 03rd and 04th November, 2012, Colombo, Sri Lanka.

34. Pal, A. K., & Katiyar, V. (2016). Nanoamphiphilic chitosan dispersed poly (lactic acid) bionanocomposite films with improved thermal, mechanical, and gas barrier properties. *Biomacromolecules*, 17(8), 2603–2618.

35. Borkotoky, S. S., Ghosh, T., Patwa, R., & Katiyar, V. (2021). Silk nanocrystal (SNC) reinforced poly (lactic acid) based microcellular foam: impact on porous structure, crystallinity, thermomechanical and surface property. *Materials Today Communications*, 27, 102258.

36. Dhar, P., Tarafder, D., Kumar, A., & Katiyar, V. (2015). Effect of cellulose nanocrystal polymorphs on mechanical, barrier and thermal properties of poly (lactic acid) based bionanocomposites. *RSC Advances*, 5(74), 60426–60440.

37. Williams, P. A., & Phillips, G. O. (2021). Gum arabic. In Phillips, G. O., & Williams, P. A. (eds.), *Handbook of Hydrocolloids* (3rd Edition) (pp. 627–652). Woodhead Publishing. Series in Food Science, Technology and Nutrition Book, Cambridge.

38. Borkotoky, S. S., Ghosh, T., Bhagabati, P., & Katiyar, V. (2019). Poly (lactic acid)/ modified gum arabic (MG) based microcellular composite foam: effect of MG on foam properties, thermal and crystallization behavior. *International Journal of Biological Macromolecules*, 125, 159–170.

39. Orellana, J. L., Wichhart, D., & Kitchens, C. L. (2018). Mechanical and optical properties of polylactic acid films containing surfactant-modified cellulose nanocrystals. *Journal of Nanomaterials*, 2018, Article ID 7124260, 12 pages, https://doi.org/10.1155/2018/7124260.

40. Tripathi, N., & Katiyar, V. (2016). PLA/functionalized-gum arabic based bionanocomposite films for high gas barrier applications. *Journal of Applied Polymer Science*, 133(21).

41. Fathima, P. E., Panda, S. K., Ashraf, P. M., Varghese, T. O., & Bindu, J. (2018). Polylactic acid/chitosan films for packaging of Indian white prawn (Fenneropenaeus indicus). *International Journal of Biological Macromolecules*, 117, 1002–1010, https://doi.org/10.1016/j.ijbiomac.2018.05.214.

42. Wang, F., Wu, H., Venkataraman, V., & Hu, X. (2019). Silk fibroin-poly(lactic acid) biocomposites: effect of protein-synthetic polymer interactions and miscibility on material properties and biological responses. *Materials Science and Engineering: C*, 104, 109890, ISSN 0928-4931, https://doi.org/10.1016/j.msec.2019.109890.

43. Arriagada, P., Palza, H., Palma, P., Flores, M., & Caviedes, P. (2017). Poly(lactic acid) composites based on graphene oxide particles with antibacterial behavior enhanced by electrical stimulus and biocompatibility. *Journal of Biomedical Materials Research Part A*, 106, doi: 10.1002/jbm.a.36307.

44. Teymoorzadeh, H., & Rodrigue, D. (2014). Biocomposites of flax fiber and polylactic acid: processing and properties. *Journal of Renewable Materials*, 2(4), 270–277.

45. Huda, M., Drzal, L., Misra, M., & Mohanty, A. (2006). Wood-fiber-reinforced poly(lactic acid) composites: Evaluation of the physicomechanical and morphological properties. *Journal of Applied Polymer Science*, 102, 4856–4869, doi: 10.1002/app.24829.

46. Wang, G., Zhang, D., Wan, G., Li, B., & Zhao, G. (2019). Glass fiber reinforced PLA composite with enhanced mechanical properties, thermal behavior, and foaming ability. *Polymer*, 181, 121803, ISSN 0032-3861, https://doi.org/10.1016/j.polymer.2019.121803.

47. Serizawa, S., Inoue, K., & Iji, M. (2006). Kenaf-fiber-reinforced poly (lactic acid) used for electronic products. *Journal of Applied Polymer Science*, 100(1), 618–624.

48. Jung, B. N., Jung, H. W., Kang, D. H., Kim, G. H., Lee, M., Shim, J. K., & Hwang, S. W. (2020). The fabrication of flexible and oxygen barrier cellulose nanofiber/ polylactic acid nanocomposites using cosolvent system. *Journal of Applied Polymer Science*, 137(47), 49536.

49. Zhou, X., Deng, J., Fang, C., Lei, W., Song, Y., Zhang, Z.,.,.. & Li, Y. (2021). Additive manufacturing of CNTs/PLA composites and the correlation between microstructure and functional properties. *Journal of Materials Science & Technology*, 60, 27–34, ISSN 1005-0302, https://doi.org/10.1016/j.jmst.2020.04.038.

50. Ramesh, P., Prasad, B. D., & Narayana, K. L. (2020). Effect of MMT Clay on mechanical, thermal and barrier properties of treated aloevera fiber/PLA-hybrid biocomposites. *Silicon*, 12(7), 1751–1760.

51. Malik, N., Shrivastava, S., & G'hosh, S. B. (2018, April). Moisture absorption behaviour of biopolymer polycapralactone (PCL)/organo modified montmorillonite clay (OMMT) biocomposite films. In *IOP Conference Series: Materials Science and Engineering* (Vol. 346, No. 1, p. 012027). IOP Publishing, Bristol.

52. Dhakal, H., Bourmaud, A., Berzin, F., Almansour, F., Zhang, Z., Shah, D. U., & Beaugrand, J. (2018). Mechanical properties of leaf sheath date palm fibre waste biomass reinforced polycaprolactone (PCL) biocomposites. *Industrial Crops and Products*, 126, 394–402.

53. Pivsa-Art, S., & Pivsa-Art, W. (2021). Eco-friendly bamboo fiber-reinforced poly (butylene succinate) biocomposites. *Polymer Composites*, 42(4), 1752–1759.

54. Abral, H., Hartono, A., Hafizulhaq, F., Handayani, D., Sugiarti, E., & Pradipta, O. (2019). Characterization of PVA/cassava starch biocomposites fabricated with and without sonication using bacterial cellulose fiber loadings. *Carbohydrate Polymers*, 206, 593–601.

55. Borkotoky, S.S., Ghosh, T. and Katiyar, V. (2020). Katiyar, V., Kumar, A., Mulchandani, N. (eds) Biodegradable nanocomposite foams: processing, structure, and properties. In *Advances in Sustainable Polymers* (pp. 271–288). Springer, Singapore.

56. Abdul Azam, F. A., Rajendran Royan, N. R., Yuhana, N. Y., Mohd Radzuan, N. A., Ahmad, S., & Sulong, A. B. (2020). Fabrication of porous recycled HDPE biocomposites foam: effect of rice husk filler contents and surface treatments on the mechanical properties. *Polymers*, 12(2), 475, https://doi.org/10.3390/polym12020475.

57. Borkotoky, S. S., Dhar, P., & Katiyar, V. (2018). Biodegradable poly (lactic acid)/ Cellulose nanocrystals (CNCs) composite microcellular foam: effect of nanofillers on foam cellular morphology, thermal and wettability behavior. *International Journal of Biological Macromolecules*, 106, 433–446.93.

58. Borkotoky, S. S., Pal, A. K., & Katiyar, V. (2019). Poly (lactic acid)/modified chitosan-based microcellular foams: thermal and crystallization behavior with wettability and porosimetric investigations. *Journal of Applied Polymer Science*, 136(12), 47236.

59. Borkotoky, S. S., Chakraborty, G., & Katiyar, V. (2018). Thermal degradation behaviour and crystallization kinetics of poly (lactic acid) and cellulose nanocrystals (CNC) based microcellular composite foams. *International Journal of Biological Macromolecules*, 118, 1518–1531.

60. Li, B., Zhao, G., Wang, G., Zhang, L., Gong, J., & Shi, Z. (2021). Biodegradable PLA/PBS open-cell foam fabricated by supercritical CO_2 foaming for selective oil-adsorption. *Separation and Purification Technology*, 257, 117949.

61. Oluwabunmi, K., D'Souza, N. A., Zhao, W., et al. (2020). Compostable, fully biobased foams using PLA and micro cellulose for zero energy buildings. *Scientific Reports*, 10, 17771, https://doi.org/10.1038/s41598-020-74478-y.

62. Kaisangsri, N., Kerdchoechuen, O., & Laohakunjit, N. (2012). Biodegradable foam tray from cassava starch blended with natural fiber and chitosan. *Industrial Crops and Products*, 37(1), 542–546, ISSN 0926-6690, https://doi.org/10.1016/j.indcrop.2011.07.034.

63. Oluwabunmi, K. E., Zhao, W., & D'Souza, N. A. (2021). Carbon capture utilization for biopolymer foam manufacture: thermal, mechanical and acoustic performance of PCL/PHBV CO_2 foams. *Polymers*, 13, 2559, https://doi.org/10.3390/polym13152559.

64. Mohammad Asim, Mohammad Jawaid, Naheed Saba, Ramengmawii, Mohammad Nasir, Mohamed Thariq Hameed Sultan. (2017). 1-Processing of hybrid polymer composites—a review. In Vijay Kumar Thakur, Manju Kumari Thakur, & Raju Kumar Gupta (eds.), *Hybrid Polymer Composite Materials* (pp. 1–22). Woodhead Publishing, ISBN 9780081007891, https://doi.org/10.1016/B978-0-08-100789-1.00001-0.

5 A Comprehensive Study of Biodegradable Composites for Food Packaging Applications

*P. Shakti Prakash, Vivek Pandey,
and Manish Kumar*
Mody University

CONTENTS

5.1 INTRODUCTION

Petrochemical-based polymers are extensively used as food packaging applications due to their easy availability and cost-effectiveness. These materials have several good characteristics such as high tensile strength, good wear and tear resistance, and heat sealability (Tharanathan 2003). Despite such good properties, one hindrance is always being encountered as its non-biodegradability which leads to serious environmental issues. Due to this, the consumer demand has shifted toward biodegradable materials obtained from natural resources such as renewable agriculture by-products and food processing industries (Vermeiren

DOI: 10.1201/9781003227908-5

et al. 1999, Álvarez 2000). Nowadays, more than 322 million tons of plastics are produced annually across the globe, which is still increasing day by day (Van den Oever et al. 2017). Currently, India generates 5.58 million tons of single-use plastic annually. Presently, nearly 99% of all plastic materials are produced by the petrochemical industries, which indicates that these plastics are being produced from non-renewable resources (Strategic 2014, Kunroo and Soni n.d.). India currently use nearly 43% of per annum synthetic polymers manufactured by packaging markets compared to the world average of 39%, which is still a bigger number (Tripathi et al. 2012, Strategic 2014). The manufacturing of the plastics is an energy-exhaustive process, which results in increasing global warming. Furthermore, on burning, plastics products release several hazardous chemical elements such as carbon monoxide (CO), furans, acetaldehyde, chlorine, styrene, dioxin, amines, nitrides, 1,3-butadiene, benzene, and hydrochloric acid. Due to this, the global environment as well as the people's health are greatly affected and deteriorated (Smith 2005).

Due their resistance to degradation, the wastes produced from the plastics create long-term problems (Van den Oever et al. 2017). India generates nearly 5.5 million tons plastic waste per year. An excessive use of plastics causes harmful effects on the environment globally, which tends to increase the requirement of extensive research in the development of eco-friendly as well as biodegradable polymers as new alternate materials (Van den Oever et al. 2017). The biodegradable polymeric materials are manufactured from renewable sources and mimic the properties of traditional polymers such as polyethylene terephthalate (PET), polyethylene (PE), and polypropylene (PP) (Kirwan et al. 2011). Hence, the biodegradable polymers can play a vital role in reducing the negative impacts on environment, which occur due to the high production of plastic and its uses. Moreover, this also leads to a greater potential use of agriculture waste products (Maharana et al. 2009).

5.2 BIODEGRADABLE COMPOSITES

Composite materials are a mixture of two or more materials having different properties; in addition, the materials should be chemically homogeneous and physically distinct in nature. The concrete, mud bricks, and fiberglass are few examples. With the increase in eco-friendly alternatives to existing food plastic (composite) packaging, development within the biodegradable polymer (composites) shows some potential. These polymers produced from natural composites materials such as polylactic acid (PLA) or other substances may be very useful for food packaging due to their biodegradability. Biodegradable polymers (composites) still have some issues such as decrease in mechanical ability and more complexity in the production methods, due to which this type of polymers (composites) have low functionality and are costly in manufacturing. By reducing such difficulties, biodegradable polymers (composites) may become a potential solution for current food packaging for environment sustainability.

5.3   PRODUCTION OF BIODEGRADABLE COMPOSITES

The market volume of biocomposites was estimated at USD 16.46 billion in 2016 and is predicted to reach USD 36.76 billion by 2022. The CAGR is estimated to be 14.44% from 2017 to 2022 and projected to grow further by huge folds (Auras et al. 2005). GreenBlue is a non-profit organization, which is dedicated to the sustainable use of eco-friendly materials in the society. The bioplastics-based packaging and films production increased from 400,000 tons in 2015 to 814,000 tons in 2020. However, the common expansion will take place in the inflexible packaging area where the 1,000,000 tons production in 2015 will increase about 7 times to a massive number, i.e., 6,897,000 tons by 2020. Within the growth of bioplastic inflexible packaging, the biodegradable composite will rise from less than 200,000 tons of production volume in 2015 to more than 1,000,000 tons in 2020. The parts of bio-based biodegradable and non-degradable composites were 36.3% and 63.7%, respectively. The most common biodegradable composites consist of PLA (10.9%), biodegradable starch blends (9.4%), biodegradable polyesters (10.8%), and polyhydroxyalkanoates (PHAs) (3.6%). In the production of bioplastics, Asia contributes 55%, Europe 19%, North America 16%, South America 9%, and Australia 1%. Mostly, non-degradable bioplastics are used in rigid packaging and biodegradable bioplastics are used for flexible packaging of the products. The upcoming scenario of bioplastics focuses on the market for biodegradable and compostable materials that are used for consumer and industrial applications (Mangaraj et al. 2019). Biodegradable polymers are used for modified atmospheric storage (MAP) of perishable items in place of traditional polymers. The MAP is required to maintain an environment lower in O_2 and higher in CO_2, which affects the metabolism of the food items to be packaged and the process of decay because microbes increases their shelf life and allows storage for long period. Moreover, MAP vastly enhances the moisture retention during change in atmosphere, which can have more effects on maintaining the quality than O_2 and CO_2 levels (Mangaraj et al. 2019).

5.4   BIOPOLYMERS AND THEIR POTENTIAL
AS PACKAGING MATERIAL

Packaging of food items has always been a challenge in food processing industries. Packaging of food is just not an art, but also includes new technology and research, which is an important aspect for maintaining the quality of food during transportation and distribution in all atmospheric conditions to the consumer at minimum cost (Robertson 2005). There are several types of traditional packaging materials made up of petrochemical elements such as PET, polystyrene (PS), PVC, PP, and polyamide (PA) (Mangaraj et al. 2019). The important characteristics of these types of materials which is may be suitable exclusively in food packaging are cost effective, attractive also includes their physical properties (molecular weight, density) mechanical properties (tensile strength) and transmission properties. With these unique properties, the packaging improves the

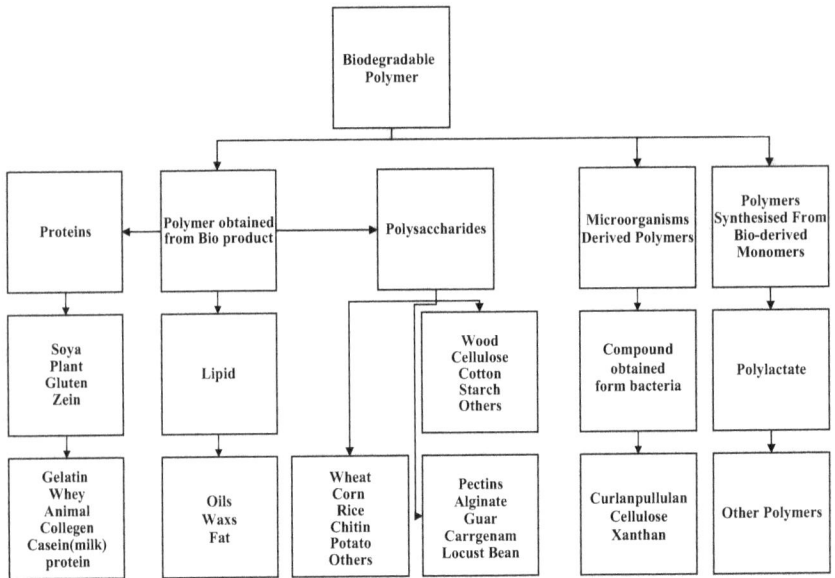

FIGURE 5.1   Types of bio-based materials.

quality of food items (Gu et al. 2011). The main challenge with synthetic materials is their non-degradation behavior in the atmosphere (Imam et al. 2008). As per ASTM standards (D-5488-94d), a biodegradable material is defined as any material capable of undergoing decomposition into natural elements such as methane, water, carbon dioxide, biomass, and inorganic compounds. With increasing demand and cognizance on sustainability, the packaging market looking for biodegradable composite materials as the substitutes of synthetic polymer around the globe. The biopolymers are materials that are decomposed by the enzymatic reaction of microbes. In previous three decades, extensive studies have been carried out on biodegradable composites for food packaging uses (Malathi et al. 2014). The biodegradable polymers-based packaging may be categorized into three important groups based on their production and origin, which is depicted in Figure 5.1.

Group 1: These polymers are obtained directly from biomass. Some polysaccharides such as cellulose and starch, and milk proteins such as gluten and casein are the elements present in this category. All these elements are naturally occurring and are hydrophilic in nature, and some are crystalline and have issues while processing. In addition, their sustainability and performance are not good, especially as a packaging material for moist food items. However, their exceptional gas resistance characteristics provide them an edge for their application in food packaging industry (Weber 2000).

Group 2: In this category, the polymeric materials manufactured by a conventional polymerization procedure are presented. These include aliphatic copolymer

(CPLA), aliphatic polyesters, aliphatic aromatic copolymers, and polylactide, using biodegradable made up of monomers such as PLA and oil-based monomers such as polycaprolactone (PCL), a polymer products made by conventional chemical method using biopolymer is PLA. It is a bio-polyester polymerized from lactic acid monomers. This type of monomer might be manufactured using fermentation of several carbohydrate feedstocks. The PLA might be changed into plasticized form of its monomers; otherwise, it can be changed into oligomeric lactic acid. The PLA can be prepared into driven film by the process of coating and injected mold objects. Hence, it shows the important role of PLA as the novel biodegradable material manufactured at commercial scale (Flieger et al. 2003, Leja and Lewandowicz 2010, Tripathi et al. 2014).

Group 3: This category represents the polymers which are generated by genetically modified bacteria. Currently, this category of biodegradable polymers is composed especially of polyhydroxyalkanoates, but still development with some polysaccharides and bacterial cellulose is required (Weber 2000, Leja and Lewandowicz 2010, Tripathi et al. 2014).

5.4.1 STARCH-BASED BIOPOLYMERS

Starch comes under the category of cheapest polysaccharides of biodegradable polymers and is available in plenty amount. It is also called hydrocolloid biopolymer just having its waterproof nature. It consists of amylopectin, which is a branched and amorphous polymer, and amylose, a linear and crystalline polymer. Depending upon the origin, the amylopectin and amylose consist of starch varying from 80% to 90% and from 10% to 20%, respectively (Mangaraj et al. 2019). The amylose is dispersible in water and generates a helical structure (Leja and Lewandowicz 2010). Biodegradable polymers can be prepared by using different kinds of starch, such as corn, potato, rice, tapioca, and cassava (Balakrishnan et al. 2017). Starch is generally used as a thermoplastic. It can be plasticized by heat and little amount of water or plasticizers for destruction and then extruded. Due to this, the thermoplastic made from starch has a high sensitivity to humidity. This sensitivity to humidity and their mechanical characteristics limit the applications of starch. To enhance the above-said properties of starch, it can be mixed with some additives and different biopolymers.

5.4.2 PROTEIN-BASED BIOPOLYMERS

Proteins such as casein, gelatin, and keratin possess characteristics similar to a polymer, such as tensile modulus, flexural and shear strength, and they also have similar material properties such as strength, toughness, and elasticity. Hence, biodegradable composites developed by such proteins are the starting materials for commercial applications. This type of composites find wide range of uses in the area of food formation and its packaging. They can also be used as biodegradable materials in which the tissue engineering and reconstructive surgery can also be a functional part. Hence, the products made up of protein can be applied in polymer reinforcement. By mixing protein polymer with other proteins and non-protein elements, the mechanical properties can also be further improved. Now, this mixing technology provides a chance to grow to the next level of biodegradable polymer which can be a substitute for the traditional plastics from the industries globally. In food packaging market, the films comprise of protein polymers such as wheat gluten, gelatin, milk proteins, corn, egg white, and soya protein. They are used as edible material, and hence, they can be eaten with the food itself. A few studies have reported plant proteins and their products made up of wheat and corn. The application of packaging materials made up from protein-based film also limits its use due to its low mechanical strength. Therefore, this concept is facilitating the effective amalgamation of bioactive constituents and their functions such as water vapor, oxygen barrier, and tampering resistance. In non-food packaging, polymers such as keratin, casein, and soy protein could play an essential role in developing different market products such as mulch film, shopping bags, and flushable sanitary product. The mixture of protein and non-protein, natural elements like chitosan, starch cellulose, and with synthetic polymer such as PVC, PE, and PP, etc., were synthesized to enhance the characteristics of plastic composed of protein-based polymer which have potential application in food packaging (Mangaraj et al. 2019).

5.4.2.1   Polylactic Acid (PLA)

Recently, the PLA has gained a high interest of various researchers due to its cost-effectiveness and commercial sustainability during processing (Mangaraj et al. 2019). The biopolymers related to the category of aliphatic polyesters are composed of alpha hydroxyl acids, which include poly(mandelic acid) or polyglycolic acid (Mangaraj et al. 2019). The PLA developed from a controlled de-polymerization of the lactic acid monomer extracted from the starch fermentation such as sugar, corn, feedstock, etc., which are easily biodegrade (Valdés et al. 2014). PLA is a stable substitute for petrochemical-derived yields, because lactides are formed by the fermentation of microorganisms in agricultural by-products, mainly the starch-rich substances (Cabedo et al. 2006). PLA is usually obtained from hydrolysis and polycondensation of lactic acid. Moreover, it can also be manufactured by processing of lactide, which is a cyclic dimer of lactic acid (Cabedo et al. 2006). The high molecular weight, water resistance, and manufacturability make PLA one of the good alternatives for food packaging materials. It is easy to prepare by biodegradability and thermoforming (John et al. 2006). PLA can also be prepared by various different approaches such as sheet formation, injection molding, film forming, thermoforming, and blow molding. The prepared PLA is available in the form of films, coatings, and containers for paper and paper boards. In addition, the PLA has a unique property that it can be chemically recycled back to lactic acid by re-polymerization. However, PLA appears to be a possible biodegradable polymer and has the potential to be applied in packaging of several food items. However, it also shows few barriers in its features such as brittleness and degradability at high temperatures.

5.4.2.2 Polyhydroxyalkanoates (PHAs)

The PHAs belongs to the family of bacterial polyesters which is manufactured and stored by different types of bacterial species under unstable evolution environments. This type of polymer can be manufactured by bacterial fermentation of lipids and sugar naturally. The structure of PHAs consists of simple large molecules consisting of 3-hydroxy fatty acid monomers. The PHAs market shares approximately 500 tons per year, which is very low volume in biopolymer. The PHAs are thermomechanically similar to synthetic polymers such as polypropylene (Mangaraj et al. 2019). This type of biodegradable polymers is obtained from traditional resources and well suited to the environment (Mangaraj et al. 2019). As they have biodegradable nature, this type of polymers can help to minimize the global warming-related gases against petroleum products displacement (Leja and Lewandowicz 2010). Poly(3-hydroxybutyrate) (PHB) is a kind of PHA that has naturally available biodegradable molecules, i.e., a chain of β-hydroxy acid linear polyester (Trainer and Charles 2006). In naturally available PHAs, the most common structure which formed by repeating units of monomers, i.e., -(CH2) n-CH3. The final product also depends upon the kind of bacteria and its feed used in this process. The PHAs find their applications as a biodegradable packaging in coatings, films, laminates, bottles, sheets, and fibers. PHAs can be helpful in the production of more than 100 monomers and copolymers. Polyhydroxybutyrate, poly(3-hydroxybutyrate-co-3-hydroxyvalerate), poly(3-hydroxybutyrate-co-3-hydroxyoctanoate), and polyvinyl butyral are some of the examples of PHA polymer. These polymers can also be applied in food industries due to their properties such as temperature stability, good mechanical strength, heat sealability, lubrication, odorlessness, flavorlessness, and ability to be easily dyed (Leja and Lewandowicz 2010).

5.5   PROPERTIES OF BIOPOLYMERS

The most important factor regarding the packaging material is its ability to protect the shelf life of the food items, and it primarily depends on the barrier properties. The barriers properties is being in a gas phase which is permeable through a material (non-porous) occurs adsorption process at the main interface. In addition, the release of an adsorbed substance from a surface of the permeable phase becomes the initial interface and it frequently measured by three factors: transmission rate, permeability, and permeance. Moreover, the barrier properties are typically estimated within equilibrium wetness circumstances under precise atmosphere (FI927 n.d.). In packaging applications, the main factors primarily measured are oxygen, carbon dioxide, and water vapor, because these factors very much affect the quality of the product if there are some changes in weight loss, color, or pH of products. One unavoidable factor also badly affects the quality of product, i.e., increase in microbial growth (Tripathi et al. 2014). Specifically, two gases, i.e., O_2 and CO_2, are always necessary for ideal storage for various supplies. To improve the quality of food products, the recommended packaging materials should have an optimum level of penetrability to gases and for water vapor. Because of these two factors are highly affects the packing situation as if the porousness of O_2 level is increase then vapor pressure also be increase which ultimately oxidize the food products which is already a perishable item. However, for some bakery products and food powders, the level of water vapor barrier property of film is still essential and to be kept at low humidity (Tripathi et al. 2014).

5.6 BIODEGRADATION OF BIOPOLYMERS

Biodegradation is the breakdown of organic matter by using the biological elements, such as enzymatic reaction with the help of bacteria and fungi. The biodegradation of biopolymers is the processed by various features that contains microorganism used polymer types, its characteristics, and nature of pre-treatment. In addition, various other factors also play a very important role in polymer degradation, such as molecular weight, polymer crystallinity, and functional groups added and their mobility (FI927 n.d.). The following reactions take place during the degradation of polymers:

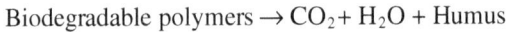

$$\text{Biodegradable polymers} \rightarrow CO_2 + H_2O + \text{Humus}$$

Petroleum based polymers

$$\rightarrow \text{monomers} + \text{polymer cross linked precipitation} + CO_2$$

5.7 BIODEGRADATION TEST

There are mainly three processes to examine the polymer's biodegradability, as shown in Figure 5.2.

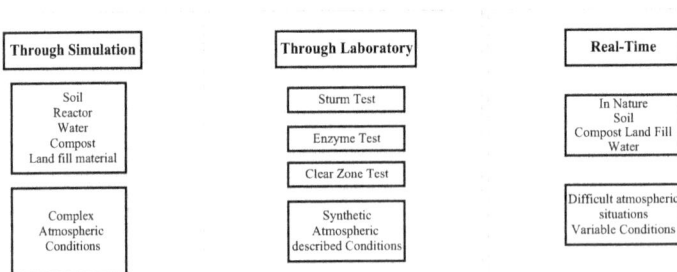

FIGURE 5.2   Different tests for biodegradation.

5.8 CONCLUSIONS

The effect of plastic production and its processing on environment can be reduced by using the biopolymers for sustainable development toward ecosystem. One of the renewable materials, i.e., biodegradable films, is manufactured with very basic by-products obtained from agricultural waste and feedstocks. It gives great interest and scope to various researchers who continuously deal with harnessing this economical prospect. However, currently the biodegradable material only reduces about 1% of the plastics conventionally used. In comparison with plastics, since last three decades biopolymers have been not up to the mark with conventional polymers due to some issues associated with them, such as their high manufacturing cost, processing, durability, and stability. Hence, there are still some unfolded areas of research in the development of different performances and cost-effective techniques for manufacturing of biopolymers. In food packaging and other applications, the use of biodegradable polymers has increased. One of the biggest advantages of biodegradable packaging is that it can be very effectively used as modified atmosphere packaging for storable products such as organic foods. In addition, there should be studies on the effect of biopolymer interaction with the food elements and their proper processing when used as a packaging material for food. Value addition research on bio-based polymers will be the futuristic approach for using them as packaging material. This can also be possible by using nanotechnology and smart sensor, which help in the enhancement of biodegradable polymers. Moreover, the biodegradable polymers show a greater possibility in solving the issues of scarcity of conventional fuel, controlling the solid waste, proper use of agricultural by-products, and environment-related issues related to plastics. Therefore, the biodegradable polymers and their composites will have a huge potential for economical and sustainable environmental changes.

REFERENCES

Álvarez, M.F., 2000. Revisión: Envasado activo de los alimentos/Review: Active food packaging. *Food Science and Technology International*, 6 (2), 97–108.

Auras, R.A., Singh, S.P., and Singh, J.J., 2005. Evaluation of oriented poly (lactide) polymers vs. existing PET and oriented PS for fresh food service containers. *Packaging Technology and Science: An International Journal*, 18 (4), 207–216.

Balakrishnan, P., Sreekala, M.S., Kunaver, M., Huskić, M., and Thomas, S., 2017. Morphology, transport characteristics and viscoelastic polymer chain confinement in nanocomposites based on thermoplastic potato starch and cellulose nanofibers from pineapple leaf. *Carbohydrate Polymers*, 169, 176–188.

Cabedo, L., Luis Feijoo, J., Pilar Villanueva, M., Lagarón, J.M., and Giménez, E., 2006. Optimization of biodegradable nanocomposites based on aPLA/PCL blends for food packaging applications. In: *Macromolecular Symposia*. Wiley Online Library, 191–197.

FI927, A., n.d. Standard test method for determination of oxygen gas transmission rate, permeability and permeance at controlled relative humidity through barrier materials using a coulometric detector.

Flieger, M., Kantorova, M., Prell, A., Řezanka, T., and Votruba, J., 2003. Biodegradable plastics from renewable sources. *Folia Microbiologica*, 48 (1), 27–44.

Gu, J.D., Ford, T.E., and Mitchell, R., 2011. Microbial degradation of materials: General processes, In: *Uhlig's Corrosion Handbook*. John Wiley and Sons, USA.

Imam, S., Glenn, G., Chiou, B.-S., Shey, J., Narayan, R., and Orts, W., 2008. Types, production and assessment of biobased food packaging materials. In: *Environmentally Compatible Food Packaging*. Wood head Publisher, UK. Elsevier, 29–62.

John, R.P., Nampoothiri, K.M., and Pandey, A., 2006. Solid-state fermentation for L-lactic acid production from agro wastes using Lactobacillus delbrueckii. *Process Biochemistry*, 41 (4), 759–763.

Kirwan, M.J., Plant, S., and Strawbridge, J.W., 2011. Plastics in food packaging. In: *Food and Beverage Packaging Technology*. Blackwell Publishing Ltd., 157–212.

Kunroo, M.H. and Soni, K., 2018. Petrochemical industry in India: Determinants, challenges and opportunities. *Progress Petrochem Science*, 1, 1–5.

Leja, K. and Lewandowicz, G., 2010. Polymer biodegradation and biodegradable polymers-a review. *Polish Journal of Environmental Studies*, 19, 255–266.

Maharana, T., Mohanty, B., and Negi, Y.S., 2009. Melt–solid polycondensation of lactic acid and its biodegradability. *Progress in Polymer Science*, 34 (1), 99–124.

Malathi, A.N., Santhosh, K.S., and Nidoni, U., 2014. Recent trends of biodegradable polymer: Biodegradable films for food packaging and application of nanotechnology in biodegradable food packaging. *Current Trends in Technology and Science*, 3 (2), 73–79.

Mangaraj, S., Yadav, A., Bal, L.M., Dash, S.K., and Mahanti, N.K., 2019. Application of biodegradable polymers in food packaging industry: A comprehensive review. *Journal of Packaging Technology and Research*, 3 (1), 77–96.

Robertson, G.L., 2005. *Food Packaging: Principles and Practice*. CRC Press. Florida, USA.

Smith, R., 2005. *Biodegradable Polymers for Industrial Applications*. CRC Press. Florida, USA.

Strategic, T., 2014. Potential of plastics industry in Northern India with special focus onplasticulture and food processing-2014. *A Report on Plastics Industry*. Federation of Indian Chambers of Commerce and Industry, New Delhi.

Tharanathan, R.N., 2003. Biodegradable films and composite coatings: Past, present and future. *Trends in Food Science & Technology*, 14 (3), 71–78.

Trainer, M.A. and Charles, T.C., 2006. The role of PHB metabolism in the symbiosis of rhizobia with legumes. *Applied Microbiology and Biotechnology*, 71 (4), 377–386.

Tripathi, A.D., Srivastava, S.K., and Yadav, A., 2014. Biopolymers potential biodegradable packaging material for food industry. *Polymers for Packaging Applications*.

Tripathi, A.D., Yadav, A., Jha, A., and Srivastava, S.K., 2012. Utilizing of sugar refinery waste (cane molasses) for production of bio-plastic under submerged fermentation process. *Journal of Polymers and the Environment*, 20 (2), 446–453.

Valdés, A., Mellinas, A.C., Ramos, M., Garrigós, M.C., and Jiménez, A., 2014. Natural additives and agricultural wastes in biopolymer formulations for food packaging. *Frontiers in Chemistry*, 2, 6.

Van den Oever, M., Molenveld, K., van der Zee, M., and Bos, H., 2017. *Bio-Based and Biodegradable Plastics: Facts and Figures: Focus on Food Packaging in the Netherlands*. Wageningen Food & Biobased Research. Netherland.

Vermeiren, L., Devlieghere, F., van Beest, M., de Kruijf, N., and Debevere, J., 1999. Developments in the active packaging of foods. *Trends in Food Science & Technology*, 10 (3), 77–86.

Weber, C.J., 2000. *Biobased Packaging Materials for the Food Industry: Status and Perspectives, A European Concerted Action*. KVL. UK.

6 Biodegradable Composites for Commodities Packaging Applications and Toxicity

V. Gayathri and B. Sabulal
KSCSTE-Jawaharlal Nehru Tropical Botanic
Garden and Research Institute Palode

CONTENTS

6.1 INTRODUCTION

The combination of two materials both differing in their physical and chemical properties, to obtain a superior material for the intended use, is called a composite material. The rapid growth of composite materials from bio-acquired materials such as recycled materials and waste resources are of current interest.

DOI: 10.1201/9781003227908-6

Biocomposites from plants with high bio-based content are constantly being approached for commercialization. Biodegradable composites have shown potential uses in sustainable packaging. Green chemistry has led to sustainable utilization of biodegradable materials and recycled plastic materials, consequently reducing the use of petroleum-based materials.

Appropriately, one can design an efficient and economical packaging material for commodities focussing on the material toxicity, deterioration rate and degradation property. Mentioned below are few points addressing the selection of packing material.

   a. The material ought to possess good physicochemical characteristics.
   b. The material must blend with the environmental factors such as temperature and humidity.
   c. It should possess resistant or repellent properties against insects and pests.

Highly perishable commodities such as fresh-cut vegetables and fruits and other cut agricultural or poultry products require efficient packaging to preserve their shelf life. Customarily, fresh fruits and vegetables are packed using petroleum-based films, which are non-biodegradable and non-renewable resources; nevertheless, they impose ecological problems. To overcome these issues, a greener approach is employed by combining biodegradable polymers with plant fibres so as to produce biodegradable composite materials. The sources for biopolymers are mainly from biomass of plant, microorganisms and polymers of biodegradable capability. The packaging films made out of the biopolymers are usually poor in mechanical property, and their barrier property is also inferior in comparison with conventional synthetic materials. The combination of the materials to form an efficient composite can cover this drawback, which is an aspect to be explored on studies related to compatibility of biopolymer composites. Our review focusses on different natural biocomposites and their toxicity profile for packaging food commodities.

6.2 COMPOSITES MATERIALS AND TOXICITY

Composite materials can be defined as the combination of two or more constituent materials with enhanced properties compared to the individual constituents. Basically, composites are classified into two main groups: synthetic (polymeric) and natural (biopolymers/biocomposites). The synthetic composites possess greater mechanical strength and controlled degradation rate, but may possess intrinsic toxicity.

6.2.1 SYNTHETIC COMPOSITE MATERIALS AND RELATED TOXICITY

Poly(lactic acid) is sorted a biomaterial due to its notable properties such as low cost, rapid degradation and commercial availability for different applications.

When combined with additives such as gelatin, PEG and NaCl, the porosity and elasticity are increased; however, leachability is decreased. It has diverse applications in textile, upholstery and various other industries [1]. PLA is widely used in sustainable slow release of drugs synthesized by microencapsulation and nanoencapsulation techniques. One of the prime factors making PLA unfit for short-life packaging of food is that more gas and vapour penetrate inside, thus restraining its use. Hence, it is suitable as a bio-package material for foods with high breathing products such as bakery items [2]. Although PLA is biologically suitable for packaging, the immunogenicity of cells when the food comes in contact with PLA is not yet studied.

Polyvinyl chloride (PVC) is highly flexible, strong and chemically inert, making it an ideal material for food packaging [1]. For packaging of food materials, PVC is generally combined with resins and plasticizers. The main concern is that leaching of resins and plasticizers from the packaging material to foods can occur, which may be harmful. Phthalates are prominent chemical compounds found in plasticizers [3]. Phthalates and other resins are noted as carcinogenic materials.

Polypropylene (PP) is prepared by the polymerization of propylene and is a resin. PP is suitable for packing items such as biscuits, snack foods and dried foods [1,4] as it has low density and high melting point, which make it moisture resistant. However, the leachant toxicity of PP for long-term storage is not yet understood.

Polymerization of ethylene yields two polymers: high-density polyethylene (HDPE) and low-density polyethylene (LDPE). HDPE films are suitable for long-term storage of food as they are durable, are highly permeable to oxygen and carbon dioxide and act as water vapour barrier. Equally, low-density polyethylene (LDPE) is less expensive, is highly water vapour resistant and is a good heat sealant. LDPE is widely used for bags manufacturing and as laminates for coating papers or boards [4]. Yet the release of leachants and storage material toxicity is not documented.

Polyvinyl alcohol (PVA) coated on polyethylene (PE) yields poly(ethylene-co-vinyl acetate) (EVA). It is flexible, heat sealable, highly adhesive and non-toxic, and it possesses low water vapour permeability. These characteristics make EVA suitable for commodity packaging [1]. Mainly, it is used as shrink wrappers and stretch wrap for refrigerated meat and poultry and also as ice bags [5].

Poly(ethylene terephthalate) (PET) possesses high mechanical strength, is considerably tough and has a high crystalline melting temperature (270°C), and it is easy to process. These properties make it ideal for bottles and food packages. Although it is BPA-free, it can leach out antimony upon heating. Antimony is a known toxic metalloid, which once in contact with food and beverages, can cause vomiting, diarrhoea and stomach ulcers. Moreover, PET deposits acetaldehyde residues, which is highly toxic [1].

The contact toxicity of these commonly used packing materials with the commodities is a challenge as they cause the accumulation of material in the form of leachants which act as a xenobiotic, when the food commodity is consumed. These xenobiotics are either conjugated to get eliminated from the system and/ or get accumulated over time in the circulation causing foreign body interaction with the immune cells. Nevertheless, a toxicological screening and the *in vivo* toxicity are again a challenge as the by-products of these materials get converted to xenobiotics triggering cellular immune response.

Recently, more advanced composites have been synthesized by natural product materials. Natural product materials are proved to be less toxic from time immemorial and are eco-friendly too.

6.2.2 BIOCOMPOSITE MATERIALS AND RELATED TOXICITY

A biocomposite material is defined as a material wherein each constituent is biocompatible. Efficient applications of biocomposites require good fabrication and modifications in their properties. This involves increasing their porosity and mechanical strength. Their applications are widespread, such as nanocomposites in drug delivery systems, packaging and in industries. Some important fabricated biocomposites and their uses as well as the possible toxicity are discussed below.

6.2.2.1 Chitosan Composite Materials

Chitosan with poly(lactic acid) (PLA) or poly(lactic-co-glycolic acid) (PLGA) is either surface coated onto the food items as edible coating or to the surface of the packing material [6]. Chitosan is one such edible coating that is reported to enhance the storage of food products, fruits and vegetables due to its intrinsic antimicrobial properties and air exchange retardance. Coating delays the ripening process and reduces water loss in fruits and vegetables. This helps in the delay of decay process. In case of meat products, coating can prevent moisture loss, thus reducing discoloration and lipid peroxidation and thereby improving their quality [7]. It also aids in controlled release of the substrate/product. However, contact material toxicity for food with these biocomposites is still in the nascent stage [8,9].

6.2.2.2 Cellulose Composites

Cellulose fillers are abundant and can be easily obtained from wood flour, lignin, wastes from olive pit, wine production, coffee, and many more. Combination of PLA with cellulose can improve the mechanical strength and reduce gas or moisture permeability, making it suitable for food and commodity packaging. PLA degradation is slow in the soil; however, this can be sped up by incorporation of natural biodegradable fillers. A merger addition of cellulose in PLA increases the pore size and water uptake capacity favouring diffusion. These factors help in increasing the shelf life of pasteurized products, sterilized vegetables and ready-to-use products. The bio-based materials in food packaging need to be assessed for their compatibility with different food products and contact toxicity [6].

6.2.2.3 Cellulose–Chitosan Composite Materials

Chitosan and cellulose composites are miscible. Their physical, chemical, mechanical and biological properties can be improved by inserting amino group to cellulose [6]. These biocomposites have a large surface area and well interconnected porous network efficient for packaging of commodity. This was reported by cell adhesion studies showing the biocompatible nature of chitosan when added to bacterial cellulose [10].

6.2.2.4 Chitosan–Collagen Composite Materials

Collagen is an ideal biomaterial for applications in tissue and scaffold engineering and drug delivery systems. Its properties are enhanced in combination with polymers such as PLGA and chitosan [5]. The two material composites are biodegradable and biocompatible, and their blending gives high porosity and good mechanical strength, which are widely used in the packaging of fruits and vegetables. Additionally, chitosan provides antimicrobial activity to increase the storage properties [11].

As mentioned in previous sections, adherent, immune and contact material toxicity for contact of food with these biocomposites is still in the emerging stage and needs to be addressed properly.

6.2.3 Micro-/Nanocomposites for Commodities Packaging: an Emerging Trend

Biopolymers are eco-friendly, biodegradable and non-toxic in nature; however, they possess poor mechanical properties and are hydrophilic in nature. Nanocomposites with polymers and natural materials have attracted great interest in food packaging industry in recent years due to their enhanced mechanical, thermal and gas barrier properties, together with non-toxic and biodegradable nature.

Incorporation of nano-sized fillers to the natural biomaterials improves their potholes. Nanofiller geometry and the interface area between the filler and the material are the criteria to be taken care of before the incorporation.

Coated films of montmorillonite (MMT), zinc oxide (ZnO-NPs)-coated silicate, kaolinite, silver NPs (AgNPs) titanium dioxide (TiO_2NPs), nanoclay and titanium nitride NPs (nano-TiN) are some of the important nanomaterials and additives used in food packaging. They retard oxygen and carbon dioxide permeation and seal the flavour of the compounds [12,13]. These nanomaterials are oxygen scavengers, antimicrobials and temperature resistant. However, efforts are taken to increase their mechanical stability, compatibility, degradability and effectiveness in conjugation with polymer matrix [12,13].

AgNPs possess antimicrobial activity, and their large surface area per mass makes it superior to other counterparts. Incorporation of AgNPs into plastic polymers involves integration of silver ions with the porous matrix. Reports show that AgNPs and ZnO-NPs can prolong the shelf life of commodities such as juices [12].

Nanoclay has been the pioneer nanocomposite used in food packaging for years. It has an advantage of barrier to gas and moisture over the plastics. Nanopaper is a common example of nanoclay composite that possesses materials that are sustainable and biodegradable. The ingredients are as follows: MMT, nanoclay, nanofibres of cellulose and chitosan. This nanocomposite material has strong modulus, strength and toughness; also, it is fire resistant. Nevertheless, it holds back oxygen exchange, making it suitable for medical and food packaging. Commercially available materials such as Durethan® KU2–2601 (trade name) are extensively used for beverage packaging that is composed of clay as pellets over the polymer matrix. This makes it excellent for exchange of liquids. Another commercial material Imperm® (trade name), which is composed of nanoclay and MXD6 resin (nylon 6 polymer manufactured by Mitsubishi Gas Chemical Company, Inc.), is produced as mono- or multi-layered films used for packaging of beverages such as soda, beer, 100% juices and flavoured alcohols, eventually are filled into the PET (poly(ethylene terephthalate) bottles [12].

Although AgNPs, nanoclays and layered silicates can be used as antimicrobial agents to protect the commodity from decaying, ZnO-NPs used for packing are supercilious because of their cost and low toxicity to animals and humans.

Catalysts such as ZnO-NPs and TiO_2-NPs function in the presence of light, which aids in the degradation of organic molecules and microorganisms thus generating reactive oxygen species (ROS) which thereby cause cell death of bacteria by oxidizing cytoplasm of bacteria whereby increasing the shelf like of food. The indicators of the rate of decay of food items such as whole fresh vegetables and cut fruits are the ROS indices such as malondialdehyde (MDA), ethylene, polyphenol oxidase activity and pyrogallol peroxidase (POD). ZnO-NPs aid in the oxidation of ethylene into H_2O and CO_2 and thus reduce MDA, polyphenol oxidase and peroxidase activity in the presence of UV irradiation. Additionally, ZnO acts as a bacteriostatic agent under UV irradiation producing hydrogen peroxide, which can eventually cause the bacterial death by generating oxidative stress burden on the organism extending the shelf life of commodities. This property makes it an effective additive for polymer composites such as PLA/ZnO-NP films used for effective packaging of commodity [12].

Titanium NPs (TiO_2-NPs/TiN-NPs) are a USFDA-approved food additive and whitener. They are thermostable and inert. They are used in water and air purification and antimicrobial cleaning and as food contact material packing. TiO_2 due to its inertness and non-toxic nature is used in packaging industry. Microbial cell membrane possesses unsaturated poly-phospholipid component, which is oxidized by TiO_2. Also, it possesses biocidal effect in the presence of sunlight or ultraviolet light on microorganisms by generating free radicals such as superoxide anions, hydrogen peroxide and hydroxyl radicals, which damage the cell wall of microorganisms. This property of TiO_2 is utilized for packaging of pasteurized milk in PET and HDPE bottles as TiO_2NPs. TiO_2 is a pigment and coating additive, which in turn acts as a whitener and brightener to the wrappers [12].

Nano-TiN is an approved material for packaging by the European Food Safety Authority (EFSA 2012). It is used together with polyethylene terephthalate in packaging due to its high mechanical strength [12,13].

Nano-starch is a natural, renewable, biodegradable and non-toxic polysaccharide, which is a commonly used material in food packaging, paper making, pharmaceuticals, rubber and plastic industries due to its eco-friendly nature, low cost and copious availability. Nano-starch possesses good mechanical and barrier properties and aids in hydration. It is suitable for use as fillers in food packaging material for improving its flexibility. They are replacing resins wet-strength resins [12].

Cellulose nanofibres are prepared from delignified coconut husk fibres. They are biodegradable and thus replace the traditional petroleum-based materials [12]. Nano-sized cellulose such as nanofibres, microfibrils or nano-whiskers are used in polymer matrix to increase its effectiveness and rigidity.

Carbon nanotubes (CNTs) in combination with various polymers such as PVA, PP, nylon and polylactic acid (PLA) are used as antimicrobial and intelligent sensors for packaging. An innovative technique designed in Germany wherein the packages are embedded with wireless chips and sprayed with CNT-based gas sensors to detect any food spoilage especially in meat packaging has made a tremendous novel venture in packaging methodology [12].

Nano-silica is a non-adhesive material used as a hydrophobic coating, which makes the food free-flowing inside containers or jars. It is mainly used for packaging of beer, wine and powdered soup. Commercially available silica NP products showed water-resistant property highly essential for powdered food commodity packaging [12].

6.2.4 Essential Oil-Based Nanocomposites for Food Packaging

Food materials, *viz.* fruits and vegetables (approx. 40%–50%), fish (35%), cereals (30%) and dairy and meat products (20%), are lost due to spoilage or alteration caused by shelf life expiration or microorganisms [13,14]. Essential oils are volatile mixtures of plant-derived secondary metabolites; they display antimicrobial, antioxidant, anti-inflammatory and other activities [15–17]. Due to these biological effects, they have potential roles in food packaging and preservation. Moreover, essential oils are categorized as generally recognized as safe (GRAS) by the USFDA [18]. Biopolymer-based nanocomposite packaging materials show good thermal, mechanical, chemical resistance, gas barrier and biodegradable properties. Essential oils embedded into these nanomatrices enhance their antimicrobial properties.

Recently, several groups have reported the development of essential oil-embedded nanocomposite materials. Motelica and co-workers fabricated films based on chitosan as the biodegradable polymer, with ZnO and Ag nanoparticles as fillers/antimicrobial agents. The nanoparticles were loaded with citronella essential oil, and the synergistic effects of chitosan, citronella essential oil and ZnO and Ag nanoparticles displayed promising antimicrobial activity. This membrane can be used as coatings for fruits [13]. Marjoram, cinnamon and clove essential oils were incorporated into alginate/clay nanocomposite films, and they retained their antimicrobial activity within the films. Films with marjoram essential oil (1.5%) showed potential activities against gram-positive bacteria, *Listeria monocytogenes* and *Staphylococcus aureus* [19]. da Costa and co-workers fabricated polymer films based on poly(hydroxybutyrate-co-hydroxyvalerate) nanocomposites for food packaging with oregano essential oil as the antimicrobial agent. Both gram-positive and gram-negative bacteria were inhibited by films containing oregano essential oil and, therefore, can be optimized as packaging films ensuring the shelf life and quality of foods [20]. In another recent study, gelatin nanofibres with encapsulated angelica essential oil were fabricated *via* electrospinning and they displayed notable inhibitory effect against both gram-positive and gram-negative bacteria. These gelatin/angelica essential oil nanofibres have great potential as food packaging materials [21]. Briefly, nanocomposite materials impregnated with antimicrobial essential oils have great prospects in developing food packaging materials.

6.3 REGULATORY CHALLENGES AND TOXICITY ASSESSMENT OF FOOD PACKAGING BIOCOMPOSITES

Till date, no standardized assessment protocol has been available for raw material testing for toxicological aspects. Nevertheless, their active components interact with the environment and cellular membranes.

The leachant transfer from the packaging material to food matrix, be it nanomaterials, rises its safety concerns. Nevertheless, some of these materials' characteristics can lead to toxicological results during these material interactions with ells, tissues and organs through food contact. This may pose risks associated with inhalation, ingestion and skin absorption. These engineered biocomposites are to be evaluated for toxicity. Risk assessment and safety of nano-food packaging is in high demand [22]. Table 6.1 shows the list of composite materials, applications, advantages and disadvantages or toxicity.

The lack of a proper guidelines to evaluate material toxicity is a handicap. Various tests are recommended by bodies such as International Organization for Standardization (ISO) to evaluate the biocompatibility of materials. Deciding on the factors such as nature of body contact of food material to the biocomposite, duration of contact and biological reactions becomes important. Every biocomposite needs to be assessed for environment and clinical safety, which in turn needs to comply with the guidelines of ISO, Current Good Manufacturing Practice (cGMP), Organisation for Economic Co-operation and Development (OECD), International Council for Harmonisation (ICH), United States Environmental Protection Agency (USEPA), USFDA, European Environment Agency (EEA), which is the agency of the European Union (EU) [50–52]. Regulatory authorities with these guidelines determine the technical requirements for biocomposites of packaging and the end user toxicity, i.e., animals and humans.

The major five norms for toxicity assessment of the biocomposites to the food and end user are listed below:

   a. Compatibility: Biocomposites must be compatible with cells, wherein it should be ideally inert to the food material. Even though they are cellular inert to the cells, they elicit severe inflammatory response in tissues in case of long storage [53].

TABLE 6.1
A List of Composite Materials and Their Properties

S. No.	Biodegradable Composite Material	Applications	Advantages	Disadvantages and Toxicity (if any)
1a	Poly(lactic acid) [1,2]	Industrial applications	Biocompatible, high Young's modulus and high tensile strength	Brittle polymer with poor toughness, low biodegradability and hydrophobicity
1b	PLA/CNT composites [23]	Medical and industrial applications	High mechanical, thermal and electrical properties	CNT may induce carcinogenicity
1c	PLA composites (fillers, such as starch and nanocellulose) [24,25]	Commodity packaging	Good mechanical, thermal and thermomechanical properties. Waste by-products are eco-friendly	–
2a	Polyvinyl chloride (PVC) [1,3]	Building construction, health care, electronics, automobile	Mechanical strength, highly tough and resistant to abrasion	Phthalates, lead, cadmium and/or organotins used as additives leach out or evaporate into atmosphere over time, posing concern to humans health
2b	Polyvinyl chloride (PVC)–vitamin C (VC)–TiO$_2$ nanocomposite film [26]	Packaging	Environment-friendly, photodegradable	–
3a	Polypropylene (PP) [1,4]	Thin-walled containers are commonly used for food packaging	Chemically resistant and weldable	Fine particles upon inhalation may cause respiratory irritation, pulmonary oedema and asthma. Biologically inert and non-toxic to environment
3b	Nanosilver-applied polypropylene [27]	Coated films for food packaging	Antimicrobial effect (*E. coli* and *S. aureus*)	–
3c	Protein-coated polypropylene films (e.g. soy protein isolate; whey protein isolate; corn zein) [28]	Food packaging	Inhibition against bacterial growth (*Lactobacillus plantarum*)	–

(Continued)

TABLE 6.1 (*Continued*)
A List of Composite Materials and Their Properties

S. No.	Biodegradable Composite Material	Applications	Advantages	Disadvantages and Toxicity (if any)
3d	Corn zein nanocomposite (CZNC) coatings on polypropylene (PP) [29]	Food packaging	Excellent barrier (oxygen and water vapour) and biodegradability	–
4a	Polyethylene (e.g. low-density polyethylene (LDPE) and high-density polyethylene (HDPE)) [4]	Food packaging	Heat resistant, inert, odour-free, good moisture barrier	Permeable to oxygen. Leach out monomers when exposed to UV and heat and over time
4b	Nanoclay-loaded low-density polyethylene (LDPE) composite [30]	Food packaging	Increased stability and improved thermal and barrier properties	Cytotoxic to A594. Noted to cause adenocarcinomic human alveolar basal epithelial cells
4c	Low-density polyethylene (LDPE) biocomposite LDPE/corn flour [31]	Food packaging	Greater interfacial adhesion between the filler and the matrix, higher mechanical properties, good biodegradation	–
5a	Poly(ethylene-co-vinyl acetate) (EVA) [5]	Food packaging	Lower heat-sealing temperature, good barrier properties and good stretch properties	Ethylene vinyl acetate (EVA) is not hazardous by ingestion
5b	Cellulose microfibres (CMFs)/poly(ethylene-co-vinyl acetate) (EVA) composites [32]	Food packaging	Improvement in biodegradation when tested on *Aspergillus niger*	–
5c	Nisin (a commercial formulation) in poly(ethylene-co-vinyl acetate) films [33]	Food packaging	Antimicrobial activity	–
5d	Bacterial cellulose nanofibres (BCNs) and poly(ethylene-co-vinyl acetate) (EVA) nanocomposite [34]	Food packaging	Good storage and swelling property, less wetting property	–

(*Continued*)

TABLE 6.1 (*Continued*)
A List of Composite Materials and Their Properties

S. No.	Biodegradable Composite Material	Applications	Advantages	Disadvantages and Toxicity (if any)
6a	Poly(ethylene terephthalate) (PET) [1]	Food packaging	High strength and stiffness, good gas (oxygen, carbon dioxide) and moisture barrier properties	Upon contact with heat, it can leach out antimony into food and beverages
6b	Moringa oleifera fruit fibres/polyethylene terephthalate composites [35]	Food packaging	Excellent mechanical and thermal properties, biodegradable	–
6c	Cellulose-based composite films with polyethylene terephthalate [40]	Food packaging	Thermally stable, improved thermal decomposition, biodegradable	–
6d	Polyethylene terephthalate/silica nanocomposites [36,37]	Food packaging	Good physical, mechanical and barrier properties	Environment hazards such as formation of dust, fumes and toxic gases
7a	Chitosan blended with glycerol (GLY), ethylene glycol (EG), polyethylene glycol (PEG) and propylene glycol (PG)[38]	Food packaging	Ductility, flexibility and storage stability, high antimicrobial and antioxidant properties	–
7b	Chitosan nanofibre and cellulose nanofibre blend composite [39]	Food packaging	Excellent mechanical and antioxidant properties	–
7c	PLA/chitosan [6]		High mechanical strength, good homogeneity	–
8a	Cellulose-reinforced poly(lactic acid) [40]		Good biodegradability and high strength	–
8b	Bacterial cellulose, poly(3-hydroxybutyrate) blended with essential oil composites [41]	Food packaging material as wrappers	Safe, environmentally friendly, antibacterial agent, biocompatible, biodegradable	–

(*Continued*)

TABLE 6.1 (*Continued*)
A List of Composite Materials and Their Properties

S. No.	Biodegradable Composite Material	Applications	Advantages	Disadvantages and Toxicity (if any)
8c	Chitosan–cellulose nanobiocomposite blends [42]	Food packaging, water treatment	Biodegradability, renewability, antimicrobial activity	–
9a	Collagen blended with sodium alginate, starch and sodium carboxymethyl cellulose [43].	Food packaging films	Mechanical properties	–
9b	Gelatin-based films and coatings [44]	Fish and meat, fruits and vegetables packaging	Antimicrobial, antioxidant properties, retard degradation due to less exchange of gases (O_2 and CO_2) and water vapour	–
9c	Collagen/chitosan and collagen/soy protein composite films [45]	Packaging material	Biodegradable, eco-friendly	–
9d	Collagen/sodium alginate blend films [46]	Food packaging	Enhanced physical, mechanical and barrier properties	–
10	Starch-based biodegradable materials [47,48]	Edible films and medicinal capsules	Starch is a hydrophilic polymer; water is a plasticizer for starch-based materials; renewable, biodegradable, abundantly available and inexpensive	–
11	Essential oils composites [21,49]	Antimicrobial agent	USFDA certified as GRAS (generally recognized as safe)	Immunotoxicity and dose-related carcinogenicity

b. Shelf life and storage properties: Leaching and degradation play a pivot role in biocomposite efficacy. By-products generated during degradation process may be non-toxic and must be eliminated from the body without intrusion with other organs. Nevertheless, biocomposites need to be economical and need to follow cGMP standard while its production. It must also be non-hazardous to the environment. They must also possess antimicrobial activity to sustain the product to be packaged. The criteria determine the shelf life of the biocomposite before it is marketed. Hence, degradation and their subsequent by-products must elicit low or negligible inflammatory response and controlled melange into the cells when in contact [53].

c. Physical properties: An ideal biocomposite needs to be mechanically strong and highly porous for gaseous exchange of air. At the same time, the moisture retention property of the biocomposite needs to be kept at minimum so as to check the contamination rate of the commodities packed [53].

d. Structure of the biocomposite: Flat films and fibres are suitable as they have large surface area. Electrospun fibres are highly recommended due to their uniformity [54].

e. Selection of materials: Raw materials used in the fabrication of a biocomposite are a key factor for the toxicity. Raw materials used in the biocomposite preparation need to be assessed for toxicity before the fabrication [54].

Biomaterials such as biocomposites are recommended for a test on animal models before marketing, and hence, predictive *in vitro* systems are also recommended. It goes without saying that raw components, final material or finished biocomposites are also needed to be tested. Biochemical screening assays, chronic, local and systemic toxicity assays are the parameters for the routine toxicity evaluation. Phase I comprises of acute/subacute toxicity phase as well as mutagenic and carcinogenic activity of the biocomposites. By this, the biological and functional ability of the material is determined. In phase II, the test biocomposite material is implanted in animal model before being used together with the available information from phase I. In phase III, the biocomposite material with quality and foolproof toxicity will be tested in humans and then marketed. The key factors influencing the toxicity of composite materials or chemicals and their respective tests are depicted schematically in Figure 6.1.

Regulatory bodies such as ISO, OECD, USFDA and US-EPA have suggested some systemic biocompatibility tests for the toxicity assessment. Below are some of them in brief.

a. Cellular toxicity test (ISO Standard 10993-5): The cytotoxicity and cytocompatibility toxicity direct or indirect contact assay is tested in an in vitro cell line culture system by MTT and LDH assays; herein, the biocomposite material interacts with the cells of interest [55].

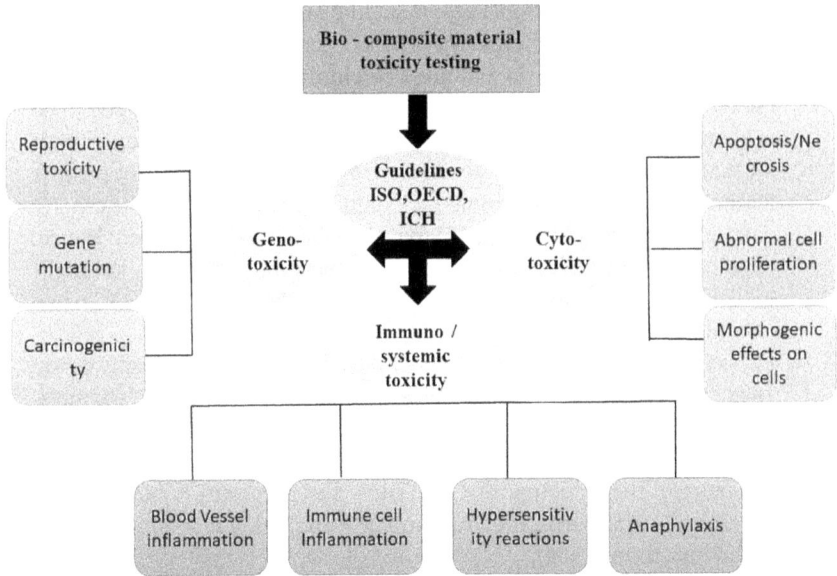

FIGURE 6.1   Key factors triggering the toxicity of composite material or chemicals and the recommended tests before the application.

b. Skin irritation and sensitization test (ISO 10993-10): Sensitization tests such as adjuvant-mediated guinea pig maximization test (GPMT) involves skin sensitization that may be triggered by biocomposites or their constituent materials. Visual manifestations of this test are necrosis in epidermis and dermis ulceration, or they may be manifested as pruritus, erythema, oedema, papules, vesicles or bullae. Immunologically mediated allergic contact dermatitis and cutaneous hypersensitive reaction are also other manifestations. Irritation tests such as patch assay also involve localized non-specific inflammatory response to biocomposites in the form of erythema (redness) of the skin and necrosis. Another type of test is the intracutaneous reactivity test; it is done mainly to detect irritant chemicals in biocomposites or that can leach out from biocomposites made up of polymers, ceramics, plastics, etc. It is a commonly used test [56].

c. Systemic toxicity test (ISO 10993-11, OECD and WHO guidelines): Here, the biocomposite material is injected in systemic circulation of rodent model, which gets distributed and absorbed from the entry route to the circulation where deleterious effects are manifested. Tests for pyrogenicity also falls in this category [57].

d. Acute, subacute and chronic toxicity tests (OECD 423, 407, 452, OECD and WHO guidelines): Serum biochemical and haematological parameters of the biocomposite material is assessed; no other documented toxicology information is available [58].

    e. Genotoxicity and carcinogenicity test (ISO 10993-3): Assessment of gene mutations, changes in chromosome structure and DNA/gene changes are caused by the test biocomposite material. It is the prime criteria of genotoxicity test. Mammalian cells, bacteria, yeasts, fungi or whole animal is used to test the genotoxicity. Carcinogenicity determines the carcinogenic potential of biocomposite materials or their by-products using multiple exposures to test animal over its life span [59].

    f. Developmental and reproductive toxicity test (OECD 421): Teratogenicity involves the development of foetus after exposure to the biocomposite material, and reproductive test describes the effects of male and female reproductive performance when exposed to the scaffold materials. Gross pathology of necropsied tissues and histopathology are evaluated [59].

    g. Haemocompatibility test (ISO 10993-4): Blood compatibility of biocomposite materials or their by-products for the possible thrombosis, coagulation and platelet function is evaluated [60].

    h. Immunotoxicology test (ISO 10993-20): Nanocomposite and biocomposite materials are tested for hypersensitivity, photosensitivity, induced autoimmunity, developmental immunotoxicity, immune suppression and stimulation on cultured cells [61,62].

The above are the novel methods suitable for material testing. However, the approach is not foolproof.

6.4   CONCLUSIONS

This chapter focussed on composites, namely synthetic biopolymers (PLA and PLGA) and natural biopolymers (chitosan, cellulose and collagen) together with their composites (PLA, PLGA, CNT, chitosan, cellulose, collagen, nanoclay, etc.). Each biopolymer molecule possesses a unique primary structure. The unique property of the composite for packaging is due to its primary structure and fabrication techniques. Properties of the fabricated composites, such as porosity, mechanical strength and water/moisture resistance, may vary depending on the material, processing method and conditions employed during processing. In view of augmenting the choice of bio-based composite applications, it is vital to take full benefit of exceptional properties of biopolymers and their composites to develop novel materials with safety and efficacy issues. However, biocompatibility, efficacy and toxicity evaluation of composites has crucial significance in the development of composites of preferred choice and hence a range of tests are utilized to do so. Regulatory bodies such as ISO, OECD and ICH provide tests/guidelines suitable to test the toxicity of chemicals and materials. *In vitro* tests on composites can be determined by direct or indirect contact toxicity assay. The toxicity of leachants on packed commodity and on the end user thereupon is determined by irritation test, sensitization, haemocompatibility, carcinogenicity, genotoxicity, etc. Biofunctional assays determine cell immune response. Acute, subacute and systemic toxicity evaluations (serum biochemical and haematological parameters) are primary parameters of toxicity evaluation for any material.

6.5 FUTURE PERSPECTIVES

Food packaging using materials composed of blends of renewable, biodegradable and sustainable synthetic and natural polymer composites is state of the art to replace the conventional non-renewable products from petrochemical industry. In addition, PLA or PLGA composites with abundant cellulose or chitosan matrix compared to other synthetic polymers with green protocol have attracted more attention as food packaging material with a revolution. Future research work on biopolymer reinforced with synthetic material should be encouraged to give eco-friendly materials. The feasible controlled and sustainable release effect of antimicrobial agents and kinetics of these materials together with the biodegradation rate and mechanism would be very useful to study the potential toxicity of the composite used in packing. This chapter mainly highlights the immense importance of biodegradable packing composites and their inherent toxicity associated with food packaging applications. Nanocomposites will pave the way for the solutions to the hurdles faced by the conventional packing materials in food industry as well as improve the material in an eco-friendly way.

REFERENCES

1. Huang, T., Qian, Y., Wei, J., Zhou, C. (2019). Polymeric antimicrobial food packaging and its applications. *Polymers*, 11(3), 560. https://doi.org/10.3390/polym11030560
2. Malinconico, M., Vink, E. T. H., Cain, A. (2018). Applications of poly (lactic acid) in commodities and specialties. *Advances in Polymer Science*, 35–50. doi:10.1007/12_2017_29.
3. Carlos, K. S., de Jager, L. S., Begley, T. H. (2018). Investigation of the primary plasticisers present in polyvinyl chloride (PVC) products currently authorised as food contact materials. *Food Addit Contam Part A Chem Anal Control Expo Risk Assess*, 35(6), 1214–1222. doi: 10.1080/19440049.2018.1447695.
4. Allahvaisi, S. (2012). Polypropylene in the Industry of Food Packaging, Polypropylene, Dr. Fatih Dogan (Ed.), ISBN: 978-953-51-0636-4, InTech, Available from http://www.intechopen.com/books/polypropylene/polypropylene-in-the-industry-of-food-packaging.
5. Ashok Kumar, S., Priya, K. D. (2016). Feasibility studies of cellulose microfiber (CMF) reinforced poly(ethylene-co-vinyl acetate) (EVA) composites for food packaging applications. *Science and Engineering of Composite Materials*, 23(5), 489–494. https://doi.org/10.1515/secm-2014-0252.
6. Sampath, U., Ching, Y., Chuah, C., Sabariah, J., Lin, P. C. (2016). Fabrication of porous materials from natural/synthetic biopolymers and their composites. *Materials*, 9(12), 991. doi:10.3390/ma9120991.
7. Souza, V., Pires, J., Rodrigues, C., Coelhoso, I. M., Fernando, A. L. (2020). Chitosan composites in packaging industry-current trends and future challenges. *Polymers*, 12(2), 417. https://doi.org/10.3390/polym12020417.

8. Hooda, R., Batra, B., Kalra, V., Rana, J. S., Sharma, M. (2018). Chitosan-based nanocomposites in food packaging. *Bio-Based Materials for Food Packaging*, 269–285. doi:10.1007/978-981-13-1909-9_12.
9. Mlalila, N., Hilonga, A., Swai, H., Devlieghere, F., Ragaert, P. (2018). Antimicrobial packaging based on starch, poly(3-hydroxybutyrate) and poly(lactic-co-glycolide) materials and application challenges. *Trends in Food Science & Technology*, 74, 1–11. doi:10.1016/j.tifs.2018.01.015.
10. Elisabeta, E. P., Maria, R., Ovidiu, P., Gabriel, M., Vlad, I. P., Amalia, C. M., Mona, E. P. (2017). Polylactic acid/cellulose fibres based composites for food packaging applications. *Materiale Plastice*, 54(4). https://doi.org/10.37358/MP.17.4.4923.
11. Wang, L. F., Rhim, J. W. (2015). Preparation and application of agar/alginate/collagen ternary blend functional food packaging films. *International Journal of Biological Macromolecules*, 80, 460–468. doi:10.1016/j.ijbiomac.2015.07.007.
12. Chaudhary, P., Fatima, F., Kumar, A. (2020). Relevance of nanomaterials in food packaging and its advanced future prospects. *Journal of Inorganica and Organometallic Polymers*, 30, 5180–5192. https://doi.org/10.1007/s10904-020-01674-8.
13. Sabulal, B., George, V. (2008). Essential oils and new antimicrobial strategies. In Ahmad, I., Aqil, F. (Eds.), *New Strategies Combating Bacterial Infection*. Wiley-VCH Verlag GmbH: Weinheim, Germany, 165–203. doi 10.1002/9783527622931.ch7.
14. Sabulal, B., George, V., Dan, M., Pradeep, N. S. (2007). Chemical composition and antimicrobial activities of the essential oils from the rhizomes of four Hedychium species from south India. *Journal of Essential Oil Research*, 19, 93–97. doi: 10.1080/10412905.2007.9699237.
15. Nikitha, S., Sabulal, B. (2021). Chemistry of Amomum essential oils. *Journal of Essential Oil Research*. doi: 10.1080/10412905.2021.1899065.
16. Jasim Ahmed, J., Hiremath, N., Jacob, H. (2017). Antimicrobial efficacies of essential oils/nanoparticles incorporated polylactide films against L. monocytogenes and S. typhimurium on contaminated cheese. *International Journal of Food Properties*, 20, 53–67. doi: 10.1080/10942912.2015.1131165.
17. Sharma, R., Mahdi Jafarib, S., Sharma, S. (2020). Antimicrobial bio-nanocomposites and their potential applications in food packaging. *Food Control*, 112, 107086. doi: 10.1016/j.foodcont.2020.107086.
18. Motelica, L., Ficai, D., Ficai, A., Trușcă, R. D., Ilie, C. I., Oprea, O. C., Andronescu, E. (2020). Innovative antimicrobial chitosan/ZnO/Ag NPs/citronella essential oil nanocomposite - Potential coating for grapes. *Foods*, 9, 1801. doi: 10.3390/foods9121801.
19. Alboofetileh, M., Rezaei, M., Hosseini, H., Abdollahi, M. (2014). Antimicrobial activity of alginate/clay nanocomposite films enriched with essential oils against three common foodborne pathogens. *Food Control*, 36, 1–7. doi: 10.1016/j.foodcont.2013.07.037.
20. da Costa, R. C., Daitx, T. S., Mauler, R. S., da Silva, N. M., Miotto, M., Crespo, J. S., Carli, L. N. (2020). Poly(hydroxybutyrate-co-hydroxyvalerate)-based nanocomposites for antimicrobial active food packaging containing oregano essential oil. *Food Packaging and Shelf Life*, 2, 100602. doi: 10.1016/j.fpsl.2020.100602.
21. Zhou, Y., Miao, X., Lan, X., Luo, J., Luo, T., Zhong, Z., Gao, X., Mafang, Z., Ji, J., Wang, H., Tang, Y. (2020). Angelica essential oil loaded electrospun gelatin nanofibers for active food packaging application. *Polymers*, 12, 299. doi: 10.3390/polym12020299.

22. Kuswandi, B. (2017). Environmental friendly food nano-packaging. *Environmental Chemistry Letters*, 15(2), 205–221. doi:10.1007/s10311-017-0613-7.
23. Kaseem, M. (2019). Poly(Lactic Acid) composites. *Materials* (Basel, Switzerland), 12(21), 3586. https://doi.org/10.3390/ma12213586.
24. Chiu, W., Chang, Y., Kuo, H., Lin, M., Wen, H. (2008). A study of carbon nanotubes/biodegradable plastic polylactic acid composites. *Journal of Applied Polymer Science*, 108, 3024–3030.
25. Yang, C., Gong, C., Peng, T., Deng, K., Zan, L. (2010). High photocatalytic degradation activity of the polyvinyl chloride (PVC)–vitamin C (VC)–TiO_2 nanocomposite film. *Journal of Hazardous Materials*, 178(1–3), 152–156. doi:10.1016/j.jhazmat.2010.01.056.
26. Jo, Y., Garcia, C. V., Ko, S., Lee, W., Shin, G. H., Choi, J. C., Park, S. J., Kim, J. T. (2018). Characterization and antibacterial properties of nanosilver-applied polyethylene and polypropylene composite films for food packaging applications. *Food Bioscience*, 23, 83–90. doi:10.1016/j.fbio.2018.03.008.
27. Lee, J. W., Son, S. M., Hong, S. I. (2008). Characterization of protein-coated polypropylene films as a novel composite structure for active food packaging application. *Journal of Food Engineering*, 86(4), 484–493. doi:10.1016/j.jfoodeng.2007.10.025.
28. Ozcalik, O., Tihminlioglu, F. (2013). Barrier properties of corn zein nanocomposite coated polypropylene films for food packaging applications. *Journal of Food Engineering*, 114(4), 505–513. doi:10.1016/j.jfoodeng.2012.09.005.
29. Han, C., Zhao, A., Varughese, E., Sahle-Demessie, E. (2018). Evaluating weathering of food packaging polyethylene-nano-clay composites: Release of nanoparticles and their impacts. *NanoImpact*, 9, 61–71. doi:10.1016/j.impact.2017.10.005.
30. Sahi, S., Djidjelli, H., Boukerrou, A. (2020). Study of the properties and biodegradability of the native and plasticized corn flour-filled low density polyethylene composites for food packaging applications. *Materials Today: Proceedings*. doi:10.1016/j.matpr.2020.05.317.
31. Sonia, A., Priya Dasan, K. (2013). Celluloses microfibers (CMF)/poly (ethylene-co-vinyl acetate) (EVA) composites for food packaging applications: A study based on barrier and biodegradation behavior. *Journal of Food Engineering*, 118(1), 78–89. doi:10.1016/j.jfoodeng.2013.03.020.
32. Scaffaro, R., Botta, L., Marineo, S., Puglia, A. M. (2011). Incorporation of Nisin in poly (ethylene-co-vinyl acetate) films by melt processing: A study on the antimicrobial properties. *Journal of Food Protection*, 74(7), 1137–1143. doi:10.4315/0362-028x.jfp-10-383.
33. Ghadikolaei, S. S., Omrani, A., Ehsani, M. (2018). Influences of modified bacterial cellulose nanofibers (BCNs) on structural, thermophysical, optical, and barrier properties of poly ethylene-co-vinyl acetate (EVA) nanocomposite. *International Journal of Biological Macromolecules*, 115, 266–272. doi:10.1016/j.ijbiomac.2018.04.071.
34. Nayak, S., Khuntia, S. K. (2019). Development and study of properties of Moringa oleifera fruit fibers/ polyethylene terephthalate composites for packaging applications. *Composites Communications*, 15, 113–119. doi:10.1016/j.coco.2019.07.008.
35. Xu, A., Wang, Y., Xu, X., Xiao, Z., Liu, R. (2020). A clean and sustainable cellulose-based composite film reinforced with waste plastic polyethylene terephthalate. *Advances in Materials Science and Engineering*, 2020, 1–7. doi:10.1155/2020/7323521.
36. Parvinzadeh, M., Moradian, S., Rashidi, A., Yazdanshenas, M. E. (2010). Surface characterization of polyethylene terephthalate/silica nanocomposites. *Applied Surface Science*, 256(9), 2792–2802. doi:10.1016/j.apsusc.2009.11.030.

37. Rusu, M. A., Leordean, D., Cosma, C., Filip, M., Moldovan, M., Silaghi-Dumitrescu, L., Hilda Orasan, O. (2017). Characterization polyethylene terephthalate nanocomposites mixing with nano-silica and titanium oxide. *MATEC Web of Conferences*, 137, 08005. doi:10.1051/matecconf/201713708005.

38. Sahraee, S., Milani, J. M. (2020). Chitin and chitosan-based blends, composites, and nanocomposites for packaging applications. *Handbook of Chitin and Chitosan*, 247–271. doi:10.1016/b978-0-12-817968-0.00008-1.

39. Hai, Le V., Lindong, Z., Hyun, C. K., Pooja, S. P., Duc, H. P., Jaehwan, K. (2020). Chitosan nanofiber and cellulose nanofiber blended composite applicable for active food packaging. *Nanomaterials*, 10(9), 1752. https://doi.org/10.3390/nano10091752.

40. Khosravi, A., Fereidoon, A., Khorasani, M. M., Naderi, G., Ganjali, M. R., Zarrintaj, P., Saeb, M. R., Gutiérrez, T. J. (2020). Soft and hard sections from cellulose-reinforced poly (lactic acid)-based food packaging films: A critical review. *Food Packaging and Shelf Life*, 23, 100429. https://doi.org/10.1016/j.fpsl.2019.100429.

41. Claro, P. I. C., Neto, A. R. S., Bibbo, A. C. C., Mattoso, L. H. C., Bastos, M. S. R., Marconcini, J. M. (2016). Biodegradable blends with potential use in packaging: A comparison of PLA/chitosan and PLA/cellulose acetate films. *Journal of Polymers and the Environment*, 24(4), 363–371. doi:10.1007/s10924-016-0785-4.

42. Albuquerque, R. M., Meira, H. M., Silva, I. D., Silva, C. J. G., Almeida, F. C. G., Amorim, J. D. P., Vinhas, G. M., Costa, A. F. S., Sarubbo, L. A. (2020). Production of a bacterial cellulose/poly(3-hydroxybutyrate) blend activated with clove essential oil for food packaging. *Polymers and Polymer Composites*, 096739112091209. doi:10.1177/0967391120912098.

43. Khalil, A. H. P. S., Saurabh, C. K., Adnan, A. S., Nurul Fazita, M. R., Syakir, M. I., Davoudpour, Y., Rafatullaha, M., Abdullaha, C. K., Haafiz, M. K. M., Dungani, R. (2016). A review on chitosan-cellulose blends and nanocellulose reinforced chitosan biocomposites: Properties and their applications. *Carbohydrate Polymers*, 150, 216–226. doi:10.1016/j.carbpol.2016.05.028.

44. Yang, H., Guo, X., Chen, X., Shu, Z. (2014). Preparation and characterization of collagen food packaging film. *Journal of Chemical and Pharmaceutical Research*, 6, 740–745.

45. Ramos, M., Valdés, A., Beltrán, A., Garrigós, M. (2016). Gelatin-based films and coatings for food packaging applications. *Coatings*, 6(4), 41. doi:10.3390/coatings6040041.

46. Ahmad, M., Nirmal, N. P., Danish, M., Chuprom, J., Jafarzedeh, S. (2016). Characterisation of composite films fabricated from collagen/chitosan and collagen/soy protein isolate for food packaging applications. *RSC Advances*, 6(85), 82191–82204. doi:10.1039/c6ra13043g.

47. Wang, Z., Hu, S., Wang, H. (2017). Scale-up preparation and characterization of collagen/sodium alginate blend films. *Journal of Food Quality*, 2017, 1–10. doi:10.1155/2017/4954259.

48. Jiang, T., Duan, Q., Zhu, J., Liu, H., Yu, L. (2019). Starch-based biodegradable materials: Challenges and opportunities. *Advanced Industrial and Engineering Polymer Research*. doi:10.1016/j.aiepr.2019.11.003.

49. Cuba, R. (2001). Toxicity myths essential oils and their carcinogenic potential. *International Journal of Aromatherapy*, 11(2), 76–83. doi:10.1016/s0962-4562(01)80021-7

50. Cirillo, G., Kozlowski, M. A., Spizzirri, U. G. (Eds.). (2018). *Composites Materials for Food Packaging*. doi:10.1002/9781119160243.

51. International Organization for Standardization (ISO) 10993, ISO/TC 194. (2018). Biological and clinical evaluation of medical devices—Part 1: Evaluation and testing within a risk management process. ISO 10993-1, 2018.

52. OECD guidelines for testing of chemicals 2018, 2019, 2020.

53. ICH Topic M 3 (R2) Non-Clinical Safety Studies for the Conduct of Human Clinical Trials and Marketing Authorization for Pharmaceuticals, 2008.

54. Chen, F. M., Liu, X. (2016). Advancing biomaterials of human origin for tissue engineering. *Progress in Polymer Science*, 53, 86–168. doi:10.1016/j.progpolymsci.2015.02.004.

55. Basu, B. (2016). Probing toxicity of biomaterials and biocompatibility assessment. *Indian Institute of Metals Series*, 291–351. doi:10.1007/978-981-10–3059-8_9.

56. ISO 10993-5:2009(en). (2009). Biological evaluation of medical devices — Part 5: Tests for in vitro cytotoxicity.

57. ISO 10993-10:2010. (2010). Biological evaluation of medical devices — Part 10: Tests for irritation and skin sensitization.

58. ISO 10993-11:2017. (2017). Biological evaluation of medical devices — Part 11: Tests for systemic toxicity.

59. OECD guidelines for testing of chemicals, Test Guidelines, 2020.

60. ISO 10993-3:2014. (2014). Biological evaluation of medical devices — Part 3: Tests for genotoxicity, carcinogenicity and reproductive toxicity.

61. ISO 10993-4:2017. (2017). Biological evaluation of medical devices — Part 4: Selection of tests for interactions with blood.

62. ISO/TS 10993-20:2006. (2006). Biological evaluation of medical devices — Part 20: Principles and methods for immunotoxicology testing of medical devices.

7 Biodegradable Composites for Conductive and Sensor Applications

V. Andal
KCG College of Technology

Karthik Kannan
Kumoh National Institute of Technology

Z. Edward Kennedy
KCG College of Technology

CONTENTS

DOI: 10.1201/9781003227908-7

7.1   INTRODUCTION

Many researchers have concentrated on developing smart eco-friendly materials, such as biodegradable composites, because they contribute to a sustainable global environment by significantly reducing non-recyclable plastic waste and are capable of transforming information while reducing its environmental impact. There is currently an increasing demand for commodities manufactured from renewable resources that can degrade in the environment. Consumers are also showing an interest in environmentally friendly materials possessing the same qualities as synthetic materials. Biodegradable composites that are not reliant on fossil fuels are in high demand around the world. The advantages and limitations of biodegradable composite make it an excellent material for a wide range of applications that benefit society. Biodegradable composites are widely used in various fields including packaging, automotive, medical, and sensing [1]. There is a demand for alternative materials as the use of non-biodegradable polymers has become ubiquitous. As a result, the development of biodegradable composites is a viable solution [2]. Because of their potential to be disintegrated by environmental agents, biopolymer composites are rapidly being recognized as better alternatives to typical non-biodegradable materials.

Biodegradable composites are made up of two-phase hybrid materials such as matrix and filler, each of which can be biodegradable on its own or both together. If the matrix and reinforcement or filler were created from biopolymers or renewable resources, the composites would be completely biodegradable. Water absorption, which affects mechanical characteristics, is one of the drawbacks of biodegradable composites made with natural polymers as a matrix. Biopolymers have been chemically and physically modified to increase their resistance to moisture to overcome these limitations, but success in creating biodegradable composites with desirable qualities has been limited. Matrix-based and reinforcement-based biodegradable composites are the two types. Our focus is on matrix-based systems. Figure 7.1 shows the classification of matrix-based biodegradable composites.

The use of biodegradable composites is not new to humanity; it has been practised since ancient times [3]. The base material, binder, and plasticizer are mixed together to make a biodegradable matrix [4]. Polymers as a matrix in natural and man-made biodegradable composites include polysaccharides, proteins, polyhydroxyalkonates, polylactides, polyglycolides, polylactones, and other biodegradable polymers [5].

High sensitivity, high selectivity, short response time, normal working circumstances, long time stability, reversibility, low cost, low maintenance, and ease of use are all the basic requirements for an ideal sensor. In the past, inorganic and organic materials were used to make the sensing layer, but biopolymers, as sensing materials in transducers, were a new focus in sensor research.

Biodegradable matrices containing composites are extensively applied as a sensor, which can naturally disappear in the environment. Biodegradable composite mitigates the e-waste and medical waste in the environment. Biodegradable

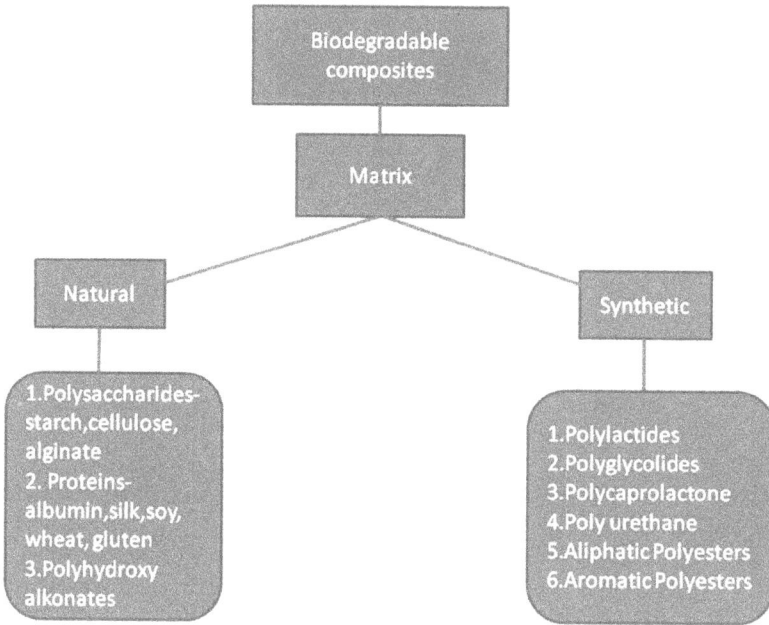

FIGURE 7.1 The classification of matrix-based biodegradable composites.

composites incorporating biodegradable polymers and fillers have recently been employed to build cost-effective, simple, selective, sensitive, and biocompatible sensors. Sensors are electronic devices, which can disappear in the environment after functioning; hence, they are referred to as biodegradable electronics [3] Biopolymer-based sensors are less expensive and have easy manufacturing procedures. Because of their outstanding biocompatibility, sensitivity, and selectivity, combining polymers with carbon nanomaterials for the manufacture of chemical sensors opens up exciting new research fields.

Chemical and biosensors are cost-effective alternatives to expensive analytical instruments because they are simple, of low cost, sturdy, portable, and easy to operate [4]. Biodegradable composites for flexible sensors with outstanding mechanical properties and sensing ability have recently been the subject of research [5,6].

This chapter presents an overview of recent reports on biodegradable composites as sensors in environmental, diagnostic, and therapeutic applications (Figure 7.2). The objective of this chapter is to provide a systematic review of the existing literature over the past two decades regarding environmental and medical sensing applications of biodegradable composites. Most of these studies focus on the removal of dyes and heavy metals, as well as dozens of studies are on antibiotics and other pollutants. A future perspective on the application of alginate-based composites for environmental remediation is presented.

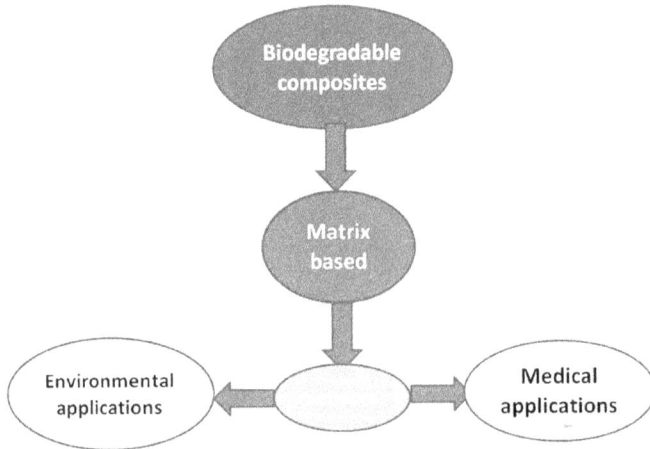

FIGURE 7.2 Schematic representation of sensing application of matrix-based biodegradable composites.

7.2 BIODEGRADABLE POLYMERS AS MATRICES IN A BIODEGRADABLE COMPOSITE

In general, biodegradable polymers are divided into two categories: natural biodegradable polymers and synthetic biodegradable polymers. Natural biodegradable polymers are made from natural resources such as plants and animals. As a result, there are two types of natural biodegradable polymers: protein-derived polymers (e.g. collagen and gelatin) and polysaccharides (e.g. alginates, chitosan, and starch) [4].

7.2.1 Natural Biodegradable Polymers

The most studied natural biodegradable polymers are natural rubber, lignin, and polysaccharides. Cellulose, chitin, chitosan, amylase, sodium alginate, and other polysaccharides are examples. We merely go over a few key points and use of common polysaccharides (such as starch, cellulose, and alginate), proteins (albumin, silk, soy, wheat, and gluten), and polyhydroxyalkonates as examples to show how natural polymers can be used as matrices.

7.2.2 Polysaccharides

7.2.2.1 Starch

Starch is a biodegradable polymer generated from plants whose potential restrictions, such as short stability and low moisture resistance, can be regulated through modification. Starch-based composites have recently been developed and explored for a variety of applications. Ibrahim et al. created a starch-based composite with the Nile rose fibre reinforcement for a variety of applications [7]. Starch carbonaceous composites are extensively investigated as sensors [8].

7.2.2.2 Cellulose

Cellulose is the most abundant biodegradable polymer. Sensing materials made of cellulose-based composites have been utilized for a long time and are gaining popularity [9]. Because of its characteristics, cellulose has great chemical and thermal stability and can be modified and used for sensors. It is generally known that cellulose has become commercially available in composites as reinforcement with both petrochemical- and bio-based polymer matrices. Mulinari et al. [10] reported a hybrid composite of bleached cellulose and hydrous zirconium oxide. The authors demonstrated that cellulose composites made from crushed sugar-cane and inorganic materials have inherent benefits such as low cost, biodegradability, and ease of preparation and handling.

7.2.2.3 Alginate

Alginate is an anionic polysaccharide made up of two different types of 1,4-linked hexuronic acid residues: b-d-mannuronopyranosyl (M) and a-l-guluronopyranosyl (G) residues. Alginate's surface functional groups make it an effective material for adsorption and other environmental applications. Shane et al. studied alginate-based composites as a biomarker for glucose [11,12].

7.2.2.4 Proteins

In tissue engineering, a protein matrix is commonly employed. Recently in biomedical research, protein-based composites have been studied for various applications. Protein-based composites possess tunable properties; hence, they meet the needs of the medical field by developing novel medical devices. Protein matrix composite also plays a key role in sensing and other applications, which are given below. Jianzhang Li et al. [13] investigated a soy protein-based composite film with excellent UV-blocking capacity [13]. Similarly, wheat gluten is also used as a filler to detect CO_2 and relative humidity in food packaging applications [14,15].

7.2.2.5 Polyhydroxyalkonates

PHAs are a type of biodegradable and biocompatible plastic made from R-hydroxyalkanoic acid polyesters. PHA biocomposites have demonstrated their adaptability and enormous potential as lightweight, long-lasting composites. Their thermoplastic and elastomeric characteristics are expensive. Several microorganisms produce them. These materials are helping to pave the way for a more sustainable future [16–18]. Polyhydroxybutyrate belongs to the class of polyhydroxyalkonates. Polyhydroxybutyrate is a crystalline polymer having a glass transition temperature (T_g) of 5°C. Grande and colleagues created a biocomposite of PHB nanofibres and ZnO nanoparticles and discovered that the ZnO nanoparticles changed the conductivity, viscosity, and thermal stability of the material. Furthermore, the biocomposite fibres' antibacterial properties were discovered against *E. coli* and *S. aureus* [19].

7.2.3 Synthetic Biodegradable Polymers

In recent years, several synthetic biodegradable polymers such as poly(lactic acid) (PLA), polyglycolic acid (PGA), polyurethane, and poly(ε-caprolactone) (PCL) have been explored as matrices for composites [20]. Aliphatic polyesters, such as PCL and poly(L-lactic acid) (PLLA), have intensively been investigated over the last two decades due to their capacity to undergo hydrolysis in the human body and in natural conditions [21–23]. PLA is a translucent thermoplastic biopolymer created by polymerizing lactic acid. It is the best alternative to petroleum-based polymer, but its use is limited due to its low biodegradability and hydrophobicity [24]. Polycaprolactone is a biodegradable polymer obtained by the ring-opening polymerization of caprolactone. In comparison with petroleum-based polymers, polycaprolactone has hydrophobicity, making it easily degradable [25]. Polyurethane resin is derived from castor oil and possesses a quality that makes it a good substitute for glass matrix [26].

7.3 SENSORS

Sensors play an important role in our daily lives; thus, they must be efficiently recycled and have a minimal environmental impact. Sensors are composed of biodegradable materials, either natural or synthetic, to ensure their long-term viability. Some materials can decay in hostile environments, and others can degrade in the environment without leaving harmful components. As a result, we will look at biodegradable composites that are used as sensors in environmental and medical applications [27].

7.3.1 Biodegradable Composites as Sensors for Environmental Applications

Biodegradable composites are made up of degradable polymers as a matrix and degradable or non-degradable reinforcing. Sensors made using biodegradable composites are used to monitor the environment. The explanation below depicts the numerous biodegradable matrix sensors and their applications in the environment.

Environmental monitoring with sensor technology is considered as one of the most important enablers for managing the environment and natural resources in a sustainable manner. Disposable sensors are inexpensive and simple to use.

Khachatryan et al. [28] reported starch-based biodegradable nanocomposite containing ZnS quantum dots capped with L-cysteine as a sensor for heavy metal detection of Cu^{2+} and Pb^{2+} ions. The ZnS quantum dots of 10–20 nm were synthesized on a potato starch matrix. The prepared biofilm shows a decrease in the photoluminescent intensity when exposed to heavy metal ions such as Cu^{2+} and Pb^{2+} [28].

The development of simple, reusable, portable, extremely sensitive, real-time detection, and low-cost poisonous gas sensor devices is critical in both industrial processing and environmental monitoring. Tawfik et al. developed a CMC–BCP (carboxymethyl cellulose–bromocresol purple) vapochromic xerogel for real-time naked eye ammonia detection with a detection limit of 9.0×10^{-2} ppb. The microporous structure is reusable, and the direct recognition procedure for trace target ammonia is reversible [29]. In gas sensing, cellulose-based composites with conducting polymers play an important role. Yun et al. [30], for example, created a multiwalled carbon nanotube–cellulose paper to detect vaporized analyte molecules such as methanol, 1-ethanol, 1-butanol, and 1-propanol. Similarly, cellulose and conducting polymer are used to make humidity and temperature sensors. Mahadeva et al. [31] used in situ polymerization to create a flexible cellulose polypyrrole nanocomposite and examined its humidity and temperature-sensing properties.

Chitosan was one of the conducting biopolymers that gained a lot of interest when it came to being employed as a sensor. Several sensors are investigated utilizing chitosan as a matrix. Li et al. [32] produced polyaniline nanofibres embedded in chitosan matrix, in which polyaniline acts as transducer, while chitosan matrix interacts with analyte hydrogen gas. Hydrogen molecules mix with oxygen to generate water, which in turn swells the chitosan matrix. As a result, this sensor has a strong selectivity for hydrogen gas. By using the sol–gel process, Abou Hammad et al. [33] generated a hybrid modified chitosan/calcium silicate nanocomposite membrane with Al_2O_3 doping. The CO_2 gas sensing selectivity of the produced nanocomposite was good. Bijouy Sankar and his colleague [34] created a low-cost, environmentally friendly sensor for detecting dangerous lead ions in aqueous media. As a sensor probe, they used a U-shaped optical fibre coated in chitosan–PVA matrix and modified with gold nanoparticles functionalized with glutathione. With a sensitivity of $0.23 \, mV \, ppb^{-1}$, the sensor probe proved effective in detecting lead ions. Chitosan's polymeric chain may easily be changed with different types of nanoparticles, making it ideal for the fabrication of nanocomposites films. Chitosan nanocomposites are frequently employed in sensor construction due to their huge specific surface area and strong charge transfer of the nanoparticles. Using a method of in situ chemical reduction, Wang and Gao [35] created a CS/Ag nanocomposite film. They used it to detect Al^{3+} concentrations in aqueous solutions. Adeosun et al. [36] created an electrochemical sensor for detecting sulphite that used chitosan to help produce the film and polypyrrole to increase the transfer charge. The sulphite sensor that was proposed had a low LOD and high sensitivity. Xu et al. [37] successfully prepared a chitosan–polypyrrole nanocomposite and used it for the determination of Pb(II) in wastewater. Using the drop casting process of chitosan, polypyrrole, and carboxyl graphene solution on the surface of the glassy carbon electrode, an electrochemical sensor for nitrite detection was developed. For NO_2 detection, the sensor demonstrated good sensitivity and selectivity [38]. In the same context, Shen et al. [39] fabricated an electrochemical sensor for the identification of dopamine by including the use of chitosan to aid film formation, poly(3,4-ethylenedioxythiophene) as a conducting polymer, and graphene to improve the electronic transfer. The resulting sensor was very sensitive to dopamine, with a detection limit of 0.29 M.

A strain sensor is a force-sensitive sensor that works by measuring the resistance-strain effect caused by an object's stress deformation (compression, bending, or stretching). The so-called resistance-strain effect describes how the resistance of a material changes as it deforms. Various conductive materials (carbon black, carbon nanotube, graphene, nanowires, etc.) are utilized to construct sensors based on the resistance-strain effect, which involves sensing resistance changes as things deform [40–44]. Zhang et al. [45] created a 3D hierarchical conductive structure of CB/natural rubber composites using latex assembly technology and cellulose nano-whiskers. The effect of a hierarchical conductive structure on liquid sensing was examined. Considering their electronic conductivity and mechanical flexibility, porous CB-based composite sponges with low density and a rapid recovery rate are attractive for the development of strain sensors. Liu et al. [46] used a solution freeze-drying process to create a CS/CB composite porous sponge that can detect various human actions (such as pronouncing, breathing, and joint bending), making it a viable strain sensor. Maximiano and his colleagues used ultrasonic mixing to create a CB/PMMA (carbon black/PMMA) conductive composite for sensing gas vapours [47]. Heavy metal ions, organic vapours, various liquid analytes, humidity, temperature, and gases such as hydrogen, CO_2, and NO_2 have all been detected using biodegradable composite-based sensors to date. Table 7.1 shows a summary of the applications of biodegradable composite-based sensors for environmental applications.

TABLE 7.1
Environmental Applications of Biodegradable Composite-Based Sensors

Biodegradable Composite	Analyte	Reference
CB/PLA (carbon black/polylactic acid)	Methanol, ethanol, chloroform, n-hexane, xylene, ethyl acetate	[48]
Polyethylenimine/carbon black	Volatile carboxylic acids	[49]
Waterborne polyurethane/carbon black	Organic vapours	[50]
Conductive polypyrrole/polyurethane composite	Chloroform	[51]
Polylactic acid/carbon black composite	Solvents – dichloromethane, ethanol, acetone, tetrahydrofuran, chloroform, and ethyl acetate	[52]
Polylactic acid/carbon nanotube composite	Humidity	[53]
Chitosan–gold nanocomposite	Caffeine	[54]
Chitosan–gold nanocomposite	Gallic acid	[55]
Amino-functionalized graphene–chitosan composite	Cu^{2+}	[56]
Carbon nanotube/cellulose composite	Water	[57]
Graphene oxide/chitosan nanocomposite	Amine vapour detection	[58]
Polyaniline-grafted chitosan sensor	Ammonia and ethanol vapour	[59]
Ag nanoparticles/alginate composite	H_2O_2	[60]
Laccase immobilized on a nanocomposite chitosan–ZnO nanoparticles	Chlorophenol	[61]

7.3.2 Biodegradable Composites as Sensors for Medical Applications

Biodegradable composites offer a fantastic chance for healthcare technology to disintegrate naturally after use as sensors. They stay away from chronic inflammations as well as technological and medical trash. Here are some of the most recent reports on body diagnostic, monitoring, and therapeutic applications. Researchers are interested in flexible sensors because of their widespread use in wearable electronics. Pollution in the environment is increasing day by day because of their use. Hence, degradable polymers such as silk, paper, and polylactic acid are used for developing flexible sensors. Liu and colleagues developed a multimodal sensor by transferring laser-induced porous carbon from polyimide to a starch layer. The sensor had good sensitivity to strain, temperature, and pressure, and it degraded in water, indicating that it was pollutant-free. The sensor can detect strain with a GF of 134.2, a response time of 130 ms, and good stability over 1000 times. It can also detect temperature changes between 25°C and 90°C with a sensitivity of $S = 1.08$°C and pressure changes between 0 and 250 kPa with a sensitivity of 0.096 kPa. Wearable technologies for sports and health monitoring, human–machine interfaces, and artificial intelligence equipment might all use flexible sensors to suit the needs of a future sustainable society [62].

Chitosan is frequently mixed with nanoparticles such as graphene, multiwalled carbon nanotubes, and conducting polymers such as polypyrrole and polyaniline to improve its electrical characteristics for sensing applications [63–66]. Mohamed Imran and his co-workers reported on the synthesis of positively charged chitosan-stabilized silver nanoparticles for the electrochemical detection of lipopolysaccharide. Since the outer covering of several gram-negative bacteria is negatively charged lipopolysaccharide, the nanoparticles were effective in detecting the pathogens. The biosensor was able to detect low concentrations of *E. coli* also [67].

Using a porous nickel molybdate nanosheets–chitosan nanocomposite-modified glassy carbon electrode, Lou et al. [68] created an electrochemical sensor for amlodipine detection. When compared to some of the electrochemical sensors created for the determination of amlodipine, such as multiwalled carbon nanotube sensors and graphite-modified paste electrodes, this device displays enhanced electrochemical performance. A sensor based on chitosan–zeolitic imidazolate framework-8 (ZIF-8)–acetylene black nanocomposites for rutin measurement had a similar case [69]. ZIF-8 and acetylene black were dispersed with the help of chitosan. The dispersion was then put over the glassy carbon electrode's surface and cured under an infrared lamp. The sensor had exceptional sensitivity, stability, and repeatability when it comes to rutin, according to the authors.

The substrate had a significant influence on the creation of e-waste in sensor production. The sensor substrate is composed of biodegradable composites to avoid this. Numerous protein-based biopolymers, such as silk, cellulose, and chitosan, have recently attracted a lot of interest as substrate materials for biodegradable devices due to their ubiquitous availability, outstanding biocompatibility, flexibility, and environmental sustainability [70].

Silk has also been utilized to facilitate the biotransfer of peptide-modified graphene nanosensors onto biomaterials and the bioselective detection of germs using dental enamel. Recently, silk has been coupled with graphene to act both as a substrate and as a passage that is electrically conducting for the development of self-healing, skin-mounted electronic tattoos that are sensitive to changes in strain, temperature and humidity [71]. Apart from silk, cellulose, the most abundant natural polysaccharide on the planet, is a promising substrate for biodegradable sensors in healthcare monitoring because of its appealing degradation behaviour in physiological environments, as well as its flexibility, transparency, high-temperature stability, and excellent biocompatibility [72,73]. In resource-constrained situations where specialized equipment and medical professionals are not always readily available, cellulose paper-based diagnostic devices can provide cost-effective and portable diagnostic options [74]. Paper was chosen as a suitable substrate to support the large-area growth of $NiSe_2$ for the fabrication of a pH sensor for non-invasive oral health monitoring, a breath sensor to monitor breath-related diseases, and a physical strain sensor for gesture recognition to assist deaf, mute, and aurally challenged people (Figure 7.3) [75].

In the same context, disposable, lightweight, and low-cost paper-based, "calibration-free" electrochemical wearable sensors have also been developed for real-time, continuous, and on-site detection of inhaled hydrogen peroxide (H_2O_2) in artificial breath [76]. A "paper watch" has also been devised to measure important body conditions in real time, such as blood pressure, heart rate, body temperature, and skin hydration, utilizing recyclable and non-functionalized materials. As a structural support, use Post-it paper [77]. A wearable paper-based chemiresistor for monitoring both sweat rate and sweat loss in the human body has also been constructed by incorporating a nanocomposite of single-walled carbon nanotubes (SWCNTs) and sodium dodecylbenzenesulphonate surfactant into the

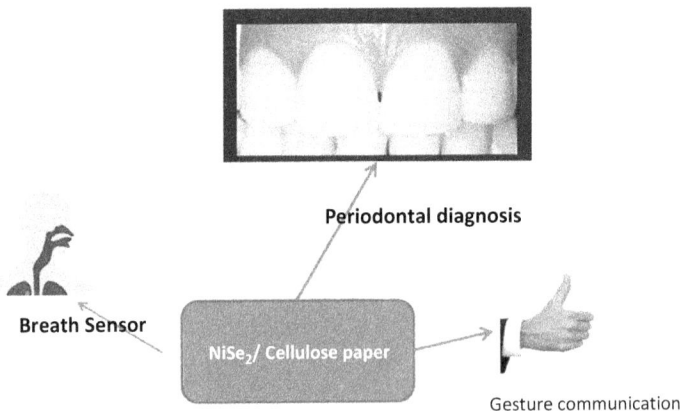

FIGURE 7.3 Schematic diagram exhibiting the use of $NiSe_2$-modified cellulose paper to make a non-invasive periodontal diagnostic sensor, a breath analyser, and a gesture sensor.

TABLE 7.2

A Summary of the Applications of Biodegradable Composite-Based Sensors for Medical Applications

Biodegradable Composite	Analyte	Reference
MWCNTs–chitosan–Co	Paracetamol	[82]
MWCNTs–chitosan nanocomposite-incorporated nano-hydroxyapatite	Nitrofurantoin	[83]
PEDOT/GPS/AuNPs	Micro-RNA	[84]
CuO/chitosan nanocomposites	Glucose non-enzymatic	[85]
Chitosan/Au nanoparticles film on GCE (glassy carbon electrode)	Glucose non-enzymatic	[86]
Chitosan/graphene oxide nanocomposite	Melamine	[87]
Chitosan–Y$_2$O$_3$ nanocomposite	Norfloxacin	[88]
Chitosan, polypyrrole, multiwalled carbon nanotubes with gold nanoparticles	*Escherichia coli*	[89]

cellulose fibres of commercial filter paper [78]. The resulting wearable device can be used for a number of on-body biofluid analysis applications and offers simple and affordable real-time perspiration measurements. Chitosan is combined with starch or glycine to provide a low-cost, biodegradable, transparent substrate for pressure sensing and other applications [79,80].

Proteins are employed as substrates for electrochemical biosensors, which are seen as a viable alternative to synthetic materials. Pal and his colleague used photolithography to create an electrochemical biosensor out of silk and conducting polymer. The designed sensor has been demonstrated to be stable and capable of detecting both particular and non-specific chemicals in vitro [81] (Table 7.2).

7.4 CONCLUSIONS

The reviewed biodegradable composites in this chapter are playing a major role for improving people's health and environment for achieving the aim of long-term healthcare and environment. Due to the extended time required for degradation, contemporary literature works progressively highlight the need to substitute synthetic polymers. This drive to create novel materials with adequate mechanical, thermal, and degrading qualities has resulted in materials that are environmentally friendly when degraded. The requirement to improve the properties of biodegradable polymeric matrices for a specific application has been a significant issue for materials manufacturers. In this chapter, the primary matrices used in the literature in recent years for biodegradable composite production for sensors were discussed: starch, chitosan, cellulose, PLA, PCL, etc. To be used as biodegradable sensors, the composites must have a high electrical conductivity. To address these critical issues and ensure the continued development of biodegradable composites, more widespread distribution of these innovative technologies is required to boost productivity and lower final material costs.

The already-in-use nanotechnology can also play a role in these advancements; the higher surface area of nanostructures opens up new possibilities for matrix and reinforcement interaction, resulting in superior qualities or even new ones that need less reinforcement in the manufacturing process.

Finally, in order to develop sensors for clinical and environmental use, more knowledge of the environment, biological systems, and immune responses is required to build sensor devices that are specifically suited to the in vivo environment. In order to do this, more interactions and collaborations between chemists, engineers, and biologists is required.

ACKNOWLEDGEMENT

The authors would like to express their gratitude to the management of KCG College of Technology for their assistance.

REFERENCES

1. M. Abhilash, D. Thomas, *Biopolymers for Biocomposites and Chemical Sensor Applications*, Elsevier, 2017, 405–435, ISBN 9780128092613.
2. S.-W. Hwang, H. Tao, D.-H. Kim, H. Cheng, J.-K. Song, E. Rill, M.A. Brenckle, B. Panilaitis, S.M. Won, Y.-S. Kim, Y.M. Song, K.J. Yu, A. Ameen, R. Li, Y. Su, M. Yang, D.L. Kaplan, M.R. Zakin, M.J. Slepian, Y. Huang, F.G. Omenetto, J.A. Rogers, A physically transient form of silicon electronics, *Science* 337 (2012) 1640.
3. D. Zimmerman ICE case studies, The Great Wall of China [http://www.american.edu/ted/ice/wall.htm; accessed in September 2008].
4. M.S. Hasnain, S.A. Ahmad, N. Chaudhary, M.N. Hoda, A.K. Nayak, C. M. Boutry, Y. Kaizawa, B.C. Schroeder et al., A stretchable and biodegradable strain and pressure sensor for orthopaedic application, *Nature Electronics* 1:5 (2018) 314–321; R. Feig, H. Tran, Z. N. Bao, Biodegradable polymeric materials in degradable electronic devices, *ACS Central Science* 4:3 (2018) 337–348.
5. A.B. Balaji, H. Pakalapati, M. Khalid, R. Walvekar, H. Siddiqui, Natural and synthetic biocompatible and biodegradable polymers, *Biodegradable and Biocompatible Polymer Composites*, Woodhead Publishing, 2018, 3–32, ISBN 9780081009703.
6. L. Wang, D. Chen, K. Jiang, G. Shen, New insights and perspectives into biological materials for flexible electronics, *Chemical Society Reviews* 46:22 (2017) 6764–6815.
7. M.M. Ibrahim, H. Moustafa, E.N. Abd, E.L. Rahman, S. Mehanny, M.H. Hemida, E. El-Kashif, Reinforcement of starch based biodegradable composite using Nile rose residues, *Journal of Materials Research and Technology* 9 (2020) 6160–6171.
8. V. Gautam, K.P. Singh, V.L. Yadav, Polyaniline/MWCNTs/starch modified carbon paste electrode for non-enzymatic detection of cholesterol: Application to real sample (cow milk), *Analytical and Bioanalytical Chemistry* 410:8 (2018 Mar) 2173–2181. doi: 10.1007/s00216-018-0880-6.
9. K. Watanabe, M. Tabuchi, Y. Morinaga, F. Yoshinaga, Structural features and properties of bacterial cellulose produced in agitated culture, *Cellulose* 5 (1998) 187–200.

10. Preparation and characterization of the Cellulose/ $ZrO_2.nH_2O$ Composites prepared by the methods of Conventional precipitation and homogeneous solution precipitation. Brazillian Polymer Congress, Águas de Lindóia, 245–246 (2005).
11. L.R. Bornhoeft, A. Biswas, M.J. Mc Shane, Composite hydrogels with engineered microdomains for optical glucose sensing at low oxygen conditions, *Biosensors (Basel)* 7:1 (2017) 8. doi: 10.3390/bios7010008
12. B. Wang, Y. Wan, Y. Zheng, X. Lee, T. Liu, Z. Yu, J. Huang, Y.S. Ok, J. Chen, B. Gao, Alginate-based composites for environmental applications: A critical review, *Critical Reviews in Environmental Science and Technology* 49:4 (2018) 318–356. doi: 10.1080/10643389.2018.1547621
13. J. Li, S. Jiang, Y. Wei, X. Li, S.Q. Shi, W. Zhang, J. Li, Facile fabrication of tough, strong, and biodegradable soy protein-based composite films with excellent UV-blocking performance, *Composites Part B: Engineering* 211 (2021) 108645.
14. F. Bibi, C. Guillaume, N. Gontard, B. Sorli, Wheat gluten, a bio-polymer to monitor carbon dioxide in food packaging: Electric and dielectric characterization, *Sensors and Actuators B: Chemical* 250 (2017) 76–84. doi: 10.1016/j.snb.2017.03.164
15. F. Bibi, C. Guillaume, A. Vena, N. Gontard, B. Sorli, Wheat gluten, a bio-polymer layer to monitor relative humidity in food packaging: Electric and dielectric characterization, *Sensors and Actuators A: Physical* 247 (2016) 355–367. doi: 10.1016/j.sna.2016.06.017
16. Z. Li, J. Yang, X. Loh, Polyhydroxyalkanoates: Opening doors for a sustainable future, *NPG Asia Materials* 8 (2016) e265. doi: 10.1038/am.2016.48
17. R. Nigmatullin, P. Thomas, B. Lukasiewicz, H. Puthussery, I. Roy, Polyhydroxyalkanoates, a family of natural polymers, and their applications in drug delivery, *Journal of Chemical Technology & Biotechnology* 90 (2015) 1209–1221. doi: 10.1002/jctb.4685
18. P. Poltronieri, P. Kumar, Polyhydroxyalkanoates (PHAs) in industrial applications. In L. Martínez, O. Kharissova, B. Kharisov (eds.), *Handbook of Ecomaterials*, Springer, Cham, 2019. doi: 10.1007/978-3-319-68255-6_70
19. H. Rodríguez-Tobías, G. Morales, A. Ledezma, J. Romero, R. Saldívar, V. Langlois, et al., Electrospinning and electrospraying techniques for designing novel antibacterial poly (3- hydroxybutyrate)/zinc oxide nanofibrous composites, *Journal of Materials Science* 51:18 (2016) 8593609
20. G.E. Luckachan, C.K.S. Pillai, Biodegradable polymers-A review on recent trends and emerging perspectives, *Journal of Polymers and the Environment*, 19 (2011) 637–676.
21. D.S. Rosa, R. Pântano Filho, Biodegradação: um ensaio com polímeros. Moara (Ed.), Itatiba, S.P; Editora Universitária São Francisco (Ed.), Bragança Paulista, S.P. (2003) 1–14.
22. N. Vogelsanger, M.C. Formolo, A.P.T. Pezzin, A.L.S. Schneider, A.A. Furlan, H.P. Bernardo, S.H. Pezzin, A.T.N Pires, E.A.R. Duek, Blendas biodegradáveis de poli(3-hidroxibutirato)/poli(-caprolactona): obtenção e estudo da miscibilidade, *Materials Research* 6:3 (2003) 359–365.
23. A.D. Raghavan, Characterization of biodegradable plastics, *Polymer-Plastics Technology and Engineering* 34:1 (1995) 41–63.
24. J.M. Raquez, Y. Habibi, M. Murariu, P. Doubois, Polylactide (PLA)-based nanocomposites, *Progress in Polymer Science* 38 (2013) 1504–1542. doi: 10.101 6/j.progpolymsci.2013.05.014
25. K. Jha, Y.K. Tyagi, A.S. Yadav, Mechanical and thermal behaviour of biodegradable composites based on polycaprolactone with pine cone particle, *Sådhanå* 43 (2018) 135. doi: 10.1007/s12046-018-0822-1

26. A. Olszewski, P. Kosmela, A. Mielewczyk-Gryń, Ł. Piszczyk, Bio-based polyurethane composites and hybrid composites containing a new type of bio-polyol and addition of natural and synthetic fibers, *Materials (Basel)* 13:9 (2020) 2028. Published 2020 April 26. doi: 10.3390/ma13092028

27. Y. Cao, K.E. Uhrich, Biodegradable and biocompatible polymers for electronic applications, *A Review Journal of Bioactive and Compatible Polymers* 34:1 (2019) 3–15.

28. G. Khachatryan, K. Khachatryan, Starch based nanocomposites as sensors for heavy metals – detection of Cu^{2+} and Pb^{2+} ions, *International Agrophysics* 33 (2019) 121–126.

29. T.A. Khattab, S. Dacrory, H. Abou-Yousef, S. Kamel, Development of microporous cellulose-based smart xerogel reversible sensor via freeze drying for naked-eye detection of ammonia gas, *Carbohydrate Polymers* 210 (2019) 196–203.

30. S. Yun, J. Kim, Multi-walled carbon nanotubes-cellulose paper for a chemical vapor sensor, *Sensors and Actuators B: Chemical* 150 (2010) 308–313.

31. S.K. Mahadeva, S. Yun, J. Kim, Flexible humidity and temperature sensor based on cellulose–polypyrrole nanocomposite, *Sensors and Actuators A: Physical* 165 (2011) 194–199.

32. W. Li, D.M. Jang, S.Y. An, D. Kim, S.-K. Hong, H. Kim, Polyaniline– chitosan nanocomposite: High performance hydrogen sensor from new principle, *Sensors and Actuators B: Chemical* 160 (2011) 1020–1025.

33. A.B. Abou Hammad, A.M. Elnahrawy, A.M. Youssef, A.M. Youssef, Sol gel synthesis of hybrid chitosan/calcium aluminosilicatenanocomposite membranes and its application as support for CO_2 sensor, *International Journal of Biological Macromolecules* 125 (2019) 503–509.

34. B.S. Boruah, R. Biswas, In-situ sensing of hazardous heavy metal ions through an ecofriendly scheme, *Optics & Laser Technology* 137 (2021) 106813. doi: 10.1016/j. optlastec.2020.106813.

35. S. Wang, Y. Gao, Fabrication of chitosan/silver nanocomposite films and their fluorescence sensing of aluminium ions, *Materials Technology* (2017). doi: 10.1080/10667857.2017.1321276.

36. W.A. Adeosun, A.M. Asiri, H.M. Marwani, Fabrication of conductive polypyrrole doped chitosan thin film for sensitive detection of sulfite in real food and biological samples, *Electroanalysis* (2020) 1–35. doi: 10.1002/elan.201900765.

37. T. Xu, H. Dai, Y. Jin, Electrochemical sensing of lead (II) by differential pulse voltammetry using conductive polypyrrole nanoparticles, *Microchimica Acta* 187:23 (2020) 1–7. doi: 10.1007/s00604-019-4027-z.

38. Q. Xiao, M. Feng, Y. Liu, S. Lu, Y. He, S. Huang, The graphene/polypyrrole/chitosan-modified glassy carbon electrode for electrochemical nitrite detection, *Ionics (Kiel)* 24 (2018) 845–859. doi: 10.1007/s11581-017-2247-y.

39. X. Shen, F. Ju, G. Li, L. Ma, Smartphone-based electrochemical potentiostat detection system using pedot: Pss/chitosan/graphene modified screen-printed electrodes for dopamine detection, *Sensors (Switzerland)* (2020) 20. doi: 10.3390/s20102781.

40. X.D. Wu, Y.Y. Han, X.X. Zhang, Z.H. Zhou, C.H. Lu, Large-area compliant, low-cost, and versatile pressure-sensing platform based on microcrack-designed carbon black@ polyurethane sponge for human–machine interfacing, *Advanced Functional Materials* 26 (2016) 6246–6256.

41. H. Liu, J.C. Gao, W.J. Huang, K. Dai, G.Q. Zheng, C.T. Liu, C.Y. Shen, X.R. Yan, J. Guo, Z.H. Guo, Electrically conductive strain sensing polyurethane nanocomposites with synergistic carbon nanotubes and graphene bifillers, *Nanoscale* 8 (2016) 12977–12989.

42. X. Yuan, Y. Wei, S. Chen, P.P. Wang, L. Liu, Bio-based graphene/sodium alginate aerogels for strain sensors, *RSC Advances* 6 (2016) 64056–64064.

43. X.D. Wu, Y.Y. Han, X.X. Zhang, C.H. Lu, Highly sensitive, stretchable, and wash-durable strain sensor based on ultrathin conductive layer@ polyurethane yarn for tiny motion monitoring, *ACS Applied Materials & Interfaces* 8 (2016) 9936–9945.

44. Y. Lin, S.Q. Liu, S. Chen, Y. Wei, X.H. Dong, L. Liu, A highly stretchable and sensitive strain sensor based on graphene–elastomer composites with a novel double-interconnected network, *Journal of Materials Chemistry C* 4 (2016) 6345–6352.

45. X.D. Wu, C.H. Lu, Y.Y. Han, Z.H. Zhou, G.P. Yuan, X.X. Zhang, Cellulose nanowhisker modulated 3D hierarchical conductive structure of carbon black/natural rubber nanocomposites for liquid and strain sensing application, *Composites Science and Technology* 124 (2016) 44–51.

46. Y. Liu, H. Zheng, M. Liu, High performance strain sensors based on chitosan/carbon black composite sponges, *Materials & Design* 141 (2018) 276–285.

47. M. Ramos, A.M. Al-Jumaily, V. Sreenivas, Conductive Polymer-Composite Sensor for Gas Detection, *1st International Conference on Sensing Technology* November 21–23 (2005).

48. K. Li, K. Dai, X. Xu, G. Zheng, C. Liu, J. Chen, C. Shen, Organic vapor sensing behaviors of carbon black/poly(lactic acid) conductive biopolymer composite, *Colloid and Polymer Science* 291 (2013) 2871–2878.

49. E.S. Tillman, M.E. Koscho, R.H. Grubbs, N.S. Lewis, *Analytical Chemistry* 75:7 (2003) 1748–1753.

50. B. Zhao, R.W. Fu, M.Q. Zhang, H. Yang, M.Z. Rong, Q. Zheng, Effect of soft segments of waterborne polyurethane on organic vapor sensitivity of carbon black filled waterborne polyurethane composites, *Polymer Journal* 38:8 (2006) 799–806.

51. X. Liu, Z. Qin, X.L. Zhang, L. Chen, M.F. Zhu, *Advanced Materials Research* 750–752 (2013) 55–58.

52. T. Sathies, P. Senthil, C. Prakash, *Materials Research Express* 6 (2019) 115349.

53. E. Devaux, C. Aubry, C. Campagne, M. Rochery, PLA/carbon nanotubes multifilament yarns for relative humidity textile sensor, *Journal of Engineered Fibers and Fabrics* 6 (2011) 13–24.

54. A. Trani, R. Petrucci, G. Marrosu, D. Zane, A. Curulli, Selective electrochemical determination of caffeine at a gold-chitosan nanocomposite sensor: May little change on nanocomposites synthesis affect selectivity, *Journal of Electroanalytical Chemistry* 788 (2017) 99–106. doi: 10.1016/j.jelechem.2017.01.049.

55. F. Nazari, S.M. Ghoreishi, A. Khoobi, Bio-based Fe_3O_4/chitosan nanocomposite sensor for response surface methodology and sensitive determination of gallic acid, *International Journal of Biological Macromolecules* (2020). doi: 10.1016/j.ijbiomac.2020.05.205

56. Z. Mo, H. Liu, R. Hu, et al., Amino-functionalized graphene/chitosan composite as an enhanced sensing platform for highly selective detection of Cu^{2+}, *Ionics* 24 (2018) 1505–1513. doi: 10.1007/s11581-017-2309-1.

57. H. Qi, E. Mäder, J. Liu, Unique water sensors based on carbon nanotube–cellulose composites, *Sensors and Actuators B: Chemical* 185 (2013) 225–230.

58. K. Zhang, R. Hu, G. Fan, G. Li, Graphene oxide/chitosan nanocomposite coated quartz crystal microbalance sensor for detection of amine vapors, *Sensors and Actuators B: Chemical* 243 (2017) 721–730.

59. S. Pandharipande, S.S. Bankar, Development of polyaniline grafted chitosan sensor for detection of ammonia & ethanol vapour, *International Research Journal of Engineering and Technology (IRJET)* 04 (2017) 534–540.

60. S. Bhagyaraj, I. Krupa, Alginate-mediated synthesis of hetero-shaped silver nanoparticles and their hydrogen peroxide sensing ability, *Molecules* 25:3 (2020) 435.

61. R.K. Mendes, B.S. Arruda, E.F. De Souza, A.B. Nogueira, O. Teschke, L.O. Bonugli, A. Etchegaray, Determination of chlorophenol in environmental samples using a voltammetric biosensor based on hybrid nanocomposite, *Journal of the Brazilian Chemical Society* 28 (2017) 1212–1219. doi: 10.21577/0103-5053.20160282.

62. H. Liu, H. Xiang, Z. Li, Q. Meng, P. Li, Y. Ma, H. Zhou, W. Huang, Flexible and degradable multimodal sensor fabricated by transferring laser-induced porous carbon on starch film, *ACS Sustainable Chemistry & Engineering* (2019).

63. W.T. Wahyuni, Z. Arif, N.S. Maryam, Development of graphene-chitosan composite modified screen-printed carbon electrode and its application for detection of rutin, *IOP Conference Series: Earth Environmental Science* 299 (2019). doi: 10.1088/1755-1315/299/1/012005.

64. W. Yu, Y. Tang, Y. Sang, W. Liu, S. Wang, X. Wang, Preparation of a carboxylated single-walled carbon-nanotube-chitosan functional layer and its application to a molecularly imprinted electrochemical sensor to quantify semicarbazide, *Food Chemistry* 333 (2020) 127524. doi: 10.1016/j.foodchem.2020.127524.

65. W.A. Adeosun, A.M. Asiri, H.M. Marwani, Fabrication of conductive polypyrrole doped chitosan thin film for sensitive detection of sulfite in real food and biological samples, *Electroanalysis* (2020) 1–35. doi: 10.1002/elan.201900765.

66. Z.R. Zad, S.S.H. Davarani, A. Taheri, Y. Bide, A yolk shell Fe_3O_4 @PA-Ni@Pd/Chitosan nanocomposite -modified carbon ionic liquid electrode as a new sensor for the sensitive determination of fluconazole in pharmaceutical preparations and biological fluids, *Journal of Molecular Liquids* 253 (2018) 233–240. doi: 10.1016/j.molliq. 2018.01.019.

67. M. Imran, C.J. Ehrhardt, M.F. Bertino, M.R. Shah, V.K. Yadavalli, Chitosan stabilized silver nanoparticles for the electrochemical detection of lipopolysaccharide: A facile biosensing approach for gram-negative bacteria, *Micromachines* 11 (2020) 413.

68. B. Lou, U. Rajaji, S. Chen, T. Chen, A simple sonochemical assisted synthesis of Porous $NiMoO_4$/chitosan nanocomposite for electrochemical sensing of amlodipine in pharmaceutical formulation and human serum, *Ultrasonics Sonochemistry* (2019) 1–27. doi: 10.1016/j.ultsonch.2019.104827.

69. Y.F. Jin, C.Y. Ge, X.B. Li, M. Zhang, G.R. Xu, D.H. Li, A sensitive electrochemical sensor based on ZIF-8-acetylene black-chitosan nanocomposites for rutin detection, *RSC Advances* 8 (2018) 32740–32746. doi: 10.1039/c8ra06452k.

70. R. Li, L. Wang, L. Yin, Materials and devices for biodegradable and soft biomedical electronics, *Materials* 11:11 (2018) 2108.

71. Q. Wang, S. Ling, X. Liang, H. Wang, H. Lu, Y. Zhang, Selfhealable multifunctional electronictattoos based on silk and grapheme, *Advanced Functional Materials* 29:16 (2019) 1808695.

72. L. Jiang, J. Zhang, 7-Biodegradable and biobased polymers. In M. Kutz (ed.), *Applied Plastics Engineering Handbook*, 2nd ed., William Andrew Publishing, pp. 127–143, 2017.

73. H. Zhu, Z. Fang, C. Preston, Y. Li, L. Hu, Transparent paper: Fabrications, properties, and device applications, *Energy & Environment Science* 7:1 (2014) 269–287.

74. A.T. Singh, D. Lantigua, A. Meka, S. Taing, M. Pandher, G. Camci-Unal, Paper-based sensors: Emerging themes and applications, *Sensors* 18:9 (2018) 2838.

75. S. Veeralingam, P. Sahatiya, A. Kadu, V. Mattela, S. Badhulika, Direct, one-step growth of $NiSe_2$ on cellulose paper: A lowcost, flexible, and wearable with smartphone enabled multifunctional sensing platform for customized noninvasive personal healthcare monitoring, *ACS Applied Electronic Materials* 1:4 (2019) 558–568.

76. D. Maier, E. Laubender, A. Basavanna, S. Schumann, F. Güder, G.A. Urban, C. Dincer, Toward continuous monitoring of breath biochemistry: A paper-based wearable sensor for real-time hydrogen peroxide measurement in simulated breath, *ACS Sensors* 4:11 (2019) 2945–2951.

77. J.M. Nassar, K. Mishra, K. Lau, A.A. Aguirre-Pablo, M.M. Hussain, Recyclable nonfunctionalized paper-based ultralowcost wearable health monitoring system, *Advanced Materials Technologies* 2:4 (2017) 1600228.

78. M. Parrilla, T. Guinovart, J. Ferre, P. Blondeau, F.J. Andrade, A wearable paper-based sweat sensor for human perspiration monitoring, *Advanced Healthcare Materials* 8:16 (2019) 1900342.

79. J. Miao, H. Liu, Y. Li, X. Zhang, Biodegradable transparent substrate based on edible starch–chitosan embedded with natureinspired three-dimensionally inter-connected conductive nanocomposites for wearable green electronics, *ACS Applied Materials & Interfaces* 10:27 (2018) 23037–23047.

80. E.S. Hosseini, L. Manjakkal, D. Shakthivel, R. Dahiya, Glycine–chitosan-based flexible biodegradable piezoelectric pressure sensor, *ACS Applied Materials & Interfaces* 12:(2020) 9008–9016.

81. R.K. Pal, V.K. Yadavalli, Silk biocomposites as flexible and biodegradable electrochemical sensors, 4, *Informatics, Electronics and Microsystems*, TechConnect Briefs (2017) 190–193.

82. S. Akhter, W.J. Basirun, Y. Alias, M.R. Johan, S. Bagheri, M. Shalauddin, M. Ladan, N.S. Anuar, Enhanced amperometric detection of paracetamol by immobilized cobalt ion on functionalized MWCNTs - Chitosan thin film, *Analytical Biochemistry* 551 (2018) 29–36. doi: 10.1016/j.ab.2018.05.004

83. S. Velmurugan, S. Palanisamy, T.C. Yang, Ultrasonic assisted functionalization of MWCNT and synergistic electrocatalytic effect of nano-hydroxyapatite incorporated MWCNT-chitosan scaffolds for sensing of nitrofurantoin a Department b Precision, *Ultrasonics Sonochemistry* (2019) 104863. doi: 10.1016/j.ultsonch.2019.104863.

84. H. Wang, H. Lü, L. Yang, Z. Song, N. Hui, Glycyrrhiza polysaccharide doped the conducting polymer PEDOT hybrid-modified biosensors for the ultrasensitive detection of microRNA, *Analytica Chimica Acta* 1139 (2020) 155–163.

85. M. Figiela, M. Wysokowski, M. Galinski, T. Jesionowski, I. Stepniak, Synthesis and characterization of novel copper oxide-chitosan nanocomposites for non-enzymatic glucose sensing, *Sensors and Actuators B: Chemical* 272 (2018) 296–307.

86. D. Feng, F. Wang, Z. Chen, Electrochemical glucose sensor based on one-step construction of gold nanoparticle–chitosan composite film, *Sensors and Actuators B: Chemical* 138:2 (2009) 539–544.

87. N. Feng, J. Zhang, W. Li, Chitosan/graphene oxide nanocomposite-based electrochemical sensor for ppb level detection of melamine, *Journal of Electrochemical Society* 166 (2019) B1364–B1369. doi: 10.1149/2.1321914jes.

88. A.K. Yadav, T.K. Dhiman, G.B.V.S. Lakshmi, A.N. Berlina, P.R. Solanki, A highly sensitive label- free amperometric biosensor for norfloxacin detection based on chitosan-yttria nanocomposite, *International Journal of Biological Macromolecules* 151 (2020) 566–575. doi: 10.1016/j.ijbiomac. 2020.02.089.

89. A. Güner, E. Çevik, M. Şenel, L. Alpsoy, An electrochemical immunosensor for sensitive detection of Escherichia coli O157:H7 by using chitosan, MWCNT, polypyrrole with gold nanoparticles hybrid sensing platform, *Food Chemistry* 229 (2017) 358–365. doi: 10.1016/j.foodchem.2017.02.083.

8 Polymers for Innovative Packaging Applications

Sonika
Rajiv Gandhi University

Sushil Kumar Verma and Vishwanath Jadhav
Deep Plast Industries

CONTENTS

DOI: 10.1201/9781003227908-8

8.1 INTRODUCTION TO PACKAGING

Packaging has become an essential activity in the current developed society and anything that is utilized to hold, security, taking care of, conveyance and show of stock, from elementary materials to completed substances, from producer to customers. Packaging is mostly isolated by the essential raw material. Packaging materials can be divided into metal, glass, polymer, paper cardboard, wood, composite, and ceramic.

Food packaging should meet various conditions, such as enactment, well-being, and numerous different conditions since it is needed to be inventive, simple to utilize, and appealing plan and has experienced an astonishing expansion, because most commercialized foodstuffs, including fresh fruits and vegetables, are being promoted inside packages. Single of the primary plants of packaging in the food commercial is to protect the result of synthetic, mechanical and microbiological effect, and moreover permits the innovation of the product and keeps all its vigorous advantage. The dynamic theme in food packaging is that the packaging is an essential part of the manufacture, conservation, storing, diffusion, and planning of food sources. The properties of the food elements are simply conceivable to possess up sincere key of suitable packaging and persistent interaction. It is possible to have the product quality and energy during the period essential for its commercialization and consumption [1,2].

However, the present inclinations contain the enhancement of packaging materials that unite with the environment and with the food, pretentious a functioning part in preservation. These novel food packaging situations have been formed as a resort to shape consumer inclinations towards slightly protected, new, delightful, and supportive food produce with a delayed time span of usability. Also, changes in retail practices, such as globalization of business sectors bringing about longer circulation distances, present significant difficulties to the food packaging industry going about as main stimuli for the development of new and improved packaging ideas that broaden timeframe of realistic usability, while keeping up the well-being and eminence of the packaged food.

8.1.1 DEVELOPMENT OF BIOPOLYMERS

Biopolymers are a new generation of materials that are immobile in progress and that have fascinated devotion as a possible replacement to conventional plastics due to an increased interest in sustainable development [2,3]. Biopolymers consist proportionately of bio-based raw materials and/or have biodegradability. According to ASTM International, biopolymers are classified as biopolymers only from a proportion of $\geq 20\%$ of renewable raw materials [4]. Bio-based in this framework means derived from a raw material that is not of petrochemical origin, but is obtained from renewable raw materials such as trees and agricultural crops based on cellulose, such as cereals, pulses, sugar cane and oil plants, and also waste [6,7]. A substance is biodegradable if it can be degraded by microorganisms such as bacteria, fungi, or protozoa by fermentation [3].

Novel biopolymers are divided into three generations of biopolymer materials. In the latest generation (third generation), the predictable mixture routes of petrochemical polymers are retained, but the crude oil content is substituted and replaced by means of biogenically produced raw materials. In this instance, biopolymers are entitled drop-in variants [4].

Recently, bioplastic has become one of the most advanced pioneering materials that are bio-based and biodegradable. Excess biomass and renewable mechanisms such as jackfruit, waste banana peels, biological waste, cultivation waste, newspaper waste [5], oil palm empty fruit cluster, sugar cane corn starch, potato stiffener, rice grass, rapeseed oil, vegetables oil, cellulose from plants, starch, cotton, bacteria, and occasionally several nano-sized elements such as carbohydrate chains (polysaccharides) are used to complete bioplastic [6]. Bioplastic can be degraded by natural microorganisms such as bacteria, algae, and fungi [7]. This chapter starts with a brief grouping of bioplastics, followed by aids and hindrances to bioplastics.

8.2   TIMELINE OF PLASTICS

8.2.1   Parkesine

In 1862, Alexander Parkes delivered first man-made plastic called Parkesine. It was an organic material obtained from cellulose that once ready and formed, kept up with its shape as it chilled off.

8.2.2   Celluloid

In 1868, John Wesley Hyatt designed celluloid (resultant from cellulose and alcoholized camphor) as an option for the ivory in billiard balls. Yet, the flexible film he created was not sufficiently able to be utilized as an option to billiard balls. The framed substance, i.e. celluloid, was not difficult to be displayed with warmth and pressing factor, and it became well known as it was the first flexible photographic film utilized for still photography and films [8].

8.2.3   Bakelite – Formaldehyde Resins

Formaldehyde was the following establishment that was developed in the line of innovation of plastics after cellulose nitrate. Around 1897, Galalith and Erinoid were the two instances of engineered plastic materials. In 1899, Arthur Smith set up British Patent for "phenol formaldehyde saps", and in 1907, Leo Hendrik Baekeland further developed phenol formaldehyde pitch and first manufactured tar, called Bakelite [9].

8.2.4   Polyvinyl Chloride (PVC)

French physicist Victor Regnault found PVC in 1912.

8.2.5 CELLOPHANE

Edwin Brandenberger utilized thick cellophane in 1913. It was his thought that he made straightforward bundling for food things.

8.2.6 POLYMETHYL METHACRYLATE (PMMA)

Barker and Skinner made PMMA. In 1924, it was sold by Rohm, under the name of Plexiglas.

8.2.7 POLYETHYLENE (PE)

It was found by E.W. Fawcett and R.O. Gibson in 1933. It is the most ordinarily utilized plastic on the planet.

8.2.8 POLYURETHANE (PUR)

In 1937, Dr Otto Bayer created PUR.

8.2.9 POLYSTYRENE (PS)

Ray McIntire created polystyrene in 1944 by chance when he was chipping away at flexible elastic.

8.2.10 POLYPROPYLENE (PP)

Working for Montedison, Giulio Natta (1963 Nobel Prize with Karl Ziegler) found a catalyst named the "Ziegler–Natta". This catalyst had the option to deliver PP with high mechanical obstruction, inertness to substance hostility, and ability to withstand temperatures above 100°C [10].

8.3 BIOPOLYMERS

Biomaterials (biopolymers) are polymers formed from endurable bases. Biopolymers are manufactured from herbal crude resources in any event. Arranged from various perspectives, for example, synthetic design, beginning, techniques for blend, cost viability, and application, polymers from inexhaustible assets are not quite the same as normal polymers on the grounds that their blend is incited purposefully. The polymers produced using usual plant materials from wheat, potato, or corn starches have atoms that are excellently microbiologically degradable. The natural polymers are for the most part innately ecological since for every protein there is a polymerase, whose aerobics produce a distinctive polymer, and there is depolymerase prepared for catalysing the decline of the polymer [11].

8.3.1 CLASSIFICATION OF BIOPOLYMERS

The conformist method of working ecological packaging (flexible and rigid) materials is classified into three groups dependent on recorded turn of proceedings. There are three essential types of bioplastics, which are biodegradable and bio-based, ecological and fossil-based, and non-biodegradable and bio-based, while non-biodegradable and oil-based are recognized as plastic.

8.3.2 PRIMARY PHASE

The primary phase of material was utilized for shopping packs, comprising of engineered polymers such as low-density polyethylene (LDPE) with an extent of 5%–15% starch fillers and favourable to reacting and autoxidative additive. Advanced materials break down or bio-fragment into further uncertain atoms that are not decomposable. Such properties have made an exceptionally awful picture of biomaterials mainly for customers who were convinced that they are biodegradable [12].

8.3.3 SECONDARY PHASE

The secondary phase of biomaterials contains a blend of pre-gelatinized starch (40%–70%) and small polyethylene (LDPE) over the increase of the hydrophilic copolymer, for instance, ethylene acrylic acidic, polyvinyl liquor, and vinyl acetic acid source. Comprehensive exploitation of starch takes 40 days, and the degradation of the comprehensive of the film keeps going beyond 2–3 years.

8.3.4 THIRD PHASE

The third phase of the material entirely consists of biomaterials and can be divided into three fundamental groups as per the beginning and manufacture strategies such as polymers extracted/remote directly from biomass, polymers produced by predictable chemical substance combination and bio-monomer, and polymers found straight from natural or genetically modified organisms [12].

8.4 ADVANTAGES AND DRAWBACKS OF BIOPLASTICS

It is recognized that plastic is one of the essential assets that are produced all day. Therefore, to reduce the ecological infection, a decision should be formed by changing the utilization of normal plastic to bioplastics. Therefore, substitution of bioplastic materials to ordinary plastic can be an advanced way for sustainable growth due to the similitude properties between regular plastic and bioplastic materials. In addition, for convinced situation bioplastic displays better properties compared to ordinary plastic like extreme in mechanical properties, oxygen porosity, gas hindrance and water emission transmission rate [13]. Table 8.1 depicts momentarily the benefits and inconveniences of bioplastics in contrast to regular plastics as revealed by earlier scientists (Figure 8.1).

TABLE 8.1

Advantages and Drawbacks of Bioplastics Compared to Conventional Plastics

Bioplastics

Advantages	Refs.	Disadvantages	Refs.
Sustainable	[14]	Expensive	[31,32]
Condensed carbon footprint	[14,16,17]	Thermal volatility	[6,16]
Condensed energy efficiency	[14,16,17]	Recycling hitches	[18]
Moderately based on usual feedstock	[14,16,18]	Crumbliness	[6,19]

Conventional Plastic

Advantages	Refs.	Disadvantages	Refs.
Cheap cost	[20]	Based on petrochemical	[20]
Moral and exceptional technical properties	[36,37]	Problematic to recycle	[20]
Save Eenergy and properties	[20,21]	Generally not biodegradable	[20]
Thermal recycling probable	[20]	Hysterical combustion can issue toxic materials	[20]

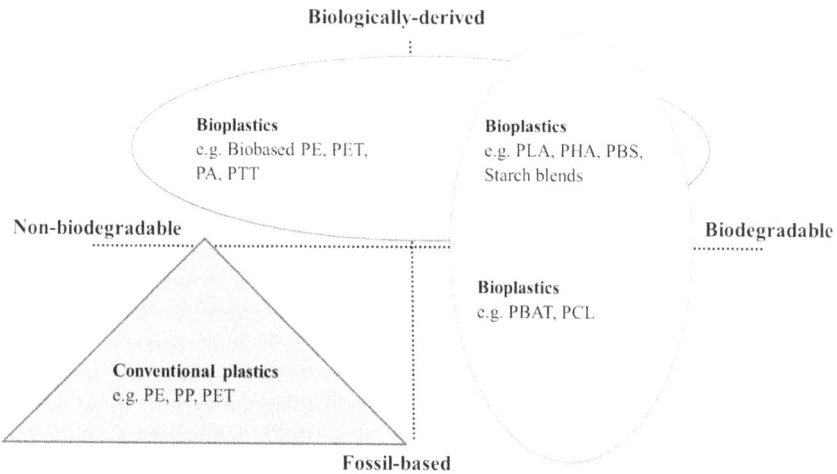

Biologically-derived

Bioplastics
e.g. Biobased PE, PET, PA, PTT

Bioplastics
e.g. PLA, PHA, PBS, Starch blends

Non-biodegradable

Biodegradable

Bioplastics
e.g. PBAT, PCL

Conventional plastics
e.g. PE, PP, PET

Fossil-based

FIGURE 8.1 Classification of bioplastic.

8.5 BIOPLASTIC EXTRACTED/ISOLATED DIRECTLY FROM BIOMASS

This group of biopolymers is abundantly available in the market. Polymers of this kind are developed from plants, marine life, and wildlife. Examples are polysaccharides, such as cellulose, chitin, and starch, whey protein, casein, collagen, soy protein, and myofibrillar proteins of animal origin, and so forth. These can be utilized alone or as a mix with industrial polyesters, for example polylactic acid (PLA). The maximum boundless category that is exploited in food packaging is cellulose-based paper.

8.5.1 CELLULOSE

Cellulose is a regular polymer and comprises of glucose monomer units that are connected together by glycosidic linkages. Cellulose is separated from its glass-like state in microfibrils. It is dissolvable in solvents such as N-methylmorpholine N-oxide. In any case, it is very difficult to use it for packaging in view of its specific properties such as helpless solvency, hydrophilic nature and profoundly glass-like structure. Because of the profoundly glass-like structure, the subsequent packaging material is fragile with helpless adaptability and elasticity [22]. Therefore, presently research has been centred around cellulose subordinates to packaging.

8.5.2 CELLULOSE DERIVATIVES

They are polysaccharides comprising of direct chains connected together by beta-(1,4)-glucosidic units. Following cellulose subsidiaries are utilized for eatable films/coating: hydroxypropyl cellulose, hydroxypropyl methylcellulose, carboxymethyl cellulose, and methylcellulose. They show thermo-gelation; that is, when suspensions are heated, they become gel-like and consequently acquire their unique consistency when cooled [23]. Such films are unfortunate water barriers due to their hydrophilic nature and show poor mechanical properties. One strategy for working on the nature of dampness obstruction can be performed, i.e. by expansion of hydrophobic mixtures, for example, unsaturated fats into cellulose lattice to foster the films [24].

8.5.3 STARCH

Bio-based crude materials are nearly inexhaustible; in view of this, starch is generally utilized. Every one of the green plants produce starch, which is a white, granular, delicate, tasteless powder and is insoluble in cooled water, liquor, or different solvents. The basic formula of starch is $(C_6H_{10}O_5)n$ [25]. Because of its low cost, accessibility, and the fact that it is derived from renewable sources, starch is possibly the most appropriate biodegradable polymer material. It is a polysaccharide comprising of amylose and amylopectin connected together by glycosidic bonds [26]. Commercially, starch and its subordinates are derived from different crude materials, for example potato and its strips, corn, wheat, and pea. It has numerous applications. In drug delivery, it is utilized as a restricting specialist. It has application in the material business, textiles, and paper and cardboard industries.

TABLE 8.2
Amylose and Amylopectin Obtainable from Various Sources

Biosource	Amylose (%)	Amylopectin (%)
Banana	20–22	80–82
Potato	18–20	83–84
Corn	28–32	70–74
Rice	35–40	65–70
Wheat	20–22	80–85

In Table 8.2, the grouping of amylose and amylopectin present in different plant parts is given. It is seen that when the grouping of amylose is high, the tensile properties of the bioplastic Increase [27]. This phase of plastics produced using starch is completely biodegradable. This efficiently substitutes PS and PE in numerous applications [3].

8.5.4 SOY PROTEINS

The illustrations formed using several soy proteins are very delicate to dampness and are not solid. To further develop elastic properties and thermal properties, for reduced dampness affectability, a stearic corrosive accumulation of around 25% is required. Further it was initiated that the soy protein along with glycerol, gellan gum, or K-carrageenan suitable for the creation of biodegradable/eatable soybean-based packaging sections plate. Incessantly in the creation of such packaging must focus on the hindrance properties to moisture in light of the fact that generally hydrophilic in flora. Collagen covers (digestive system) for the present stay the solitary generally applied item dependent on proteins. It is recognized that the film dependent on a protein revealed extensive applications as an edible film [28].

8.5.5 CHITIN AND CHITOSAN

It is the second most abundant regular polysaccharide occurring in the nature after cellulose and varies from it simply by the OH group. Chitosan is acquired from chitin. Chitosan is ready by deacetylation portion within the sight of alkali and it is a substantial misuse of fishery industry [29]. It structures films with no expansion of added chemical substances and comprises of excellent mechanical properties and microbial properties against microscopic organisms such as moulds, and yeasts. It additionally displays great oxygen and carbon dioxide permeability, yet the significant downside is that it has poor solubility in unbiased arrangements. To obtain a solvent product, the necessary level of deacetylation should be 80%–85% [30]. It structures straightforward films which works on the quality just as capacity life of the products.

They are also beneficial for making various environmentally friendly packaging films, the most common of which they utilise as a disposable coating to extend the shelf life and usage of new green foods. As a result, allowing excessive use of materials that make environmentally friendly food packaging is essential for extending the shelf life of usable food sources [31].

8.5.6 LIGNIN

Lignin is considered as quite possibly the most plentiful sustainable biopolymers close to cellulose and is significant because of its renewable nature, minimal expense, simple accessibility, and decent simplicity of chemical and mechanical modifications, which are also recognized as eco-viable [32,33]. Lignin is a potential source of bioactive parts in the food and drug industries. Lignin is polar organization polymers comprising normally of 1–2 hydroxyl groups for every monomeric unit. Lignin removed from rural waste is holding an extraordinary potential in the plan and creation of eco-viable plasticware. It is a promptly accessible material characterized by low density, low abrasive features, and low cost. These components make it intriguing as a natural filler suitable to supplanting the inorganic ones [34].

8.5.7 COLLAGEN

Collagen is a protein discovered to a great extent in well-evolved animal (25% of our absolute protein in mass) and is the significant strength supplier to tissue. A customary collagen particle comprises of three trapped protein chains that structure a helical construction. Collagen is non-harmful, delivers just an insignificant insusceptible reaction, and is astounding for connection and natural cooperation with cells. It might likewise be prepared into various configurations, such as permeable wipes, gels, and sheets, and can be cross-linked with synthetic substances to make it more grounded or to modify its debasement rate. Based upon how it is handled, collagen might possibly cause adjustment of cell conduct, have admissible mechanical properties, and show constriction or shrinkage [35].

8.5.8 GELATIN

Gelatin is perhaps the most widely recognized biopolymer. It is acquired either by incomplete corrosive hydrolysis, or by partial basic hydrolysis of animal collagen. It's a denaturized stringy protein.. Gelatin is garish and odourless. It is a glassy, fragile faintly yellow in colour. Gelatin is solvent in aqueous arrangements of polyhydric alcohols like glycerol and propylene glycol. It's insoluble in less polar natural solvents, including benzene, acetone, essential alcohols, and dimethylformamide. Gelatin's colour relies upon the extraction method and the crude materials utilized. Gelatin's two most helpful properties are gel strength and thickness. Gelatin can be utilized as an emulsifying, foaming, and wetting specialist in the food industry in medication and beauty care products. Gelatin possesses a usual pH range from 3.8 to 5.5 [36,37].

8.5.9 ALGINATE

Mostly, alginate is obtained from brown sea kelp; it's a polysaccharide. Alginate can be handled effectively in water as chitosan and is genuinely non-harmful and non-inflammatory, so it has been endorsed in certain nations for wound dressing and use in food items. It has been demonstrated that alginate is biodegradable. It has controllable porosity and might be connected to other organically dynamic atoms. Alginate can frame a strong gel under gentle preparing conditions, which permits its utilization for entangling cells into dots and different shapes. The fascinating matter is that encapsulation of certain cell types into alginate beads may really enhance cell development and endurance. Alginate has been investigated for use in liver, nerve, heart, and ligament tissue engineering. Like others, alginate has some disadvantages including mechanical softness and unfortunate cell adhesion. To overcome these impediments, the strength and cell conduct of alginate were improved by blends with different materials, including the normal polymers agarose and chitosan [38,39].

8.5.10 PECTIN

Pectin is a biodegradable polymer, a combination of polysaccharides that make up around 33% of the cell divider dry substance of higher plants. These are dissolvable in unadulterated water. The monovalent cation salts of pectinic and pectic acids are for the most part soluble in water; however, divalent and trivalent cation salts are insoluble or weakly soluble. The convergences of pectin are most maximum in the centre lamella of the cell wall, which steadily diminishes as one goes through the essential wall towards the plasma layer. 10%–15% of gelatin is found in apple pomace on a dry matter premise, while citrus strip contains 20%–30%. Citrus pectins are light cream in colour; apple pectins are frequently more obscure. The main utilization of pectin depends on its capacity to shape gels. Gels can be framed by HM pectin with sugar and corrosive. The development of gel is brought about by hydrogen bonding between free carboxyl groups on the pectin particles and furthermore between the hydroxyl groups of adjoining atoms [40].

8.5.11 OTHER PLANT PROTEINS

The two maximum usual vegetable proteins exploited in the manufacture of biodegradable packaging are chickpeas and divided soy protein. Further proteins that are employed include those obtained from wheat, pistachios, pea plant, and sunflower [41]. Further, polymers based on proteins similar to egg whites, casein, fibrinogen, silk, and elastin are used as a raw material for the creation of ecological packaging because of their biodegradability.

8.5.12 Poly-Beta-Hydroxyalkanoates (PHB)

PHB degrades in the presence of miniature life form that interacts with the polymer, discharge proteins, and break down into more modest parts. The properties of PHB are (i) 100% protection from water, (ii) 100% biodegradability, and (iii) ability to process thermoplastic [42]. The maximum extensively standard is the utilization of derivative poly(hydroxybutyrate) obvious PHB. PHB is an ecological polyester directly prepared by bacterial ageing of sugar or lipid. It tends to be utilized for food packaging, beauty care products, and drug items, just as in farming. The vigorous conditions are totally debased into water and carbon dioxide.

8.5.13 Polylactic Acid (PLA) Plastics

PLA is the most appealing material from the packaging perspective because of its astounding biocompatibility, biodegradability, and preparing capacity. It is handled by infusion forming, blow shaping, thermoforming, and expulsion. It is made out of lactic corrosive (2-hydroxy propionic corrosive) and comprises methyl group on alpha C molecules.

Commercially, it was the first bio-based polymer with a huge scope, which could be moulded into different articles and films [43]. It substitutes HDPE (high-density polyethylene), LDPE (low-density polyethylene), PET (polyethylene terephthalate), and PS (polystyrene) as a packaging material.

8.6 BIOPLASTIC FROM UNEATABLE SUBSTANCES

In today's world where food is scare, we can create bioplastics from non-consumable resources as well. Things such as orange strip, pomegranate strip, banana strip, and potato strip are utilized for the production of bioplastic. In the most recent pattern, bioplastic films from polysaccharide build-up feedstock are of incredible interest. Cellulose, hemicelluloses, starch, and gelatin make these lignocellulosic feedstocks valuable for the creation of bioplastic (Figure 8.2).

8.6.1 Pomegranate Peel

This has rich wellspring of bioactive compounds. It consists of lignin (5.7%), hemicelluloses (10.8%), cellulose (26.2%), and gelatin (27%). During corrosive hydrolysis, the polysaccharides present in the strip are changed into monosaccharides, which can be separated into cellulose, hemicelluloses, and lignin parts. These segments further are utilized to create bioplastic [44].

8.6.2 Orange Peel

The strip contains carbs which can be utilized for the production of biomolecules. Imprudent release of natural strips causes numerous ecological issues [44]. Consequently, it is prescribed to gather the waste and convert it into bioplastics.

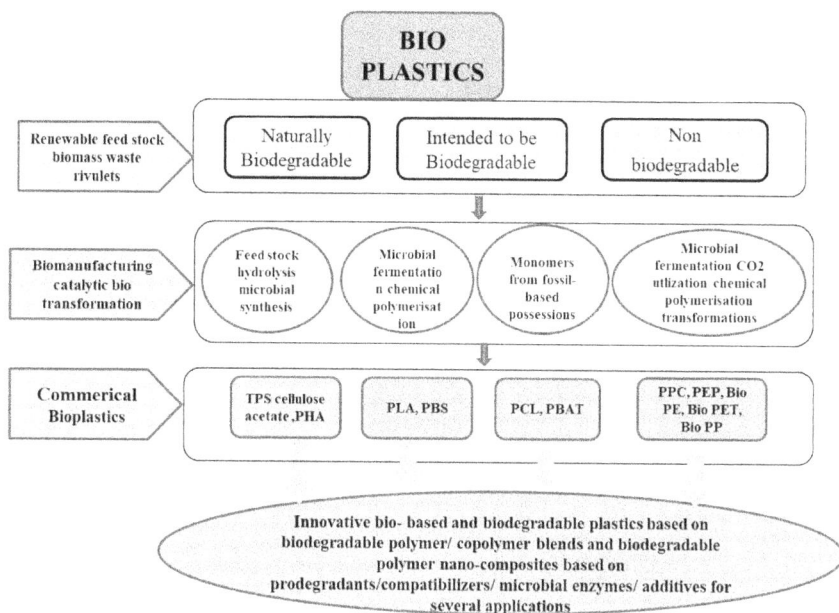

FIGURE 8.2 Schematic diagram of industrial approaches in manufacturing commercial bioplastic applications.

8.7 POLYMERS PRODUCED BY CONVENTIONAL CHEMICAL SYNTHESIS OF BIO-MONOMERS

In the category of the polymers produced from bio-based monomers, the polyesters used to be more popular. Thus, historically, the mainly studied monomers were bifunctional molecules, such as lactic acid, an α-hydroxy acid able to self-condense for the production of PLA; 1,3-propanediol (PDO) leading to Dupont's Sorona after condensation with terephthalic acid and succinic acid, expected to be a key bio-based building block and leading to polybutylene succinate (PBS) after condensation with 1,4-butanediol. The extremely popular of the biopolymers of this group is PLA. PLA is a green, thermoplastic, linear polyester has properties similar to PS. There are numerous ways to conveniently manufacturing (or high molecular weight) PLA [45]. The two major monomers are lactic acid and lactide.

PLA is obtained through direct condensation of lactic acid monomer. This technique should be carried out at a temperature lower than 200°C because at this temperature, low molecular weight supplies will develop. Subsequently, the polycondensation is carried out in solution or the melt where the short oligomeric units combine to give a high molecular weight polymer chain. Even higher molecular weight can be attained by crystallization of the basic polymer from the melt.

Aerobic degrades lactic acid completely into water and carbon dioxide, and biodegradation occurs in favourable settings for 3–4 weeks [46].

TABLE 8.3
Trade Names and Suppliers of PLA

Trade Name	Company	Country
NatureWorks@	Cargill Dow	The USA
Biomer@ L	Biomer	Germany
Lacea@	Mitsui Chem.	Japan
Lacty@	Shimadzu	Japan
Heplon@	Chronopol	The USA
Galacid@	Galactic	Belgium
Eco Plastic@	Toyota	Japan
Treofan@	Treofan	Netherlands
CPLA@	Dainippon Ink Chem.	Japan
PDLA@	Purac	Netherlands
Ecoloju@	Mitsubishi	Japan
Ecoflux@	BASF	Germany
Ecoworld@	Ecoworld	China

Therefore, PLA has primarily managed into thermoformed pads and flasks for packing and serving food, films, and other packaging materials. PLA suffers from its cost and also some weaknesses in thermomechanical properties (mainly heat resistance); however, it is considered as one of the most promising bioplastics for the substitution of the petroleum-based polymers in materials and packaging applications [47]. Companies include NatureWorks, a joint venture between Cargill and PTT Global Chemicals that has announced annual PLA production of 140, 000 metric tonnes. Diverse concerns commercialize PLA with several percentages of D/L lactide, and trade names and suppliers of different grades of PLA are listed in Table 8.3.

8.7.1 Polyglycolide (PGA)

PGA is the simplest linear aliphatic polyester. It is prepared by ring-opening polymerization of a cyclic lactone, glycolide. It is highly crystalline, with a crystallinity of 45%–55%, and hence, it is not soluble in maximum organic solvents. It has a high melting point (220°C–225°C) and a glass transition temperature of 35°C–40°C [48]. It has excellent mechanical strength. However, its biomedical applications are partial due to its low solubility and high rate of degradation in resilient acidic products. Therefore, copolymers of glycolide through caprolactone, lactide, or trimethylene carbonate have been equipped for medical devices [48].

8.7.2 Polycaprolactone (PCL)

The ε-caprolactone has a cheap cyclic monomer with semi-crystalline linear polymer is attained from ring-opening polymerization of ε-caprolactone in existence of tin octoate catalyst [49]. It is soluble in a wide range of solvents and its glass transition temperature is low, around −60°C, and its melting point is 60°C–65°C. It is a semi-rigid material at ambient temperature and has a modulus in the array of LDPE and HDPE, a low tensile strength of 23 MPa, and a high elongation at break (more than 700%). It has a low Tg and is frequently employed in polyurethane preparations as a compatibilizer or a soft block.

Enzymes and mildews simply biodegrade PCL [3]. It advances the degradation rate, and several copolymers with lactide or glycolide have been prepared [48]. PCL is commercially available under the trade names CAPA® (from Solvay, Belgium), Tone® (from Union Carbide, the USA), or Celgreen® (from Daicel, Japan). It has probable applications in the medical field.

8.7.3   POLYBUTYLENE SUCCINATE **(PBS)**

It is obtained by polycondensation reactions of glycols, such as ethylene glycol and 1,4-butanediol, with aliphatic dicarboxylic acids, such as succinic and adipic acids [50]. It was conceived in 1990 and advanced by Showa Highpolymer (Japan) under the trade name Bionolle®. EnPol® is the trade name of the similar class of polymers commercialized by Ire Chemical (Korea). Diverse poly(alkylene dicarboxylate)s have been organized: PBS, poly(ethylene succinate) (PES), and a copolymer, i.e. poly(butylene succinate-co-adipate) (PBSA). PBSA is attained by the addition of adipic acid. Their molecular weights range from numerous tens to several hundreds of thousands. The custom of a minor amount of coupling agents as chain extenders consents the molecular weight to be enlarged [51]. Alternative copolymer was organized by condensation of 1,2-ethylenediol and 1,4-butanediol with succinic and adipic acids by SK Chemicals (Korea) and marketed under the trade name Skygreen®. Lunare SE® trademark is alternative aliphatic copolyester marketed by Nippon Shokubai (Japan). The assembly of those copolymers, i.e. the nature of diacids and diols used, stimulates their properties as well as their biodegradation rates [50].

8.7.4   POLY(P-DIOXANONE) **(PPDO)**

These are well-known aliphatic polyester having good physical and chemical properties. It is produced by ring-opening polymerization of p-dioxanone. PPDO is semi-crystalline, with a low glass transition temperature in the range of −10°C to 0°C. These properties are PPDOs with a variety of molecular weights that have been investigated [52]. The evolution of molecular weight can advance the thermal stability of poly(p-dioxanone). Conferring to the consequences of rheological tests, poly(p-dioxanone) displays a shear-thinning behaviour. The tensile strength and modulus grow with molecular weights. Poly(p-dioxanone) has decisive biodegradation due to the ester's bonds in the polymer chains. Nevertheless, poly(p-dioxanone) is more expensive than poly(butylene succinate-co-adipate). Novel biodegradable polyester was organized by chain extension of poly(p-dioxanone) with poly(butylene succinate-co-adipate) [53]. Toluene di-isocyanate was used as a chain extender. Equally, these polymers have good compatibility.

8.7.5 POLYANHYDRIDES

These are interesting biodegradable materials because they have two hydro-lysable sites in the repetition element. Aromatic polyanhydrides will reduce gradually over an extensive period, while aliphatic polyanhydrides can degrade in a few days. Aliphatic homo-polyanhydrides have potential applications because of their high crystallinity and fast degradation. This is the instance of poly(sebacic anhydride). The degradation rate of polyanhydride can be achieved via regulating the hydrophobic and hydrophilic mechanisms in the copolymer. The increase in the hydrophobicity of the diacid structure blocks of the polymers resulted in slower degradation. In the copolymers with a hydrophobic aromatic comonomer such as carboxy phenoxy propane have been widely explored as biomaterials [54]. The mechanical strength and degradation rate of these cross-linked polyanhydrides are based on the nature of the monomer types.

8.8 POLYMERS OBTAINED DIRECTLY FROM NATURAL OR GENETICALLY MODIFIED ORGANISMS

Numerous microbes amass these polymers as a source of energy and as a carbon save. For the most part, in excess of 250 categories of microbes are standard for the creation of PHAs, yet just a scarce of them are suitable for the industrial production of PHAs [55]. This group contains polyhydroxyalkanoates (PHAs) and bacterial cellulose.

PHAs come in a variety of types. Alcaligenes, Azotobacter, Bacillus, Halobacterium, Rhizobium, and a variety of other bacteria can be transported in large quantities biotechnologically, on an infinite substrate, by utilising ageing and completing physical, chemical, and morphological cycles separated from biomass after synthesis. It has liable upon the microbes and the carbon basis, the PHA might be produced from rigid fragile to plastic to elastic like polymer. There are comparative properties such as propylene and PE, flexible and the thermoplastic (detained after cooling structures).

The use of minor poly(hydroxybutyrate) marked PHB is commonly used to identify them. PHB is a biodegradable polyester linear produced by microbial development of sugar or lipid. It tends to be utilized in food packaging, cosmetic products, and pharmaceutical products, as well as in agricultural industries. The aerobic conditions are completely degraded into water and CO_2. The biodegradation occurs in 5–6 months and a half [56].

Separately sustainable, biodegradable plastics can be created from engineered polymers by utilizing microorganisms. The bacterium *Pseudomonas putida* has deviations over styrene monomer in the PHA, biodegradable plastic which has a wide range of consumptions. PHA is a water-insoluble, biodegradable, and compostable material whose improvement is necessary before its commercialization [31].

8.9 PROPERTIES OF BIODEGRADABLE MATERIALS

The materials must be valued in the food packaging sector because their physical and mechanical features enable their acceptability and usage to a certain degree; nevertheless, this also applies to expenses to an excessive degree.

8.9.1 BARRIER PROPERTIES

One the very apex of critical elements in food packaging polymers is their hindrance or permeability concert against transmission of gases, water vapour, and cologne atoms. Biomaterials made of polysaccharides having poor barrier properties with regard to water vapour and other polar substances in an enormous extent of the humidity, yet at low or centre piece of humidity make good properties to O_2 and other non-polar substances such as diverse flavours and oils. Humidity vapour transmission rate was arranged from the starch substantial is 4–6 times higher than customary materials produced using synthetic polymers. The adjoining of barrier properties (Table 8.4) of the biomaterial- and oil-derived materials is represented. PHAs have comparative moisture vapour transmission rate as well as materials produced using petrochemical. PHB has improved barrier properties to oxygen compared to PET and PP and sufficient barrier properties with regard to fat and scents for products with a short time span of usability. The barrier properties to gases in most biomaterials rely upon the surrounding humidity, but PLA and PHAs are exclusions.

TABLE 8.4
Barrier Properties of Biopolymers and Oil-Derived Polymers [12].

Polymer	Temperature (°C)	Thickness (mm)	OTR (mLm^{-2}day^{-1} at 0% relative humidity)	MVTR (gm^{-2}day^{-1} at 100% relative humidity)
PLA (polylactic acid)	23	0.1	200	66
OPLA (oriented polylactic acid)	22	4.6	56.33	15.30
PHB (polyhydroxybutyrate)	30	1.0	183	1.16
PET (polyethylene terephthalate)	22	4.6	9.44	3.48
LDPE (low-density polyethylene)	38 (at 90% relative humidity)	0.75		7.9
OPS (oriented polystyrene)	22	4.6	532	5.18
LDPE+5% starch	38 (at 90% relative humidity)	0.75		36.85

8.9.2   OXYGEN SCAVENGERS/ABSORBERS

Oxygen, a necessary component for life, jeopardises the use of food due to degradative oxidation reactions and the development of moulds and aerophilic microorganisms, which cause quality degradation through the generation of flavor, off-odors, and destructive compounds When oxygen (O_2) comes into contact with an electron somewhere between one and four, it is reduced to a transitional compound that produces superoxides, hydroxyl radicals, hydrogen peroxide, and water, all of which are extremely reactive (free radicals in nature) with the exception of water, resulting in oxidative reactions [57]. The excellent loss of oxygen-sensitive nutrition products such as milk powder, packaged pasta, biscuits, fruit juices, organic product, etc., can be decreased utilizing the oxygen scavengers which reduce the oxygen atoms left after the packaging system [56]. However, nowadays oxygen scavenging components are being incorporated inside the packaging material itself utilizing proteins, mono- or multilayer substances, and receptive conclusion liners for containers and jugs.

Oxygen scavengers help to diminish the permeation of oxygen through the package during storage, transportation, and retail purchase. Effective oxygen scavengers are highly competent oxygen collectors and able to eliminate oxygen and work indefinitely till the scavengers are available. They can eliminate the oxygen to under 0.01% as announced by certain producers [58].

This is significant in processed meat products such as fermented sausages or cooked ham where quick discolouration might be caused if the packaging with indications of oxygen is presented to the lighting producing the photooxidation measures.

Generally, the oxygen scavengers work on the iron powder oxidation; however, non-metallic oxygen absorbers have been as of late produced for limiting the malicious impacts of metal-based scavengers like upcoming medical problems, causing arc while microwave heating, being perceived in the metal detectors, etc. Some microorganisms, for example *Pichia subpelliculosa* and *Kocuria varians*, have been involved in the oxygen scavengers as a substitute to chemical scavengers, which are getting the compensations of continuing the sustainability. The spores of *Bacillus amyloliquefaciens* were combined as a scavenger in the PET copolymer having 1,4-cyclohexanedimethanol, which can be activated within 48 hours under the high humidity at 30°C after which it could capably absorb the oxygen for at least 15 days [59].

For commercial applications, the oxygen scavengers are usually presented to the packaging materials to avoid the non-edible waste along with the food, reducing the uncertainty of sudden rupture or crack of the sachets in the package and feeding of their substances.

8.9.3   MOISTURE SCAVENGERS/ABSORBERS

In case of foods sources with high water activity, surplus moisture is formed inside the package, resulting in the bacterial and mould growth, which leads to the decrease in the shelf life and food quality. Therefore, moisture scavengers are essential for controlling the moisture development to prevent microorganism growth and further develop the product appearance. After oxygen absorbers, moisture scavengers are the commercially evolved category and accessible in numerous forms such as pads, sachets, sheets, or blankets. Desiccants such as activated clays, silica gel, calcium oxide (CaO), and added minerals are usually porous and tear-resistant plastic-based sachets, which are used to switch moisture and humidity instance of dried food packaging. The most popular moisture scavengers are predictable silica gels and molecular sieves such as zeolites for absorbing odours, after drying. Currently, moisture drip scavenging sheets, pads, and blankets are being manufactured by many companies to control the moisture in foods such as fruits, vegetables, meats, poultry, and fish. Such absorbers naturally involve a double layer of microporous non-woven polymer such as PE or PP, which is located onto the very absorbent compounds such as starch-based copolymers, carboxymethyl cellulose, and polyacrylate salts.

Similarly, the double-action carbon dioxide (CO_2) or oxygen absorber sachets and labels are commercially for the foil-packaged and preserved coffee in the USA and Japan [60].

8.9.4 Carbon Dioxide (CO_2) Scavengers/Emitters

Carbon dioxide (CO_2) is harmful to the development of aerobes (microbes or fungi) due to the decreased relative oxygen level and antimicrobial effects which result in the extended delay phase as well as group period in the log phase during the evolution of microorganisms. High carbon dioxide stages (nearly 10%–80%) find applications in meat conservation by hindering the growth of microorganisms and thus increasing the storage life [61]. Thus, carbon dioxide producers are viewed as a steady system to oxygen scavengers.

Aerophilic microbes such as *Pseudomonas* may be prevented by using the medium to high levels (10%–20%) of CO_2, while the lactic acid bacteria reproduction may be triggered by the CO_2. Besides, a few microorganisms such as *L. monocytogenes*, *C. botulinum*, and *C. perfringens* are partially repressed by the concentration (<50%) of carbon dioxide. A study conveyed the expanded production of *C. botulinum* at higher CO_2 levels while reducing the bacterial development rate [62]. Thus, the application of CO_2 packaging should be exactly examined relying on different meat products and CO_2 levels. In modified climate packaging (MCP), carbon dioxide usually has a microbiological inhibitory consequence, but excess levels of carbon dioxide may poorly affect the product (sometimes change the sensitivity of product). Hence, it is essential to remove the carbon dioxide in some packaging systems to ensure food conservation. Calcium/sodium/potassium hydroxide, silica gel, and calcium oxide are generally used as CO_2 absorber to prevent the package swifting [63]. On high water bustle conditions, the calcium hydroxide combines with carbon dioxide to form calcium carbonate:

$$Ca(OH)_2 + CO_2 \quad CaCO_3 + H_2O$$

CO_2 emitters are functional mostly to decrease the gas-to-product volume ratio, resulting in decreased headspace. They are commercially available as spongy pads and sachets in meat, poultry, and cheese packaging [64].

8.9.5 FLAVOUR AND ODOUR ABSORBERS OR RELEASERS

The integration of odours and flavours is essential to make the food more interesting to the clients and improve the fragrance or flavour of fresh food or managed product in initial stage of packaging. Such fragrances and spirits are generated regularly and consistently inside the stuffed product during its packing life, or their spread might be controlled while food preparation or initial the package. Measured distribution of fragrance can be used to stabilize the inherent loss of smell or taste of food during the complete storage life [65].

On the other hand, the odour exclusion from the package interior can be unfavourable and helpful also. Moreover, in the previous case, the absorption of fragrance mechanisms can withdraw the chosen constituents from the product as irregularly the aromatic mixtures are certainly collected in the interior of the package as in the instance of orange juice. In such conditions, the deficiency of desired fragrance must be prohibited from the food product, which is also one of the aims of the barrier packaging. Nevertheless, the exclusion of odour or fragrance is valuable numerous times under the domain of active packaging. Some foods such as cereal products and new poultry produce a specific odour called detention odour. The major way for their exclusion from package interior is to avoid or eliminate the potential side effects of these confinement odours. Further, probable reasons for presenting odour scavengers may be to eliminate the effect of odour shaped in the package materials.

8.9.6 ETHYLENE ABSORBERS AND ADSORBERS

Ethylene is a usual plant chemical produced by ageing. It accelerates produce inhalation, resulting in maturity and ageing. Eliminating ethylene from a package environment helps increase the shelf life of new produce. The maximum communal agent of ethylene removal is potassium permanganate, which oxidizes ethylene to acetate and ethanol [66]. Ethylene may also be taken out by physical adsorption on active surfaces such as activated carbon or zeolite. Potassium permanganate is generally provided in sachets, while other adsorptive or permeable chemicals may be provided as sachets or combined in the packaging materials.

8.9.7 MECHANICAL PROPERTIES

The mechanical properties have the most organic material and comparable to petroleum-derived materials. For example, lactic acid occurs in two forms: the optically active levo (L)-enantiomer and the racemic blend of dextro (D)- and levo-enantiomers. Polymerization of the two forms results in poly(L-lactic acid) and poly(DL-lactic acid), raised to then as PLLA which shows regularly crystalline

TABLE 8.5

Mechanical Properties of Bio-Based Polymers and Those Derived from Oil [12].

Polymer	Melting Temperature T_m (°C)	Glass Transition Temperature T_g (°C)	Tensile Strength (N mm^{-2})	Elongation at Break (%)
PLA (polylactic acid)	130–180	40–70	48–53	30–240
PHB (polyhydroxybutyrate)	140–180	0	25–40	5–8
Starch	110–115		35–80	580–820
PET (polyethylene terephthalate)	245–265	73–80	48–72	30–300
LDPE (low-density polyethylene)	110	−30	10	620
PS (polystyrene)	100	70–115	8–20	100–1000
PP (polypropylene)	176	0	3.8	400

and extremely resistant to hydrolysis and PDLLA is mainly amorphous and more liable to hydrolysis.

PLA is distinct by molecular weight of the polymer chain structure (linear with respect to the branched), the degree of crystallization, etc. Positioning of PLA improves the mechanical strength and thermal stability. The amorphous and poorly crystallized PLA has a translucent, glossy surface; a highly crystalline PLA has a dense surface. Table 8.5 represents the mechanical properties of bio-based polymers and those derived from Oil.

The physical properties of PHAs copolymer depend on the configuration and molecular structure of the copolymer. The PHB is a hard, highly crystalline, thermoplastic polymer, which mostly resembles the isotactic polypropylene because of its mechanical properties. As such polymer PHB usually rigid and brittle, the outline of the hidoksivaleratne subunit copolymers advances his mechanical properties so that it decreases the level of crystallization and melting temperature subsequent in a decrease or increase in stiffness durability and resistance to impact. It is obvious that a variability of PHAs is used in numerous applications due to its properties and melting temperature, which is 150°C–180°C. The experimental systems try to regulate which materials have superior properties and are permitted out and also the mixing of diverse categories to obtain an improved packing material.

8.10 CURRENT RESTRICTIONS

The primary issue was the most-used for food packaging are their properties, privilege and cost. It precise, brittleness, low temperatures at which generates distortion, low resistance during processing (excluding PHA polymers) and their hindrance properties, mostly to water vapour, restrictive their custom. The charge has been decreasing in recent years and will most likely continue to do so in the future, but with time, the measure development and industry's ability to produce such materials should improve. The limited availability of raw materials is still possibly the most pressing issues that hinder the expansion of such materials. However, it is unlikely that it will be sufficient to PLA and to see the needs of the food industry for some time Biodegradable packaging comes in a variety of shapes and sizes to familiarise you with the materials for packaging and storage of diverse items, including the most biodegradable gels, films, bags, boxes with lids, and trays.

8.10.1 BIOACTIVE PACKAGING

Bioactive packaging might be defined as "an innovative packaging technique where the bioactive/functional packaging materials hold the essential bioactive compounds possessing functional properties at ideal level till they are formed within the package during the packing or before consumption to advance the consumer's health". Bioactive packaging makes the packaged foods better and thus is directly associated with the consumer's health. Numerous techniques which sustain the unique characteristics of biopolymers include microencapsulation, nanoencapsulation, enzyme encapsulation, and enzyme immobilization.

Phytochemicals, vitamins, prebiotics, and nanofibres are the perfect functional mechanisms for integration in the package to encourage the health. For the effective combination of bioactive substances, the subsequent factors are essential to obtain the compulsory issue rate as soon as the package is unlocked and before consumption: manufacture technique of the films, optimal temperature/time grouping for mixing bio-based packaging materials and functional substances, suitable packaging material and production mechanism.

Proteins, fats, dextrins, starches, alginates, and several oil compounds are used as encapsulating agents. The bioactive constituents are released from the capsule using appropriate methods such as solvent initiation. They are site- and stage-specific and signalled using modification in temperature, osmotic shock, contamination, or pH [67]. The choice of suitable immobilization method and biomaterial support to production the enzymatic packages can be strongly based on the physical characteristics of biocatalyst such as sterile enzymes or whole cells derived from bacterial or fungal source, probable storage limitations, kind of packed nutrition, specific consumption of the biocatalyst.

8.10.2 NANO-PACKAGING

The nanotechnology, a science of small materials, is active in food packaging to constrain food decomposition, preserve quality, improve shelf life, and confirm the freshness of food products and drinks. It may also contribute to the customers to advance the food as per their taste necessities and nutritional demands. A very minute volume of nanoparticle is sufficient to transform the packaging materials short of significantly moving their transparency, density, packaging, and processing properties since of a large aspect ratio of nanoparticle. Nanotechnology modifies the assembly of slightly packaging substantial at molecular level.

Nanoscale-based innovation offers innovative modifications to food packaging by (i) improving barrier as well as mechanical properties, (ii) sensing the pathogens, and (iii) dynamic/intelligent packaging, thus unveiling the food eminence and protection advantages. The biodegradable films obtained from usual polymer have limited applications in packaging due to the lower barrier and mechanical properties against heat, carbon dioxide, oxygen, flavour and volatiles, moisture stability, and UV-blocking as exhibited by the natural polymers, which can be enhanced using nanocomposites. Nanocomposites lower the packaging excess linked with managed products and preserve the fresh foods ranging the shelf life. The packaging materials, bottles, and other heavy packages by addition of nanoparticles can be turned into lighter ones that have stronger mechanical and thermal properties.

8.10.3 RESPONSIVE PACKAGING

Responsive packaging is the new innovation in packaging and can be defined as "the packaging system where package links due to the explicit change in the package, food, headspace or exterior adjacent and produces condensed nutrients or energetic ingredients under specific atmosphere to increase the shelf life and value of food". Responsive packaging system responds only to incentive current inside and outside of package, where improvements can be anything which unfavourably moves the food like food-borne threats, microorganisms, moulds, impurities, pH, humidity, or gas levels in the headspace. The direct interaction between quality indicator/sensor and headspace or food product is essential for providing the response about the food eminence [68].

Responsive materials, for example, self-assembled nanoparticles, hydrogels, coated films, supramolecular materials, and surface-grafted substances, must be added into the packaging system that expresses deviations in chemical or physical characteristics in response to impetuses [69]. The responsive food packaging can decrease the instances of food-borne diseases as perceived in real time and also minimize the food waste since fresh food is simply recognized by clients and processors. Responsive food packaging can be bioresponsive, chemoresponsive, thermoresponsive, and mechanoresponsive in nature and constructed on the spurs present in food or package. Almost certainly, the responsive packaging is a kind of revolutionary innovation for packing the food in the innocuous method, but several standards must be taken into consideration before its commercialization. The concert of reactive materials must be obviously defined in standings of detection limit, sensitivity, working environments, and variety. This technology costs high, but its rewards clients in terms of enhanced safety and value of food.

8.10.4 EDIBLE PACKAGING

Edible packaging makes it possible to eat food products on the retail counter alongside their edible skins. It might decrease the food and packaging excess and migration of chemicals from the package to food. Edible packaging films and coatings are naturally composed of proteins, carbohydrates, or fats built on their use. Edible package must possess the essential and important functional characteristics to act as a humidity barrier, and gases and new blends or composites may be expressed for regulating the transport of nutrients organized with food-grade additives [70]. Currently, the edible packaging revolution lies in the following five groups in the food industry, such as food packed in an edible/biodegradable package, food contained in food, a container or cup to be expended with its drink, package that disappears, and edible packaging at quick-service cafeterias. Further nanoscale assemblies may be incorporated in edible packages to advance their applications; however, full protection assessment is required before addition.

8.11   BIODEGRADATION AND COMPOSTING

The biodegradation has the biochemical material conversion process in the water, biomass, CO_2, or CH_4 gas in relations of the achievement of microbes. This process of biodegradation of polymer contains of two steps. Primarily, the development of reducing the polymer chain breaking of C bonds in terms of the consequence of thermal (degradation rate depends on temperature), moisture and the existence of microorganisms. Next, part of the development of biodegradation initiates when shorter chains become energy causes of microorganisms (bacteria, fungi, or algae). This development has in full intellect recognized as biodegradation only when carbon compounds convert food and microbes are converted into water, biomass, or CO_2.

The composting technique has a changing of organic matter remains in the productive humus. Natural substances which further develop soil structure formed from organic excess valued support to retain humidity, the soil more breathable, growth soil microparasitic action, enriching it with nutrients and development the resistance of plants to pests and diseases.

It is a biological procedure in which with the precise conditions of higher temperature and activity of certain microbes (composting cycle), there is a degradation of polymers to bio-based as fast as the others decomposition of biological excess, subsequent in a water, CO_2 and compost. The organic compost is entirely globally neutral and, in agronomical terms, shall have the same features as other composts. The development of composting is a key part of industry with organic waste and return the relics of biodegradable resources in the innovative use [71].

8.11.1   THE BIODEGRADATION POLYMER MECHANISM

- Degradable polymers: the physical and chemical structure changes under the influence of humidity and oxygen.
- Biodegradable polymers: these are fragmented down under the stimulus of naturally occurring microbes.
- Hydrolytic degradable polymers: these are degraded by hydrolysis.
- Oxidation degradable polymers: these are fragmented down by oxidation.

- Photodegradable polymer: degradation occurs under the influence of natural sunlight and oxygen. Under the effect of photooxidation, physicochemical bonds are broken. This is a response involving radicals and chain reaction.
- Thermally degradable polymers: thermally degradable in presence of sunlight and environmental conditions.

8.12 BIODEGRADABLE PLASTIC

The biodegradation of plastic occurs if the biological structure (the body) uses organic materials as a basis of nutrients. Microbes recognize biodegradable plastics as food, feed on them, and digest them. Biodegradable plastics can be founded on renewable raw materials – biomass (e.g. starch) – or non-renewable fossil raw materials (e.g. oil) managed by chemical or biotechnological developments. Basis or interaction that produces biodegradable plastic does not affect the arrangement of biodegradable plastics.

- It is rapidity the grade of degradation, in addition to the chemical configuration of, depends on the initial material, and the outline of the domain of the end product, which can be changed by adding fillers and plasticizers to advance properties or reduce price.
- The degradation cycle of biodegradable plastics may comprise instantaneous or successive abiotic (cold and humidity) and biotic aspects (microorganisms) and must comprise of a step of biological mineralization.
- Fundamental phase – in this phase, division occurs (macroscopic decomposition and conversion to oligomers).
- Inferior phase – mineralization occurs (conversion of the organic substances into inorganic ones, under the stimulus of microorganisms).
- Chemical mechanisms: hydrolysis, oxidation, and reduction in ambient conditions.

The biodegrade and composting products are responsive alternative to ensure the situation in order to preserve petroleum product, and decrease CO_2 emissions.

8.13 COMPOSTABLE PLASTICS

The compostable plastics are biodegradable in the conditions and inside the time surround of the cycle of composting. In this during industrial composting form temperature can affect up to 70°C. Composting ascends in humid atmospheres, and the composting procedure takes place for few months. The significant has to recognize that a biodegradable plastic is not certainly compostable (can biodegrade over time or below diverse conditions), whereas still compostable plastic biodegradable. The determination of rules for compostable plastics is significant because materials that are not suitable for composting can decrease the final quality of compost [72]. The compostable plastic is characterized by several national and international standards (e.g. EN13432 and ASTM D-6900), including industrial composting. EN13432 describes has the appearances of packaging materials must chance to be standard as compostable and suitability for reprocessing of organic solid excess. EN14995: 2006 provides the opportunity for plastic that is cast off for commercialized application. These standards form the basis for various documentation systems [73].

At a low temperature in the compost stack, home composting is more diversified than the industrial. Plastics must be precisely tested to determine compostability in home and green environments.

8.14   APPLICATIONS OF BIODEGRADABLE POLYMERS

Biopolymers have an extensive scope of application in various fields, including medicine, packaging, agriculture, and automotive industries [74]. Biopolymers that are active in packaging keep on getting more updates than those used in other applications. China and Germany are approving the extensive use of biodegradable packaging materials to reduce the volume of latent materials presently being arranged in landfills, inhabiting unusual existing space. The packaging waste materials have caused significant environmental effects. The biodegradation of packaging materials has garnered increasing attention [75].

8.14.1   Applications in Medicine and Pharmacy

Recent applications of biodegradable polymers include surgical inserts in vascular or orthopaedic operation and pure membranes. Due to the good strength and variable degradation speed, biodegradable polyesters are usually active as porous structure in tissue engineering. Gelatin is a usual polymer used for coatings and microcondensing several medicines for biomedical applications and also active for preparing biodegradable hydrogels. Poly(3-hydroxybutyrate-co-3-hydroxyvalerate), usually recognized as polyhydroxybutyrate, has the novel property of being piezoelectric, which is used in various applications where electrical simulation is applied. PGA fabrics (non-woven) have been explored as scaffolding conditions for tissue revival. Chitin and its derivatives are used in drug carriers as well as anti-cholesterolemic medicines, plasma anticoagulants, anti-tumor, and cancer-related goods [76]. Collagen's cell adhesion property is revealed by PLGA; thus, it can be used for tissue engineering applications and polymeric shell in nanoparticles used as drug delivery schemes. Polyanhydrides have been explored in precise release devices for drugs treating eyes syndrome and using as limited anaesthetics, chemotherapeutic agents, anticoagulants, neuro-active drugs, and anticancer specialists [77].

8.14.2   Applications in Stuffing

The packaging has alternative significant extent where biodegradable polymers are used. It reduces the volume of waste; biodegradable polymers are regularly used. The biopolymers show properties such as air permeability and low temperature sealability. PLA has a standard permeability level to water vapour and oxygen. It is used in packaging applications such as films, plates, and flasks [75,78]. PCL is also used in easily compostable packaging. The novel tendency in food packaging is thus the use of blends of different types of biopolymers. Chitosan is used in paper-based packaging as a coating, to transport an oil hindrance packaging etc. Films based on chitosan have been recognized to be active in food conservation and can be possibly used as antimicrobic packaging [79,80].

8.14.3 APPLICATIONS IN AGRICULTURE

Insecticides and nutrients, fertilisers, and insect repellent pheromones are among the grown chemical compounds in question. In aquatic agriculture, biopolymers are used in a variety of wires and spinning nets. The agricultural films have located in the soil are disposed to degradation and ageing during their suitable lifetime, so they essential to have some specific properties. Once starch interacts with soil microbes, it degrades into non-toxic products. This is the reason starch films are used as agriculture mulch films [81].

8.14.4 APPLICATIONS IN OTHER FIELDS

Automotive: The automotive sector aims to manufacture lighter vehicles by the use of bioplastics and biocomposites.

Electronics: PLA and kenaf fibre are used as a composite in electronics applications. PLA has previously been used to make computer cases by Fujitsu.

Construction: PLA fibres have been used in stuffing and pavement. Its inflammability, which lower than that of chemical synthetic fibres, offers more security.

Infrequent applications: It has lot of applications which do not fit into any of the prior groups. Thus, combs, pens, and computer pads made of biodegradable polymers have also been considered as promotion tools. As food and feedstuff additives, chitin and chitosan are mostly used [82].

8.15 ADVANTAGES AND DISADVANTAGES OF BIO-PACKAGING

In this literature can be initiate to advices for both positive and negative characteristics of biodegradable packaging. The fact that all of the research and development has led to the statistic that the disadvantages of biodegradable packaging and the need to eliminate its manufacture continues to grow is attired. Advantages and disadvantages of biodegradable packaging are given in Table 8.6.

8.16 CONCLUSIONS

From this research and literature review, it can be determined that biodegradable packaging has a bright future in the food packaging industry. Biopolymers show a significant influence on sustainable development considering the extensive range of disposal options at the lower eco-friendly effect.

TABLE 8.6

Seven Advantages and Disadvantages of Biodegradable Packaging

Advantages	Disadvantages
Easier recycling	The cognizance of people
Non-toxic	Installations of the productions
Less energy to produce	A single immoral property
Environment-friendly	Compostability
Renewable	The lack of arable land
Reduced carbon emission (CO_2)	Short life
Reduced dependence on oil	Processing plants

Currently, it is needed to take phases to the development of biodegradable products and exploit the eco-friendly, social, and industrial benefits. It's the achievement of such very advanced products is the achievement of high quality (eco-friendly quality) values. Biodegradable polymers have previously been recognized for their ability to develop advanced, novel, and effective drug delivery systems. They are capable of delivering a wide range of bioactive materials. The natural biopolymers have a significant role in the controlled and targeted release of drugs. Polymers have received far more attention in recent years as a result of their possible applications in domains such as regular protection and actual prosperity assistance. For enhancing the properties of biodegradable polymers, extensive procedures have been created, for example random and piece copolymerization or connection. Various aspects with policy and governmental changes, as well as ecosphere demand for food and energy capitals, will certainly stimulate the growth of bio-packaging. There is no uncertainty that the manufacture of and demand for this bio-packaging will grow partly because of the better properties of biodegradable packaging and comparatively due to the decrease in its expanse, which is now improper in relation to the value of additional bio-packaging materials. Through increasing the awareness of society, training, and most large marketing chains acting as the industrial manufacturers, the consumers can rise the growth and advance of biodegradable packaging. In order to overcome this large of packaging the food industry requests to further investigate. Most scientists and researchers in this field support that the future of biodegradable packaging depends on a blend of biodegradable polymer nanocomposite materials, which will advance its concert. They approve also that the extreme upcoming of biodegradable material has PHAs whose competitiveness depends on production, and increased manufacture leads to a direct decrease in price. Such kind of policies improve both the biodegradation percentage and the mechanical properties of the polymers. Biopolymers are eco-friendly polymers. If these polymers can replace an equivalent amount of fossil fuel-based polymers, then about 192 trillion of fossil-derived fuel will be saved every year, which results in the decrease in the emission of CO_2 by 10 million tons. To avoid the negative impacts on the ecosystem, developments should be recurring without creating any chemical or biological disparity.

REFERENCES

1. L. Vermeiren, F. Devlieghere, J. Debevere, Effectiveness of some recent antimicrobial packaging concepts, *Food Addit. Contam.* 19 (2002) 163–171. https://doi.org/10.1080/02652030110104852.

2. J.F. Martucci, R.A. Ruseckaite, Biodegradable bovine gelatin/Na+-montmorillonite nanocomposite films. Structure, barrier and dynamic mechanical properties, *Polym. -Plast. Technol. Eng.* 49 (2010) 581–588. https://doi.org/10.1080/03602551003652730.

3 R. Chandra, R. Rustgi, Biodegradable Polymers. *Prog. Polym. Sci.* 23 (1998) 1273–1335. http://dx.doi.org/10.1016/S0079-6700(97)00039-7.

4. H.-J. Endres, A. Siebert-Raths, Engineering biopolymers, *Eng. Biopolym.* (2011) I–XVI. https://doi.org/10.3139/9783446430020.fm.

5. G.G. Sudhanshu Joshi, Ujjawal Sharm, Bio-plastic from waste newspaper, *Interntional J. Eng. Res. Technol.* (2016) 24–27.

6. N. Jabeen, I. Majid, G.A. Nayik, Bioplastics and food packaging: a review, *Cogent Food Agric.* 1 (2015) 1–6. https://doi.org/10.1080/23311932.2015.1117749.

7. B.L. Momani, *Digital WPI Interactive Qualifying Projects (All Years) Interactive Qualifying Projects Assessment of the Impacts of Bioplastics: Energy Usage, Fossil Fuel Usage, Pollution, Health Effects, Effects on the Food Supply, and Economic Effects Compared to Petr,* (2009). https://digitalcommons.wpi.edu/iqp-all.

8. Z. Kuruppalil, Green plastics: an emerging alternative for petroleum-based plastics, *Int. J. Eng. Res. Innov.* 3 (2011) 59–64. http://ijeri.org/IJERI-Archives/issues/spring2011/IJERIVol3N1Spring2011final1.PDF#page=61.

9. S. Pathak, C. Sneha, B.B. Mathew, Bioplastics: its timeline based scenario & challenges, *J. Polym. Biopolym. Phys. Chem.* 2 (2014) 84–90. https://doi.org/10.12691/jpbpc-2-4-5.

10. P. Chalmin, Field actions science reports the history of plastics: from the capitol to the Tarpeian rock, *F. Actions Sci. Reports; J. F. Actions.* 2019 (2019) 6–11. http://journals.openedition.org/factsreports/5071.

11. G.R. Castro, E. Bora, B. Panilaitis, D.L. Kaplan, Emulsan-alginate microspheres as a new vehicle for protein delivery, *ACS Symp. Ser.* 939 (2006) 14–29. https://doi.org/10.1021/bk-2006-0939.ch002.

12. E. Chiellini, A. Barghini, P. Cinelli, V.I. Ilieva, Overview of environmentally compatible polymeric materials for food packaging, *Environ. Compat. Food Packag.* 2010 (2008) 371–395. https://doi.org/10.1533/9781845694784.3.371; A. Ivankovic, K. Zeljko, S. Talic, A. Martinovic Bevanda, M. Lasic, Biodegradable packaging in the food industry, *J. Food Saf.* 68 (2017) 23–52.

13. A. Pandey, P. Kumar, V. Singh, Application of bioplastics in bulk packaging : a revolutionary and sustainable approach. https://vdocument.in/application-of-bio-plastics-in-bulk-comparison-for-use-in-bulk-packaging-in.html?page=1 (2010).

14. I. Muhammad Shamsuddin, Bioplastics as better alternative to petroplastics and their role in national sustainability: a review, *Adv. Biosci. Bioeng.* 5 (2017) 63. https://doi.org/10.11648/j.abb.20170504.13.

15. P. Shivam, Recent developments on biodegradable polymers and their future trends, *Int. Res. J. Sci. Eng.* 4 (2016) 17–26.

16. Y.J. Chen, Bioplastics and their role in achieving global sustainability, *J. Chem. Pharm. Res.* 6 (2014) 226–231.

17. R.L. Reddy, V.S. Reddy, G.A. Gupta, Study of bio-plastics as green & sustainable alternative to plastics, *Int. J. Emerg. Technol. Adv. Eng.* 3 (2013) 82–89. http://www.ijetae.com/files/Volume3Issue5/IJETAE_0513_13.pdf.

18. Ezgi Bezirhan Arikan, Havva Duygu Ozsoy, A review: investigation of bioplastics, *J. Civ. Eng. Archit.* 9 (2015) 188–192. https://doi.org/10.17265/1934-7359/2015.02.007.

19. R.A. Ilyas, S.M. Sapuan, M.L. Sanyang, M.R. Ishak, Nanocrystalline cellulose reinforced starch-based nanocomposite: a review, *Conference Paper.* (2016) 82–87. https://www.researchgate.net/publication/315675302_Nanocrystalline_cellulose_reinforced_starch-based_nanocomposites_A_Review.

20. M. Lackner, Bioplastics - Biobased plastics as renewable and/or biodegradable alternatives to petroplastics, 2015. https://doi.org/10.1002/0471238961.koe00006.

21. A.L. Andrady, M.A. Neal, Applications and societal benefits of plastics, *Philos. Trans. R. Soc. B Biol. Sci.* 364 (2009) 1977–1984. https://doi.org/10.1098/rstb.2008.0304.

22. K. Jamshidi, S.H. Hyon, Y. Ikada, Thermal characterization of polylactides, *Polymer (Guildf).* 29 (1988) 2229–2234. https://doi.org/10.1016/0032-3861(88)90116-4.

23. G.O. Phillips, P.A. Williams, *Handbook of Hydrocolloids*: Second Edition, 2009. https://doi.org/10.1533/9781845695873.

24. V. Morillon, F. Debeaufort, G. Blond, Critical reviews in food science and nutrition factors affecting the moisture permeability of lipid- based edible films : a review factors affecting the moisture permeability of lipid-based edible films : a review, cri, *Rev. Food Sci. Nutr.* (2002) 37–41.

25. J. Jane, Starch properties, modifications, and applications, *J. Macromol. Sci. Part A.* 32 (1995) 751–757. https://doi.org/10.1080/10601329508010286.

26. P. Jariyasakoolroj, P. Leelaphiwat, N. Harnkarnsujarit, Advances in research and development of bioplastic for food packaging, *J. Sci. Food Agric.* 100 (2020) 5032–5045. https://doi.org/10.1002/jsfa.9497.

27. L. Ceseracciu, J.A. Heredia-Guerrero, S. Dante, A. Athanassiou, I.S. Bayer, Robust and biodegradable elastomers based on corn starch and polydimethylsiloxane (PDMS), *ACS Appl. Mater. Interfaces.* 7 (2015) 3742–3753. https://doi.org/10.1021/am508515z.

28. S. Guilbert, B. Cuq, Material formed from proteins, *Handbook of Biodegradable Polymers* (2020). https://doi.org/10.1515/9781501511967-011.

29. L. Sánchez-González, C. González-Martínez, A. Chiralt, M. Cháfer, Physical and antimicrobial properties of chitosan-tea tree essential oil composite films, *J. Food Eng.* 98 (2010) 443–452. https://doi.org/10.1016/j.jfoodeng.2010.01.026.

30. H.M. Park, X. Li, C.Z. Jin, C.Y. Park, W.J. Cho, C.S. Ha, Preparation and properties of biodegradable thermoplastic starch/clay hybrids, *Macromol. Mater. Eng.* 287 (2002) 553–558. https://doi.org/10.1002/1439-2054(20020801)287:8<553::AID-MAME553>3.0.CO;2-3.

31. E. Chiellini, Environmentally compatible food packaging, 2008. https://doi.org/10.1533/9781845694784.

32. R. Pucciariello, M. D'Auria, V. Villani, G. Giammarino, G. Gorrasi, G. Shulga, Lignin/Poly(ε-Caprolactone) blends with tuneable mechanical properties prepared by high energy ball-milling, *J. Polym. Environ.* 18 (2010) 326–334. https://doi.org/10.1007/s10924-010-0212-1.

33. H. Nitz, H. Semke, R. Mülhaupt, Influence of lignin type on the mechanical properties of lignin based compounds, *Macromol. Mater. Eng.* 286 (2001) 737–743. https://doi.org/10.1002/1439-2054(20011201)286:12<737::AID-MAME737>3.0.CO;2-2.

34. P. Alexy, B. Košíková, G. Podstránska, The effect of blending lignin with polyethylene and polypropylene on physical properties, *Polymer (Guildf).* 41 (2000) 4901–4908. https://doi.org/10.1016/S0032-3861(99)00714-4.

35. G. Vaissiere, B. Chevallay, D. Herbage, O. Damour, Comparative analysis of different collagen-based biomaterials as scaffolds for long-term culture of human fibroblasts, *Med. Biol. Eng. Comput.* 38 (2000) 205–210. https://doi.org/10.1007/BF02344778.

36. A.A. Karim, R. Bhat, Gelatin alternatives for the food industry: recent developments, challenges and prospects, *Trends Food Sci. Technol.* 19 (2008) 644–656. https://doi.org/10.1016/j.tifs.2008.08.001.

37. K. Shyni, G.S. Hema, G. Ninan, S. Mathew, C.G. Joshy, P.T. Lakshmanan, Isolation and characterization of gelatin from the skins of skipjack tuna (katsuwonus pelamis), dog shark (scoliodon sorrakowah), and rohu (labeo rohita), *Food Hydrocoll.* 39 (2014) 68–76. https://doi.org/10.1016/j.foodhyd.2013.12.008.

38. R. Glicklis, L. Shapiro, R. Agbaria, J.C. Merchuk, S. Cohen, Hepatocyte behavior within three-dimensional porous alginate scaffolds, *Biotechnol. Bioeng.* 67 (2000) 344–353. https://doi.org/10.1002/(SICI)1097-0290(20000205)67:3<344::AID-BIT11>3.0.CO;2-2.

39. K. Masuda, R.L. Sah, M.J. Hejna, E.J.M.A. Thonar, A novel two-step method for the formation of tissue-engineered cartilage by mature bovine chondrocytes: the alginate-recovered-chondrocyte (ARC) method, *J. Orthop. Res.* 21 (2003) 139–148. https://doi.org/10.1016/S0736-0266(02)00109-2.

40. S. Raj, A review on Pectin: chemistry due to general properties of pectin and its pharmaceutical uses, (2012). https://doi.org/10.4172/scientificreports.550.

41. K. Dean, Yu L. Biodegradable protein-nanoparticles composites. In: *Biodegradable Polymers for Industrial Applications*, Edited by R. Smith, Woodhead Publishing Ltd, UK (2005) 289–312.

42. Y. Kumar, P. Shukla, P. Singh, P.P. Prabhakaran, V.K. Tanwar, Y. Kumar, Bio-Plastics: a perfect tool for eco-friendly food packaging: a review, *J. Food Prod. Dev. Packag.* 1 (2014) 1–06. www.jakraya.com/journal/jfpdp.

43. R.M. Rasal, A.V. Janorkar, D.E. Hirt, Poly(lactic acid) modifications, *Prog. Polym. Sci.* 35 (2010) 338–356. https://doi.org/10.1016/j.progpolymsci.2009.12.003; S. Ramesh Kumar, P. Shaiju, Kevin E.O. Connor, P. Ramesh Babu, Bio-based and biodegradable polymers - State-of-the-art, challenges and emerging trends, *Curr. Opin. Green Sustain. Chem.* 21 (2020) 75–81. https://doi.org/10.1016/j.cogsc.2019.12.005.

44. Manali Shah, Sanjukta Rajhans, Himanshu A. Pandya, Archana U. Mankad, Bioplastic for future: a review then and now, *World J. Adv. Res. Rev.* 9 (2021) 056–067. https://doi.org/10.30574/wjarr.2021.9.2.0054.

45. S.I. Woo, B.O. Kim, H.S. Jun, H.N. Chang, Polymerization of aqueous lactic acid to prepare high molecular weight poly(lactic acid) by chain-extending with hexamethylene diisocyanate, *Polym. Bull.* 35 (1995) 415–421. https://doi.org/10.1007/BF00297606.

46. N.P. Mahalik, A.N. Nambiar, Trends in food packaging and manufacturing systems and technology, *Trends Food Sci. Technol.* 21 (2010) 117–128. https://doi.org/10.1016/j.tifs.2009.12.006.

47. J.R. Dorgan, H. Lehermeier, M. Mang, Thermal and rheological properties of commercial-grade poly (lactic acid)s, *J. Polym. Environ.* 8 (2000) 1–9.

48. L.S. Nair, C.T. Laurencin, Biodegradable polymers as biomaterials, *Prog. Polym. Sci.* 32 (2007) 762–798. https://doi.org/10.1016/j.progpolymsci.2007.05.017.

49. M. Mochizuki, Structural effects on biodegradation of aliphatic polyesters, *Sen'i Gakkaishi.* 52 (1996) 203–209. https://doi.org/10.2115/fiber.52.5_P200.

50. T. Fujimaki, Processability and properties of aliphatic polyesters, "BIONOLLE", synthesized by polycondensation reaction, *Polym. Degrad. Stab.* 59 (1998) 209–214. https://doi.org/10.1016/s0141-3910(97)00220-6.

51. Y. Doi, Biodegradable plastics and polymers, *J. Pestic. Sci.* 19 (1994) 1–8. https://doi.org/10.1584/jpestics.19.S11.

52. K.K. Yang, X.L. Wang, Y.Z. Wang, H.X. Huang, Effects of molecular weights of poly(p-dioxanone) on its thermal, rheological and mechanical properties and in vitro degradability, *Mater. Chem. Phys.* 87 (2004) 218–221. https://doi.org/10.1016/j.matchemphys.2004.05.038.

53. Y.H. Zhang, X.L. Wang, Y.Z. Wang, K.K. Yang, J. Li, A novel biodegradable polyester from chain-extension of poly(p-dioxanone) with poly(butylene succinate), *Polym. Degrad. Stab.* 88 (2005) 294–299. https://doi.org/10.1016/j.polymdegradstab.2004.11.003.

54. K.W. Leong, B.C. Brott, R. Langer, Bioerodible polyanhydrides as drug-carrier matrices. I: Characterization, degradation, and release characteristics, *J. Biomed. Mater. Res.* 19 (1985) 941–955. https://doi.org/10.1002/jbm.820190806.

55. K. Jõgi, R. Bhat, Valorization of food processing wastes and by-products for bioplastic production, *Sustain. Chem. Pharm.* 18 (2020). https://doi.org/10.1016/j.scp.2020.100326.

56. A. Botana, M. Mollo, P. Eisenberg, R.M. Torres Sanchez, Effect of modified montmorillonite on biodegradable PHB nanocomposites, *Appl. Clay Sci.* 47 (2010) 263–270. https://doi.org/10.1016/j.clay.2009.11.001.

57. Dong Sun Lee, Antioxidative Packaging System, Innovations in Food Packaging (2014) 111–131. doi:10.1016/B978-0-12-394601-0.00006-0

58. L. Vermeiren, L. Heirlings, F. Devlieghere, J. Debevere, Oxygen, ethylene and other scavengers, *Novel Food Packaging Techniques* (2003) 22–49.

59. T. Anthierens, P. Ragaert, S. Verbrugghe, A. Ouchchen, B.G. De Geest, B. Noseda, J. Mertens, L. Beladjal, D. De Cuyper, W. Dierickx, F. Du Prez, F. Devlieghere, Use of endospore-forming bacteria as an active oxygen scavenger in plastic packaging materials, *Innov. Food Sci. Emerg. Technol.* 12 (2011) 594–599. https://doi.org/10.1016/j.ifset.2011.06.008.

60. M.L. Rooney, Active packaging in polymer films, *Act. Food Packag.* (1995) 74–110. https://doi.org/10.1007/978-1-4615-2175-4_4.

61. J.P. Kerry, M.N. O'Grady, S.A. Hogan, Past, current and potential utilisation of active and intelligent packaging systems for meat and muscle-based products: a review, *Meat Sci.* 74 (2006) 113–130. https://doi.org/10.1016/j.meatsci.2006.04.024.

62. I. Artin, A.T. Carter, E. Holst, M. Lövenklev, D.R. Mason, M.W. Peck, P. Rådström, Effects of carbon dioxide on neurotoxin gene expression in nonproteolytic Clostridium botulinum type E, *Appl. Environ. Microbiol.* 74 (2008) 2391–2397. https://doi.org/10.1128/AEM.02587-07.

63. R. Ahvenainen, Active and intelligent packaging: an introduction, In: *Novel Food Packaging Techniques*, Edited by R. Ahvenainen, Woodhead Publishing, CRC Press, Boca Raton (2003) 5–21 https://sci-hub.hkvisa.net/http://dx.doi.org/10.1533/9781855737020.1.5.

64. C.E. Realini, B. Marcos, Active and intelligent packaging systems for a modern society, *Meat Sci.* 98 (2014) 404–419. https://doi.org/10.1016/j.meatsci.2014.06.031.

65. E. Almenar, R. Catala, P. Hernandez-Muñoz, R. Gavara, Optimization of an active package for wild strawberries based on the release of 2-nonanone, *LWT - Food Sci. Technol.* 42 (2009) 587–593. https://doi.org/10.1016/j.lwt.2008.09.009.

66. A. López-Rubio, E. Almenar, P. Hernandez-Muñoz, J.M. Lagarón, R. Catalá, R. Gavara, Overview of active polymer-based packaging technologies for food applications, *Food Rev. Int.* 20 (2004) 357–387. https://doi.org/10.1081/FRI–200033462.

67. A. Lopez-Rubio, R. Gavara, J.M. Lagaron, Bioactive packaging: turning foods into healthier foods through biomaterials, *Trends Food Sci. Technol.* 17 (2006) 567–575. https://doi.org/10.1016/j.tifs.2006.04.012.

68. J. Brockgreitens, A. Abbas, Responsive food packaging: recent progress and technological prospects, *Compr. Rev. Food Sci. Food Saf.* 15 (2016) 3–15. https://doi.org/10.1111/1541-4337.12174.

69. M. Zelzer, S.J. Todd, A.R. Hirst, T.O. McDonald, R. V. Ulijn, Enzyme responsive materials: design strategies and future developments, *Biomater. Sci.* 1 (2013) 11–39. https://doi.org/10.1039/c2bm00041e.

70. C.A. Campos, L.N. Gerschenson, S.K. Flores, Development of edible films and coatings with antimicrobial activity, *Food Bioprocess Technol.* 4 (2011) 849–875. https://doi.org/10.1007/s11947-010-0434-1.

71. B. Xi, X. Zhao, X. He, C. Huang, W. Tan, R. Gao, H. Zhang, D. Li, Successions and diversity of humic-reducing microorganisms and their association with physical-chemical parameters during composting, *Bioresour. Technol.* 219 (2016) 204–211. https://doi.org/10.1016/j.biortech.2016.07.120.

72. D. Muscat, B. Adhikari, R. Adhikari, D.S. Chaudhary, Comparative study of film forming behaviour of low and high amylose starches using glycerol and xylitol as plasticizers, *J. Food Eng.* 109 (2012) 189–201. https://doi.org/10.1016/j.jfoodeng.2011.10.019.

73. C.J. Weber, V. Haugaard, R. Festersen, G. Bertelsen, Production and applications of biobased packaging materials for the food industry, *Food Addit. Contam.* 19:S1 (2002) 172–177. https://doi.org/10.1080/0265203011008748.

74. A.G. Andreopoulos, T. Theophanides, Degradable plastics: a smart approach to various applications, *J. Elastomers Plast.* 26 (1994) 308–326. https://doi.org/10.1177/009524439402600401.

75. K. Petersen, P. Væggemose Nielsen, G. Bertelsen, M. Lawther, M.B. Olsen, N.H. Nilsson, G. Mortensen, Potential of biobased materials for food packaging, *Trends Food Sci. Technol.* 10 (1999) 52–68. https://doi.org/10.1016/S0924-2244(99)00019-9.

76. R.A.A. Muzzarelli, Chitin and its derivatives: new trends of applied research, *Carbohydr. Polym.* 3 (1983) 53–75. https://doi.org/10.1016/0144-8617(83)90012-7.

77. S.E.M. Ibim, K.E. Uhrich, M. Attawia, V.R. Shastri, S.F. El-Amin, R. Bronson, R. Langer, C.T. Laurencin, Preliminary in vivo report on the osteocompatibility of poly(anhydride- co-imides) evaluated in a tibial model, *J. Biomed. Mater. Res.* 43 (1998) 374–379. https://doi.org/10.1002/(SICI)1097-4636(199824)43:4<374::AID-JBM5>3.0.CO;2-5.

78. R. Auras, B. Harte, S. Selke, An overview of polylactides as packaging materials, *Macromol. Biosci.* 4 (2004) 835–864. https://doi.org/10.1002/mabi.200400043.

79. F. Ham-Pichavant, G. Sèbe, P. Pardon, V. Coma, Fat resistance properties of chitosan-based paper packaging for food applications, *Carbohydr. Polym.* 61 (2005) 259–265. https://doi.org/10.1016/j.carbpol.2005.01.020.

80. P.K. Dutta, S. Tripathi, G.K. Mehrotra, J. Dutta, Perspectives for chitosan based antimicrobial films in food applications, *Food Chem.* 114 (2009) 1173–1182. https://doi.org/10.1016/j.foodchem.2008.11.047.

81. D. Briassoulis, An overview on the mechanical behaviour of biodegradable agricultural films, *J. Polym. Environ.* 12 (2004) 65–81. https://doi.org/10.1023/B:JOOE.0000010052.86786.ef.

82. E. Agulló, M.S. Rodríguez, V. Ramos, L. Albertengo, Present and future role of chitin and chitosan in food, *Macromol. Biosci.* 3 (2003) 521–530. https://doi.org/10.1002/mabi.200300010.

9 Edible Film and Coating for Food Packaging

Aishwarya Dhiman and Rajni Chopra
National Institute of Food Technology
Entrepreneurship and Management, Kundli

Meenakshi Garg
Bhaskaracharya College of Applied Sciences,
National Institute of Food Technology
Entrepreneurship and Management
University of Delhi

CONTENTS

DOI: 10.1201/9781003227908-9

9.1   INTRODUCTION

Edible films and coatings are the biopolymers that are narrowing down the frontiers between food, preservation, and packaging and are extensively being researched for the protection, maintenance of quality, and packaging of food. An environmental concern associated with the increased amount of waste caused by unsustainable and imperishable packaging matter has inspired scientists to come up with greener, eco-friendly alternatives. With the introduction of a film which is consumable and ecological, prevents migration of moisture and oxidation of lipids, preserves color, eliminates off-odors, improves shelf life and functional properties in foods, not only the notion of food, preservation, and packaging has been amalgamated, but also, by the utilization of by-products from waste generated by food industries, a solution for menace of rising environmental concerns has been achieved while meeting consumer demands for a natural, nutritious, and wholesome product. Additionally, these films can be utilized to boost the sensory, microbiological, and nutritive attributes of the food commodities (Lopez-Rubio et al., 2017). They are used for various food groups to improve their quality, e.g., fruits and vegetables (to retain moisture and prevent weight loss), meat, poultry, fish (to reduce lipid oxidation and prevent moisture loss and discoloration), bakery, snacks, dairy (to prevent hydration), and oil-fried products. The biopolymers essential in the manufacturing of edible films and coatings are generally derived from wastes or by-products from food industries, or underutilized sources of proteins, lipids, or polysaccharides that are eco-friendly, safe for consumption (e.g., corn zein from ethanol production, cheese-derived whey protein, and chitosan from crustacean shells) and can act as carriers of active agents such as antimicrobials, antioxidants, flavorings, and nutraceuticals (Umaraw et al., 2020). The use of agro- and food industry waste for the fabrication of bio-packaging can not only lead to various environmental and economic benefits, but also aid in reducing competition for food resources. Edible packaging systems were traditionally designed to be tasteless and transparent so as to not have any effect on the sensory properties of the food; however, recent novel works have shown that for certain products, such as pizza toppings and sushi wraps, the organoleptic properties contributed by the films might be necessary (Hambleton et al., 2008). Edible films and coatings are presently restricted for administration only in high-value products due to their cost, deficiency of materials with the acceptable functionality, high expense of installation of equipment for their manufacture, drawbacks in the manufacturing process, and specifications (Rojas-Graü et al., 2009; Umaraw et. al., 2020).

9.2   HISTORY AND BACKGROUND

Although the concept of edible packaging has been associated with food technology since the last 50 years, their use dates back to the time long before the chemistries behind them were known. They have been used for centuries in ways such as wax on fruits in order to avoid moisture and to improve the aesthetic appeal of the product by creating a shiny surface. Wax coatings improve the shelf life of fruits and vegetables as they have the ability to retard the rate of respiration, transpiration, and moisture migration in the coated products (Baldwin, 1994). Edible films had a negligible commercial use to as late as 1967 and were only restricted to use as a wax layer on fruits; however, the number of companies offering such products rocketed from around 10 in 1986 to 600 in 1996 as the business flourished and even today, the bio-packaging system utilization has grown rapidly with a considerable gain of more than 100 million dollars yearly to maintain the quality of various foods (Dehghani et al., 2018).

The technique has been practiced with different names in history. In Europe, the process of storing fruits in edible fats or waxes for use in future was known as "larding"; the skin of boiled soy milk was used in Japan to produce an edible film called "Yuba", which maintained the quality and appearance of the food product (Park et al., 2002). A US patent was issued in the 19th century for maintaining the quality of meat using gelatin. The commonly used commercial applications of edible films are coatings of shellac and wax on fruits and vegetables, sugar on nuts and drug pills, zein on candies, and gelatin films on soft capsules (Han and Aristippos, 2005).

9.2.1   CHARACTERISTICS

Edible films and coatings are materials formulated using a biopolymer or a combination of biopolymers. Although the terms coatings and films may be used interchangeably in some scenarios, an edible film is independently manufactured and later applied on the product, whereas a coating is straight away produced

TABLE 9.1

Comparison between Edible Films and Coatings

S. No.	Edible Coatings	Edible films
1.	Coatings are formed directly on the food surface.	Films are produced separately and then applied to food products.
2.	Coatings are applied in liquid form on food.	Films are first molded as solid sheets and then used to cover the food surface.
3.	They cannot be manufactured or used independently.	They are self-independent materials.
4.	Techniques such as dipping, spraying, and brushing are utilized for casting.	Extrusion plays a major role in casting process.

on the surface of the product (Dehghani et al., 2018). Coatings are administered on the food surface in the form of solution by immersion, spray, or brush, while films are stand-alone or self-standing materials (e.g., wraps) covering the surface of food products. The difference between films and coatings is summarized in Table 9.1.

The minimum thickness of 254 micrometers should be maintained by any material to be considered as a film (Robertson, 2012). They should not contain any toxic or non-digestible components and should cover the surface of food product uniformly with satisfactory adhesion. They must possess structural stability and have the ability to prevent migration of aroma and flavor in heterogeneous food systems and not interfere or degrade the sensory characteristics of the food. They must be cost-effective and economically feasible. Edible films and coatings have properties that can prevent UV light, provide barriers against mechanical damage, increase shelf life of the product, possess bioactive components, transport solutes such as additives, salts and pigments, avoid transport of water vapors, and organic components such as aromas and solvents, and gases between food and atmosphere, exhibit antimicrobial behavior against bacteria and fungus, and are produced from biodegradable natural substances (Díaz-Montes and Castro-Muñoz, 2021).

9.3 CLASSIFICATION

The classification of edible films and coatings is summarized in Figure 9.1. Based on their principle ingredients, they can broadly be categorized into four groups.

9.3.1 Polysaccharide-Based Edible Films and Coatings

Edible packaging materials derived from polysaccharides such as cellulose, starch, pectin, agar, exudate gums, and seaweed extracts are categorized as polysaccharide-based edible films and coatings. Polysaccharide macromolecules are

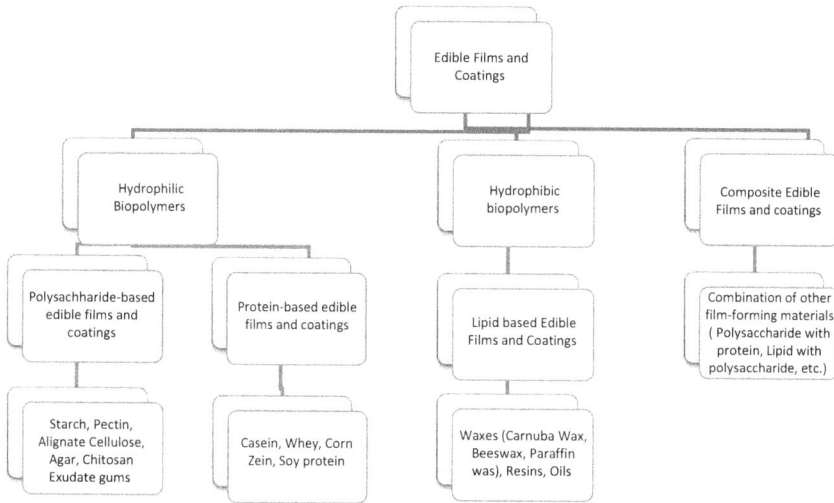

FIGURE 9.1 Classification of edible films and coatings.

generally bioactive and sourced from agricultural feedstocks or crustacean shell wastes. Due to the presence of an organized network of bonded hydrogen, they primarily act as oxygen blockers; however, they are poor moisture barriers due to their high hydrophilicity (Li et al., 2017). This makes them capable for application in fruits and vegetables where they lower the respiration rate by modifying the environment inside the product. They are generally prepared using cellulose and its derivatives, (namely starch, esters, ethers, pectin, and exudate gums). Chitin-based films delay deterioration in harvested fruits and vegetables such as cherries, mangoes, tomatoes, fresh beans, bananas, and strawberries (Prashanth and Tharanathan, 2007). Pectin-based edible coatings comprising of tea leaf extract showed improved the radical scavenging with reduced lipid oxidation when used in gamma-irradiated cooked pork patties (Kang et al., 2007).

9.3.2 PROTEIN-BASED EDIBLE FILMS AND COATINGS

Edible bio-packaging can be manufactured using proteins obtained from both animal and plant sources. Proteins, namely whey protein, corn zein, and casein, due to their abundance, cheap price, high availability, and moisture-blocking characteristics have extensively been used in the manufacture of edible coatings. Protein-based edible films exhibit excellent barrier properties against gaseous exchange as compared to lipid and polysaccharide films, but lower mechanical strength and water vapor resistance in comparison with synthetic polymers, which thereby restricts their applications in food packaging systems (Bourtoom, 2009; Kamal, 2020). The mechanical strength and moisture resistance properties can be enhanced by utilizing various methods such as chemical and enzymatic techniques to modify structure and properties of proteins, by blending proteins and hydrophobic constituents or suitable biopolymers, or by exerting a physical force (Bourtoom, 2009).

9.3.3 LIPID-BASED EDIBLE FILMS AND COATINGS

Hydrophobic substances generally included in the fabrication of edible coatings include waxes, essential oils, oils, fats, natural resins, emulsifiers, and surface-active agents. Wax coatings are applied to enhance the characteristics of various fruits such as apples and other citrus fruits. Historically, larding, a phenomenon of utilization of lards or fats to improve the shelf life of meat products, had been practiced in England. They have a non-polar structure, because of which they are employed as efficacious preventers of moisture migration (Morillon et al., 2002). Cocoa butter and cocoa-based films have extensive applications in bakery and confectionery industry to enhance the product quality. Food-grade varnishes and lacs are used for imparting color, gloss, and sheen to enhance the surface appearance and decrease surface stickiness. In fish and meat products, surface-active agents and emulsifiers are applied as barriers against gaseous and moisture exchange to improve the shelf life (Huber and Embuscado, 2009).

9.3.4 COMPOSITE EDIBLE FILMS AND COATINGS

Composite films are the films formulated by using two or more hydrocolloids in combination with the dissolution of the biopolymer molecules to achieve the benefits of each component. Recent novel innovations involve the use of polymer nanotechnology to develop polymers or copolymer matrices employing specialized nanoscale particulates. A few studies have reported the utilization of nanocomposite consumable packaging materials as a sound substitute for the packaging of food products. Sasaki et al. (2016) designed a film using nanoemulsion of clove oil and pectin, which showed an enhancement in its mechanical and antimicrobial effect against *Escherichia coli* and *Staphylococcus aureus*.

9.4 COMPONENTS OF EDIBLE FILMS

9.4.1 FILM-FORMING MATERIALS

Biopolymers, essentially polysaccharides, lipids, resins, and proteins, either independently or together with other polymers as a blend serve as primary film-forming materials for edible films and coatings. The qualities of the resulting bio-packaging systems are remarkably impacted by the chemical and physical characteristics of these polymers (Han and Aristippos, 2005). In order to control the hydrophobicity and hydrophilicity, along with maintaining the edibility of the film, the solvents are limited to ethanol and water only.

The most common materials for the formation of films are proteins, obtained from many different plants and animal origins, such as milk, eggs, grains, animal tissues, and oilseeds (Chiralt et al., 2018).

Polysaccharides such as starch, gums, non-starch carbohydrates, and fibers are another class of primary materials used for manufacturing films. The excessive amount of hydrophilic and hydroxyl groups occurring in their structure signifies that hydrogen linkages in the case of polysaccharides have a major contribution in the formation of films and may have effects on the attributes of the manufactured film. This organized network of bonded hydrogen makes them act as oxygen blockers; however, their hydrophilic nature restricts them to act as moisture barriers (Li et al., 2017).

Owing to their edibility, biodegradability, and cohesiveness, biomaterials such as lipids; waxes, e.g., carnauba wax, beeswax, rice bran wax, candelilla wax, and terpene; shellac; and resins are also used as raw materials for the manufacture of films. At room temperature, they exist as soft solids, which with the application of heat, followed by techniques such as molding and casting can be molded into any desirable physical structure due to the reversible change in phase that takes place between the fluid, soft-solid, and crystalline structure (Han and Aristippos, 2005). The films and coatings fabricated using lipids and resins are water-resistant and have low surface energy because of their hydrophobic nature (PBrez-Gago and Krochta, 2002).

In order to modify and procure desired film structures and characteristics, biopolymers can also be used as multiple composite layers, i.e., by blending two or more polymers to yield a homogeneous film layer.

9.4.2 PLASTICIZERS

Plasticizers are compounds with low atomic mass, added to the primary polymeric materials required for film formation to lower their temperature of glass transition by reducing the proportion of crystalline to amorphous space so as to improve their flexibility and processability (Krochta, 2002). They have the ability to position themselves between the interlinking polymers and hinder the extensive interactions between the polymer molecules that cause them to be brittle and stiff. Hydrophilic and hygroscopic compounds such as glycerin, sorbitol, sucrose, and polyethylene glycol act as plasticizers due to their ability to produce a hydrodynamic complex with water by attracting water molecules. Factors responsible for affecting the functioning of plasticizers include their ability to bind water, quantity of oxygen atoms and the distance between them within the plasticizer molecule, and the size and shape of molecules in the plasticizer.

9.4.3 ADDITIVES

As a means to enhance quality and safety of food products, various active agents, antioxidants, emulsifiers, nutraceuticals, antimicrobials, flavors, and colorants are employed as additives in materials for film formation. The incorporation of antioxidants and antimicrobials in the film-forming materials provides active functions by protecting them from oxidation and microbial damage, thereby improving their characteristics.

The applications of these components are summarized in Table 9.2.

TABLE 9.2
Applications of Various Components in Edible Films and Coatings

S. No.	Component of Edible Films and Coatings	Applications
1.	Film-forming materials • Polysaccharides • Proteins • Lipids • Composites	Primary raw materials provide structure and physical and chemical properties to the edible coatings and films.
2.	Solvent – water or ethanol	Formulation of solution containing all raw materials.
3.	Plasticizers – glycerin, sorbitol, sucrose, polyethylene glycol	Reduce glass transition temperature, improve elasticity/flexibility.
4.	Additives • Antioxidants • Active agents • Antimicrobial compounds • Emulsifiers • Nutraceuticals	Improve quality, safety, and stability of edible films and coatings. Antioxidants prevent oxidation, thereby reducing instances of off-flavors and odors, antimicrobial compounds improve safety, while emulsifiers improve stability. Active agents impart functional properties.

9.5   FILM FORMATION PROCESS AND MECHANISM

Dry and wet casting methods can be used for the formulation of films, and in both of these methods, a spatially rearranged gel structure with all the ingredients for film formation, including biopolymers, solvents (for wet casting), plasticizers, and additives should be formed. Film-forming solutions are first generated by the gelatinization of biopolymers, which are then dried in order to remove the residue solvent from the gelatinized network. Various techniques, namely X-ray diffraction, FTIR spectroscopy, NMR spectroscopy, polarizing microscopy, electrophoresis, and other polymer chemistry laboratory methods, can be employed to study the complete mechanism behind the formation of film from the biopolymers after gelation. In dry casting, procedures such as thermoplastic properties, gelatinization, flow profile, polymer melting, and polymer rearrangement are used to identify the film-forming mechanisms.

Techniques such as molten casting, extrusion, and heat processing are employed in the dry film formation method. In order to make the film-forming materials flow, heat is exerted on them to cause a rise in the temperature above their melting point. Plasticizers reduce the temperature of glass transition of the film-forming materials.

In the wet process, solvents are utilized for the production of dispersion of film-forming ingredients, and later, the solvent is removed by drying to form the structure of the film. The selection of solvents is one of the most crucial steps in the wet process as all of the film-forming materials including active agents, plasticizers, and biopolymers are incorporated or dispersed homogeneously in the solvent. Commonly, water, ethanol, and their mixtures are used as solvents

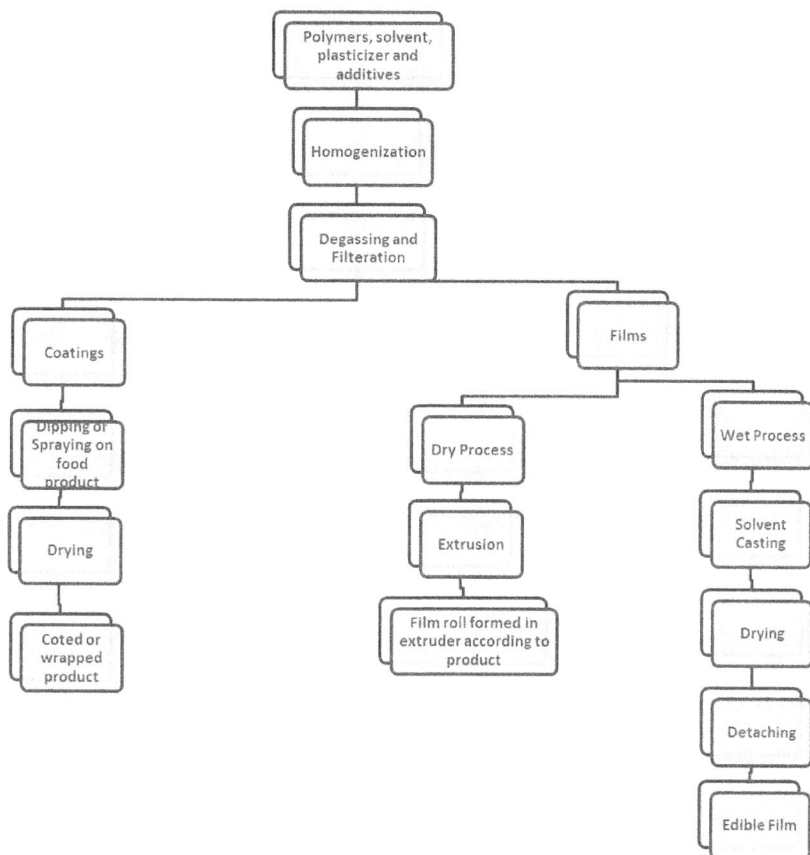

FIGURE 9.2 Film-forming process.

due to their edibility and biodegradability. The solution formed by blending of the film-forming materials and solvents is then applied onto a flat surface with a sprayer, spreader, or dipping roller, which subsequently is dried to get rid of excess solvent, and finally, film is formed. The process of formation of edible films and coatings is demonstrated in Figure 9.2.

The use of incompatible materials may result in their phase separation from the solution, which is not desirable unless a bilayer film is to be produced, in which case phase separation is intentional. Film-forming solution is added with emulsifiers to avoid phase separation and to obtain a homogenous film structure. Due to the immense amount of water or ethanol present in the film-forming solution, they possess a higher surface tension than the surface energy of dry films, which makes them difficult to be evenly coated on a flat surface with low surface energy using high-speed coating equipment; nevertheless, the surface energy of the solution is reduced during the drying process by the elimination of solvent and concentration of the solution. A low-viscosity solution is avoided to prevent the uneven coating and drizzling of solution from the surface to the floor caused by an acceleration of the separation process. Therefore, a high-viscosity film solution is used to reduce the separation rate; however, the peeling process of the low-surface-energy films after drying becomes very hard due to the high adhesion caused by a minor difference of surface energy between the film and flat surface, which contrarily is a desirable phenomenon for direct coating of films onto the food surface to avoid the peeling process.

The higher the difference between the surface energies of the coating material and uncoated product, the lower is the adhesion between the surfaces and worse is the coating performance. The difference between the surface energies can be reduced by the addition of surface-active agents such as emulsifiers and other amphiphilic chemicals, thereby improving the adhesion and coating performance (Díaz-Montes and Castro-Muñoz, 2021).

9.6 NEED FOR EDIBLE FILMS AND COATINGS/FUNCTIONS OF EDIBLE FILMS AND COATINGS/ADVANTAGES

9.6.1 ENVIRONMENTAL SAFETY AND EDIBILITY

The biopolymers required for edible biofilms and coatings are generally derived from wastes and by-products from food industries and food-grade underutilized origins of proteins, polysaccharides, and lipids (e.g., corn zein from ethanol production, cheese-derived whey protein, and chitosan from crustacean shells), which can thus help to solve the menace created by the conventional packaging materials and food wastes in the environment.

9.6.2 PROTECTION FROM PHYSICAL AND MECHANICAL DAMAGE

Physical damage to food products from pressure, mechanical impact, vibrations, or other factors can be reduced by using edible packaging systems. They are tested for their mechanical strength, tensile strength, elongation at break, elasticity, rigidity, and elasticity modulus, along with the mass transfer rates which include permeation, absorption, and diffusion. Both mechanical strength and rate of mass transfer are dependent on the type of substances and process used for the generation of these biopolymer structures.

9.6.3 BARRIER TO MIGRATION OF GASES

The deterioration of quality of majority of food products is attributed to the phenomenon of mass transfer occurring between the food and atmosphere, food product and packaging material, or among the food constituents themselves, including oxygen transfer, moisture absorption, flavor and aroma loss, leaching of components, and off-odor pickup by food packaging materials (Krochta, 2002). For instance, oxygen present in external environment can penetrate into the food and cause its oxidation; inks and other additives can migrate into food products and lead to undesirable changes; volatile flavors and aromas from food might be lost to the material used for packaging or to the surrounding environment; and some products such as biscuits and pies may lose their crispiness due to moisture absorption. Biodegradable edible packaging can avert this migration phenomenon and maintain the products' quality by covering the product or being present between the heterogeneous ingredients of the food. The barrier characteristics of the edible bio-packaging systems are greatly influenced by their composition and atmospheric factors.

9.6.3.1 Carriers of Active Substances

Edible films and coatings have the ability to carry and control the release of various active substances, including some food ingredients, antimicrobial agents, antioxidants, inks, nutraceuticals, pharmaceuticals, bioactive and agro-chemicals in the form of microcapsules, nanocapsules, soluble strips, and flexible pouches (Kester and Fennema, 1986). Edible packaging systems have the ability to control the release of bioactive ingredients to the surroundings in accordance with the type of their application, which could require a slow, immediate, specific rate or non-migration of the ingredient.

9.6.3.2 Food Quality

Edible films and coatings can reduce the rate of drying on surface, improve the oxidative stability of food components, and retard ripening/aging, moisture absorption, aroma loss, and microbial spoilage in food products. Apart from improving the physical and chemical attributes of the products, they also promote the visual or sensory qualities, including surface smoothness, edible color prints, flavor carriers, and other factors (Krochta, 2002). The oxygen-blocking property of edible packaging materials not only improves the quality by preventing oxidation of lipids, flavors, and colorants in food products such as nuts, confectionery, fried and colored products, but also helps in lowering the respiration rate of freshly harvested products (Park, 2000). Food systems with heterogeneous food ingredients such as pie fillings and crust, raisins, and baking cereals are very prone to moisture migration; therefore, their quality can be enhanced by the employment of these moisture barrier edible packaging systems.

9.6.4 Microbial Safety and Shelf Life

Edible films and coatings impart protection to the food by reducing the chances of contamination by foreign matter and lowering spoilage rate by incorporation of antimicrobial compounds, thereby extending their shelf life and improving product attributes. The functions of edible packaging are demonstrated in Figure 9.3.

9.6.5 Convenience

They can provide surface strength to delicate products, which makes them convenient to handle. Coating provides more resistance to tissue damage and bruising caused by physical impact and vibration in fruits and vegetables.

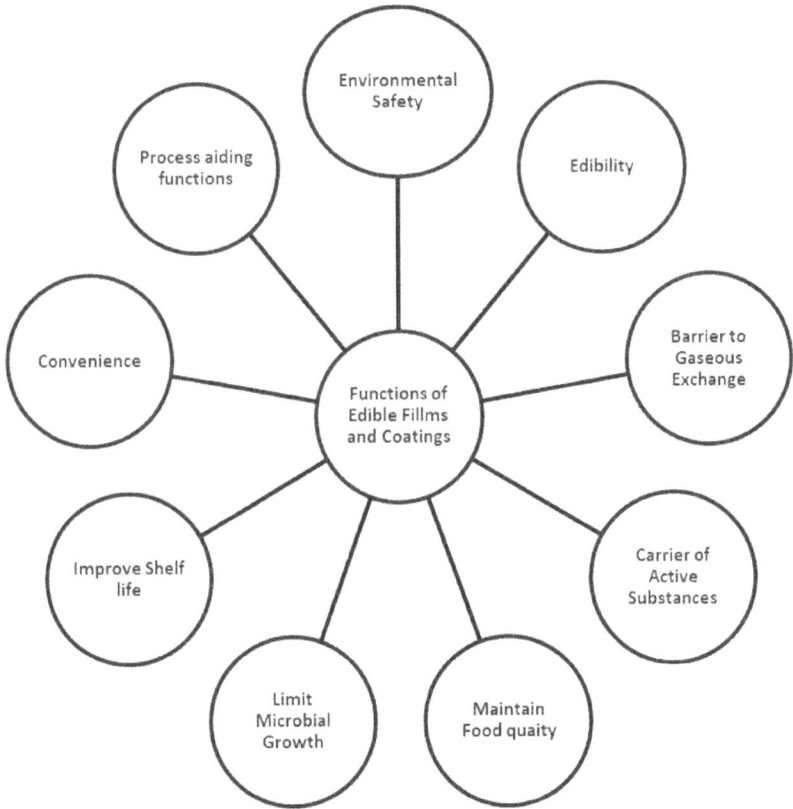

FIGURE 9.3   Functions of edible films and coatings.

9.6.6 PROCESS-AIDING FUNCTIONS

The efficacy of various unit operations in food industries can be amplified by the use of edible films and coatings. For instance, the amount of oil absorbed by potato strips and slices can be reduced by using edible coatings (Han and Aristippos, 2005); the longevity of irradiated fruits and vegetables can be further escalated through edible coatings by retarding the oxidation reactions in dried product occurring during dehydration (Lacroix and Ouattara, 2000). They can also enhance the potency of the popcorn making and act as an adhesive between various diversified components of food (Dabrowska and Lenart, 2001). The prevention of evaporation of volatile flavors and aromas during the process of freeze-drying is another benefit of moisture-permeable coatings.

9.7   APPLICATIONS IN VARIOUS FOOD GROUPS

Table 9.3 represents the applications of edible packaging systems in various food groups.

TABLE 9.3

Applications of Edible Films and Coatings in Different Food Groups

S. No.	Food Group	Applications
1.	Meat, ready-to-eat meat, poultry products, beef, chicken, pork	Inhibit lipid oxidation and discoloration; prevent moisture loss and flavor loss; retain fresh juices; and inhibit the growth of pathogenic microbes.
2.	Fish products	Limit microbial load and maintain quality and safety.
3.	Fresh produce	Improve appearance, freshness, shelf life, and safety of product.
4.	Fruits	Enhance color and mechanical strength; reduce respiration rate and weight loss; maintain firmness; reduce oxidation and moisture loss; prevent tissue disintegration; delay ripening; and improve shelf life.
5.	Vegetables	Reduce microbial load and improve shelf life.
6.	Dairy products	Act as a barrier to gaseous exchange and water vapors to reduce water loss and control maturation rate; reduce microbial growth; and improve shelf life.
7.	Bakery and confectionary	Delay stalling and off-flavors; maintain crispiness of products such as biscuits and snacks; maintain humidity in sponge cakes, pies, and other heterogonous products; retard lipid oxidation; and improve shelf life.

9.7.1 MEAT AND POULTRY PRODUCTS

The quality, shelf life, and safety of fresh and minimally processed (ready-to-eat) meat and poultry products are susceptible to temperature, moisture, light, endogenous enzymes, and microbial damage. These factors have unwanted effects on texture, color, flavor, and other sensory properties of the products, with microbial spoilage being the leading root of meat deterioration. Techniques such as larding have been in use since the 16th century for the preservation of meat pieces. Edible coatings and films inhibit lipid oxidation and discoloration, prevent loss of moisture from tissues during storage, restrict volatile flavor loss, maintain juiciness in fresh meat cuts, and ensure fresh RTE meat and poultry products to maintain standards and safety by inhibiting the growth of pathogens (Kapetanakou et al., 2014).

9.7.2 FISH PRODUCTS

Marine products are considered to be highly perishable due to the presence of a high amount of polyunsaturated fatty acids, autolytic enzymes, high water activity, and neutral pH. Studies involving the use of films such as chitosan, calcium alginate, whey proteins, and gelatin combined with chitosan, alone or with the incorporation of antimicrobial agents (sodium lactate, clove, nisin, etc.) have shown a decreased load of microorganisms (*L. monocytogenes*, TVC, *Salmonella anatum*, *Pseudomonas spp.*, Enterobacteriaceae, and LAB) on products such as cold-smoked salmon, silver carp, and cod fillets (Gómez -Estaca et al., 2010, Datta et al., 2008).

9.7.3 FRESH PRODUCE

The multiple processing stages (such as peeling, cutting, and washing pesticide residues and plant remains) involved in fresh or minimally processed perishable foods make them very susceptible to spoilage, due to which antimicrobial treatments are required to maintain their safety levels and longevity (Robson et al., 2008). Edible bio-packaging also contributes to the improved appearance and freshness of the products till their purchase.

9.7.4 FRUITS

Edible coatings can be used on whole fruits as waxes to improve their color and mechanical strength as well as on fruit pieces in fruit-based salads to enhance their organoleptic appeal. They have the ability to reduce the rate of respiration and transpiration along with avoiding weight loss while maintaining the pigmentation and firmness of fruits; however, their functionality can be improved by the addition of antimicrobial bioactive ingredients. Polysaccharide coatings applied to a variety of foods including papaya, apple, and mango with or without the addition of antioxidants (ascorbic acid) and antimicrobial agents (trans-cinnamaldehyde) have shown a reduction in oxidation, moisture loss, and tissue disintegration, thus improving physical appearance and overall quality of the fruits (Brasil et al., 2012, Kapetanakou et al., 2014). Chitosan-based edible coatings used in fruits such as strawberries, fresh-cut papaya, table grapes, and cantaloupe showed delayed ripening and reduced transpiration rates (Vu et al., 2011). A shelf life extension of two weeks was demonstrated in freshly cut apples coated with alginate and gel-based edible packaging.

9.7.5 VEGETABLES

Most studies on edible coatings for vegetables are directed on freshly harvested or minimally processed vegetables such as broccoli, carrots, and potatoes, but their use in leafy vegetables is limited due to the possible difficulties in their application on such types of vegetables. Polysaccharide films composed of alginate, chitosan, cassava starch, and blends of chitosan and casein with or without the addition of antimicrobials have been applied on potatoes, freshly cut broccoli, pumpkin cylinders, and carrots (fresh and cut), and they have been found to reduce the microbial load, thereby upgrading the quality, safety and shelf life of vegetables (Kapetanakou et al., 2014).

9.7.6 DAIRY PRODUCTS

Dairy products often have low quality and shelf life, attributing to their complex microenvironments containing salts, fats, proteins, etc., and their variety of compositions. The atmospheric conditions maintained while storing and handling the product make their surfaces susceptible to spoilage by yeast, mold, and bacteria. Edible packaging can serve as a carrier of antimicrobial compounds and, therefore, reduce microscopic organisms' load on the surface of cheese (Costa et al., 2018). Chitosan alone or in combination with antimicrobials has indicated a surge in the shelf life of different varieties of cheese, including mozzarella, Emmental, Portuguese cheese, Apulia spreadable cheese, and cheddar cheese. (Kapetanakou et al., 2014).

9.7.7 Bakery and Confectionary Products

Edible biopolymers may be applied to bakery commodities such as pizza, pies, and cakes to control the humidity of sponge cake inside the heterogeneous ingredients used in its manufacture. Cocoa butter and cocoa-based films have extensively been used in the bakery and confectionary organizations to enhance their product quality. The low moisture permeability rates of edible packaging systems can be used to maintain the crispiness of biscuits and other snacks. They can also retard lipid oxidation in bakery products and thus preserve the flavor and aroma of these products.

9.8 RECENT ADVANCES/CURRENT RESEARCH

9.8.1 Carriers of Bioactive Compounds

Bioactive compounds are the food components present in meager amounts (e.g., flavonoids, carotenoids, and phytoestrogens) that contribute to various health benefits due to their antioxidant, anti-inflammatory, antimutagenic, anti-cancer, anti-cholesterol, and apoptotic effects (Kris Etherton et al., 2002, Munoz et al., 2016, 2020, Montes et al., 2020).

9.8.2 Flavor Encapsulation

Flavoring substances are incorporated in the food products to enhance the satisfaction of the consumers. The rate of liberation of the flavoring agents incorporated these films is dependent on the process of encapsulation and the composition of the film. Aroma compounds can be encapsulated in emulsion-based edible films in the form of lipid globules. Carrageenans can encapsulate and reduce the volatility of the flavors and fragrances and can be used in the form of gels, films, and beads (Chakraborty, 2017).

9.8.3 Carrier of Probiotics

Edible films and coatings function as carriers of probiotics/living organisms and can pose a new alternative toward biopreservation and healthier food products (Guimaraes et al., 2018). Few latest researches have indicated the application of edible films as carriers of probiotics; for instance, high water vapor permeability with lower tensile strength and stability of 42 days more than the control film was exhibited by a carboxymethyl cellulosic film developed by incorporating four microbial probiotic strains, namely *Lactobacillus casei, L. rhamnosus, L. acidophilus,* and *B. bifidum* (Ebrahimi et al., 2017).

9.9 CHALLENGES AND OPPORTUNITIES

Despite the various advantages, edible packaging still faces numerous challenges for commercial application due to its unenlightenment and lack of statistics, which restrains the modification of edible films for necessary application and specifications.

9.9.1 REGULATIONS

All the constituents of consumable packaging systems must meet the specified regulations of food products and food-grade substances. They can be categorized as "food products, food additives, food ingredients, food packaging materials, or food contact substances" with respect to their application (Debeaufort et al., 1998). The primary materials used in the manufacture of edible packaging must be of food-grade quality and GRAS. The additives and other ingredients used must be within the limitations specified by the nation or FDA and in accordance with GMP. The presence of any allergens such as milk, peanuts, eggs, soy, wheat, nuts, shellfish, and fish must be labeled according to "Food Allergen Labeling and Consumer Protection Act of 2004", in order to intimate consumers with intolerances or allergies to any used food ingredients (Janjarasskul and Krochta, 2010). The regulation of the "Federal Food, Drug, and Cosmetic Act (21 USC 343)" also requires the producer to declare all the constituents used in the fabrication of film and coating on the label.

9.9.2 FEASIBILITY OF COMMERCIALIZED SYSTEMS

Currently, the manufacturing of edible packaging is limited to a laboratory scale due to its high cost of production in comparison with conventional plastic films. The feasibility of the commercialized edible packaging system is influenced by the resistance of manufacturers to the use of novel or new packaging systems, the difficulties in the operation procedure, cost of financing for coating and film manufacturing resources and equipment, and potential comparisons with traditional packaging systems (Han and Gennadios, 2005). To enhance the commercial systems' feasibility, research based on the cost reduction of edible packaging along with its large-scale production is essential.

9.9.3 CONSUMER ACCEPTANCE

Consumer acceptance issues significantly affect the potential use and commercialization of edible packaging materials. It is a consolidated indicator of the preferences of consumers including safety, organoleptic properties (flavor, taste, texture, appearance, and other sensory attributes), advertising and cultural or religious restrictions, and hesitation to the consumption of new ingredients (Han and Aristippos, 2005). Commercialization of these films and coatings is also dependent on marketing factors, which include price, and special attention to any special instructions required for opening, cooking, consuming the packaged/coated foods, or disposing of the packaging (Janjarasskul and Krochta, 2010).

9.10 CONCLUSIONS

Edible films and coatings can prove to be the up-and-coming systems for food packaging due to their positive impacts not only on food products but also on the environment. With adequate research on various aspects including cost management, new techniques for large-scale production, and better durability, they can replace the present conventional packaging techniques in the near future. The selection of film-forming materials has the primary function in determining the characteristics of the resulting edible packaging. The addition of active compounds has improved the functional properties of edible films and coatings, which have thereby impacted their antimicrobial and antioxidant activity positively and thus improved the shelf life of the perishable goods. Consumer awareness drives shall be run to make the society aware about the ill-effects of traditional packaging and the benefits of bio-packaging systems over them. Proper regulations should be framed by the government and regulatory bodies for their manufacture and commercial applications.

REFERENCES

Baldwin, E. A. (1994). Edible coatings for fresh fruits and vegetables: past, present and future. *Edible Coatings and Films to Improve Food Quality* (J. M. Krochta, E. A. Baldwin and M. Nisperos-Carriedo, eds.), pp. 25–64. Technomic Publishing, Lancaster, PA.

Bourtoom, T. (2009). Edible protein films: properties enhancement. Review Article. *International Food Research Journal* 16, 1–9.

Brasil, I. M., Gomes, C., Puerta-Gomez, A., Castell-Perez, M. E., and Moreira, R. G. (2012). Polysaccharide-based multilayered antimicrobial edible coating enhances quality of fresh-cut papaya. *LWT-Food Science and Technology* 47(1), 39–45.

Chakraborty, S. (2017). Carrageenan for encapsulation and immobilization of flavor, fragrance, probiotics, and enzymes: a review. *Journal of Carbohydrate Chemistry* 36, 1–19

Castro-Muñoz, R., Yañez-Fernandez, J., and Fila, V. (2016). Phenolic compounds recovered from agro-food by-products using membrane technologies: an overview. *Food Chemistry* 213, 753–762. https://doi.org/10.1016/j.foodchem.2016.07.030

Chiralt, A., González-Martínez, C., Vargas, M., and Atarés, L. (2018). Edible films and coatings from proteins. *Proteins in Food Processing*, 477–500. doi:10.1016/b978-0-08-100722-8.00019-x

Costa, M. J., Maciel, L. C., Teixeira, J. A., Vicente, A. A., and Cerqueira, M. A. (2018). Use of edible films and coatings in cheese preservation: opportunities and challenges. *Food Research International* 107, 84–92, ISSN 0963-9969.

Dabrowska, R., and Lenart, A. (2001). Influence of edible coatings on osmotic treatment of apples. *Osmotic Dehydration and Vacuum Impregnation: Applications in Food Industries*, CRC Press, pp. 43–49, ISBN. 97080429132216.

Datta, S., Janes, M.E., Xue, J., Losso, Q.-G., and La Peyre, J.F. (2008). Control of Listeria monocytogenes and Salmonella anatum on the surface of smoked salmon coated with calcium alginate coating containing oyster lysozyme and nisin. *Journal of Food Science* 73 (2). M67-M71

Debeaufort, F., Quezada-Gallo, J. A., and Voilley, A. (1998). Edible films and coatings: tomorrow's packagings: a review. *Critical Reviews in Food Science and Nutrition* 38, 299–313.

Dehghani, S., Vali Hosseini, S., and Regenstein, J. M. (2018). Edible films and coatings in seafood preservation: a review. *Food Chemistry* 505–513, ISSN 0308-8146.

Díaz-Montes, E., and Castro-Muñoz, R. (2021). Edible films and coatings as food-quality preservers: an overview. *Foods* 10, 249. https://doi.org/10.3390/foods10020249

Ebrahimi, B., Mohammadi, R., Rouhi, M., Mortazavian, A., Shojaee-Aliabadi, S., and Reza Koushki, M. (2017). Survival of probiotic bacteria in carboxymethyl cellulose-based edible film and assessment of quality parameters. *LWT-Food Science and Technology* 81, 54–60. doi:10.1016/j.lwt.2017.08.066

Gómez-Estaca, J., López de Lacey, A., López-Caballero, M. E., Gómez-Guillén, M. C., and Montero, P. (2010). Biodegradable gelatin-chitosan films incorporated with essential oils as antimicrobial agents for fish preservation. *Food Microbiology* 27, 889–896.

Guimaraes, A., Abrunhosa, L., Pastrana, L. M., and Cerqueira, M. A. (2018). Edible films and coatings as carriers of living microorganisms: a new strategy towards biopreservation and healthier foods. *Comprehensive Reviews in Food Science and Food Safety* 17, 594–614.

Hambleton, A., Debeaufort, F., Beney, L., Karbowiak, T., and Voilley, A. (2008). Protection of active aroma compound against moisture and oxygen by encapsulation in biopolymeric emulsion-based edible film. *Biomacromolecules* 9(3), 1058–1063. doi:10.1021/bm701230a

Han, J. H., and Aristippos, G. (2005). Edible films and coatings: a review. *Innovations in Food Packaging*, 239–262.

Huber, K., and Embuscado, M. (2009). *Edible Films and Coatings for Food Applications*. doi:10.1007/978-0-387-92824-1

Janjarasskul, T., and Krochta, J. M. (2010). Edible packaging materials. *Annual Review of Food Science and Technology* 1(1), 415–448. doi:10.1146/annurev.food.080708

Kamal, I. (2020). Edible *Films and Coatings: Classification, Preparation, Functionality and Applications-A Review*. 4, 501–509. doi:10.32474/AOICS.2019.04.000184. 30-435

Kang, H. J., Jo, C., Kwon, J. H., Kim, J. H., Chung, H. J., et al. (2007). Effect of a pectin based edible coatings containing green tea powder on the quality of irradiated pork patty. *Food Control* 18(5), 430–435.

Kapetanakou, A., Manios, S., and Skandamis, P. (2014). Application of edible films and coatings on food. *Novel Food Preservation and Microbial Assessment Techniques*, 237–273. doi:10.1201/b16758-11

Kester, J. J., and Fennema, O. R. (1986). Edible films and coatings: a review. *Food Technology* 48(12), 47–59.

Kris-Etherton, P. M., Hecker, K. D., Bonanome, A., Coval, S. M., Binkoski, A. E., Hilpert, K. F., Griel, A. E., and Etherton, T. D. (2002). Bioactive compounds in foods: Their role in the prevention of cardiovascular disease and cancer. *The American Journal of Medicine* 113(9B), 71S–88S. https://doi.org/10.1016/s0002-9343(01)00995-0

Krochta, J. M. (2002). Proteins as raw materials for films and coatings: definitions, current status, and opportunities. *Protein-Based Films and Coatings* (A. Gennadios, ed.), pp. 1–41. CRC Press, Boca Raton, FL.

Lacroix, M., and Ouattara, B. (2000). Combined industrial processes with irradiation to assure innocuity and preservation of food products - a review. *Food Research International* 33(2), 719–724.

Li, L., Sun, J., Gao, H., Shen, Y., Li, C., Yi, P., He, X., Ling, D., Sheng, J., Li, J., Liu, G., Zheng, F., Xin, M., Li, Z., and Tang, Y. (2017). Effects of polysaccharide-based edible coatings on quality and antioxidant enzyme system of strawberry during cold storage. *International Journal of Polymer Science* 2017, 8.

Lopez-Rubio, A., Fabra, M., Martinez-Sanz, M., Mendoza, S., and Vuong, Q. (2017). Biopolymer-based coatings and packaging structures for improved food quality. *Journal of Food Quality* 2017, 1–2. doi:10.1155/2017/2351832

Morillon, V., Debeaufort, F., Blond, G., Capelle, M., and Voilley, A. (2002). Factors affecting the moisture permeability of lipid-based edible films: a review. *Critical Reviews in Food Science and Nutrition* 42(1), 67–89.

Park, H. J. (2000). Development of advanced edible coatings for fruits. *Trends in Food Science and Technology* 10, 254–260.

Park, S. K., Hettiarachchy, N. S., Ju, Z. Y., and Gennadios, A. (2002). *Formation and Properties of Soy Protein Films and Coatings. Protein-based Films and Coatings* (A. Gennadios, ed.), pp. 123–137. CRC Press, Boca Raton, FL.

PBrez-Gago, M. B., and Krochta, J. M. (2002). *Formation and Properties of Whey Protein Films and Coatings. Protein-Based Films and Coatings* (A. Gennadios, ed.), pp. 159–180. CRC Press, Boca Raton, FL.

Prashanth, H., and Tharanathan, R. N. (2007). Chitin/chitosan: modifications and their unlimited application potential—An overview. *Trends in Food Science & Technology* 18, 117–131. doi:10.1016/j.tifs.2006.10.022

Robertson, G. L. (2012). Edible, biobased and biodegradable food packaging materials. (G. L. Robertson, ed.), *Food Packaging: Principles and Practice*. third ed. CRC Press, Boca Raton.

Robson, M. G., Ferreira Soares, N. F., Alvarenga Botrel, D., and de Almeida, G. L. (2008). Characterization and effect of edible coatings on minimally processed garlic quality. *Carbohydrate Polymers* 72(3), 403–409.

Rojas-Graü, M., Soliva-Fortuny, R., and Martin-Belloso, O. (2009). Edible coatings to incorporate active ingredients to fresh-cut fruits: a review. *Trends in Food Science & Technology* 20, 438–447. doi:10.1016/j.tifs.2009.05.002

Sasaki, R. S., Mattoso, L. H. C., and DE Moura, M. R. (2016). New edible bionanocomposite prepared by pectin and clove essential oil nanoemulsions. *Journal of Nanoscience and Nanotechnology* 16, 6540–6544.

Umaraw, P., Munekata, P., Verma, A., Barba, F., Singh, V. P., Kumar, P., and Lorenzo, J. M. (2020). Edible films/coating with tailored properties for active packaging of meat, fish and derived products. *Trends in Food Science & Technology* 98. doi:10.1016/j.tifs.2020.01.032

Vu, K. D., Hollingsworth, R. G., Leroux, E., Salmieri, S., and Lacroix, M. (2011). Development of edible bioactive coating based on modified chitosan for increasing the shelf-life of strawberries. *Food Research International* 44, 198–203.

10 Smart and Intelligent Packaging Based on Biodegradable Composites

Theivasanthi Thirugnanasambandan
Kalasalingam Academy of Research and
Education (Deemed University)

CONTENTS

10.1 INTRODUCTION

Intelligent packaging provides some intelligent functions to the consumer. It can measure the properties of the food environment to inform the user. Fang et al. (2017) reported that intelligent packaging provides safety to the consumers and can give warning regarding the perishable foods through some intelligent functions such as sensing, tracing and communicating. Smart devices such as labels, barcodes and tags attached with the packaging get information about the condition of food, which is transferred to the users. Smart packaging technologies include time–temperature indicators (TTIs), gas indicators, freshness indicators and pathogen indicators. Conventional information regarding food is packed date, weight and nutritional information, and other instructions are also stored in the barcodes, which are then read by smart phones.

The freshness of the package is determined by measuring the concentration of pathogens, amines, carbon dioxide and oxygen. Nowadays, indicators use vanillin to detect various types of molds, yeasts and bacteria. The ripeness of fruits can be checked by sensing ethylene gas, acetaldehyde and acetic acid. Carbon dioxide indicators play a major role in smart packaging because the processes such as ripeness, fermentation, microbial spoilage and degassing are all related to the level of carbon dioxide. The use of nanomaterials as sensors and printed electronics will lead to the commercial applications of smart packaging systems (Sand and Boz, 2018).

Mustafa Fatima and Silvana Andreescu (2018) detailed the smart and active packaging technology concept. Figure 10.1 illustrates the smart packaging technology. The quality of food is assessed in food industries by analyzing the freshness, microbial spoilage and gas or heat spoilage. The indicators exhibit color changes whenever any changes are produced inside the packaging. In conventional biosensors, the components such as probes, transducers and biorecognition elements (enzymes and aptamers) are needed. There is no need for these things in indicators where detection is possible through visual observation.

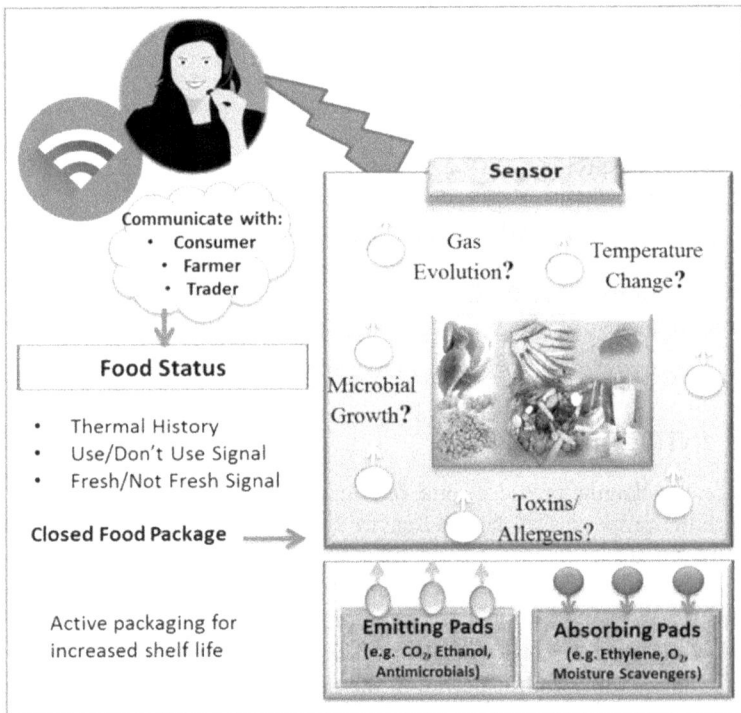

FIGURE 10.1 Schematic diagram representing the smart and active packaging technology concept (Mustafa and Andreescu, 2018).

Smart packaging can provide support for safety monitoring of food. The spoilage of food causes various issues related to food safety. The freshness of food is conventionally monitored by chromatographic techniques and fluorescent and electrochemical principles. These methods give more limitations for practical applications since they require high-end equipment for analysis. Organic amines released from the food cause color changes in the sensing materials. The conventional methods of food analysis are time-consuming, and skilled persons are needed to operate the high-end equipment.

Biochemical reactions occur when food is stored for a long time. Such reactions can release amine compounds, carbon dioxide and organic acids, resulting in a change in pH. Various freshness sensors and TTIs are constructed by observing the color change visually. In the case of smart fruit/food packaging, electrical-, electrochemical- and chemical-based sensor signals are utilized as input for calibration. Results can be read from the outputs such as change in color, fading of colors and voltage.

A radio-frequency identification (RFID) chip is attached on the packaging. RFID systems consist of two major components: transponder or tag and interrogator or reader, which create wireless data transmission. TTIs, ripeness indicators, chemical sensors, biosensors and RFID are the components in intelligent packaging. TTIs and RFID are mostly used for commercial applications. Alam Arif et al. (2021) explained the sensor system utilized in the smart fruit/food packaging technology. Figure 10.2 shows the sensor system and smart packaging.

Smart food packaging is highly sensitive for changes in pH due to the increase in time and temperature. The sensor shows the color change because of the release of chemical compounds and gases from the food. The technology will be more helpful for the users if it is available at a low cost using biodegradable polymers.

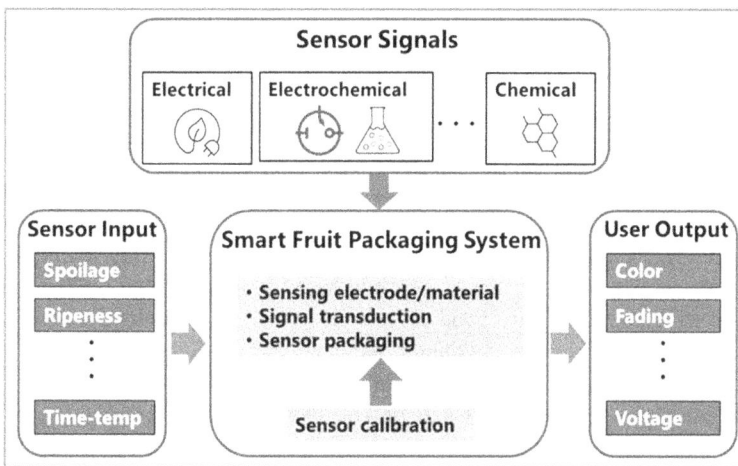

FIGURE 10.2 Schematic diagram demonstrating the details of a sensor system applied in smart fruit/food packaging technology (Alam Arif et al., 2021).

Maintaining the temperature is performed conventionally in the field of food preservation. The variations in temperature affect the quality of food products such as meat and fish (Motelica et al., 2020). Developing colorimetric sensors is emerging as a new research for monitoring the pH changes in food. The spoilage in food is indicated by the changes in pH.

Biodegradable materials such as cellulose and starch have a high market in intelligent food packaging. They are renewable, recyclable and suitable for large-scale production at low cost (Halonen et al., 2020). Biodegradable composites with sensing ability are used for detecting the freshness of food. This kind of smart and intelligent packaging not only covers the food, but can also monitor the changes in the food environment. This new advanced technology helps ensure food safety and can also decrease the wastage of food.

10.2   BIODEGRADABLE POLYMERS/COMPOSITES

Cellulose, starch and chitosan are some of the natural polymers that can be obtained from plants, animals and microbial resources (Huang et al., 2019). Polymer matrices are involved to prepare biodegradable composites. Starch, chitosan, cellulose triacetate, polylactic acid (PLA), poly(vinyl alcohol) (PVA), polyglycolic acid (PGA), polycaprolactone (PCL), poly(β-hydroxybutyrate-co-β-hydroxyvalerate) (PHBV), polyhydroxybutyrate (PHB), poly(butylene adipate-co-terephthalate) (PBAT) and nylon-2-nylon-6 are some examples for the polymer matrices utilized in the biodegradable composites.

The fillers in these composites can also be natural materials such as cellulose fibers, vegetable fibers and rice husk ash. Although polymers are non-conductive in nature, some of them have conducting property. Particularly, conducting polymers such as polypyrrole (PPy) and polyaniline are well known for their conducting property and potential biomedical applications. Barra Ana et al. (2020) discussed several biopolymers in their report. Figure 10.3 shows the chemical structures of some biopolymers.

10.3   CONDUCTIVITY OF BIODEGRADABLE POLYMERS

As already mentioned, polymers are generally non-conductive in nature. They can be made conductive by adding carbon nanomaterials and metallic nanoparticles as fillers. Biodegradable polymers are added with carbon nanomaterials to improve the electrical conductivity. Graphene is a carbon-based nanomaterial. It is a two-dimensional semi-metal. It is a zero-bandgap material with highly conductive nature.

Quantum dots are semiconductor nanomaterials. Their particle size is in between 1 and 5 nm. They have the ability to exhibit photoluminescence properties. Graphene quantum dots can be considered as an example for quantum dots. If the bandgap of graphene is improved, it will replace silicon in electronics industries. Hence, many researchers are working on the bandgap engineering of graphene. One such effort is the synthesis of graphene quantum dots in which the bandgap can be varied with respect to the particle size (Chen et al., 2019).

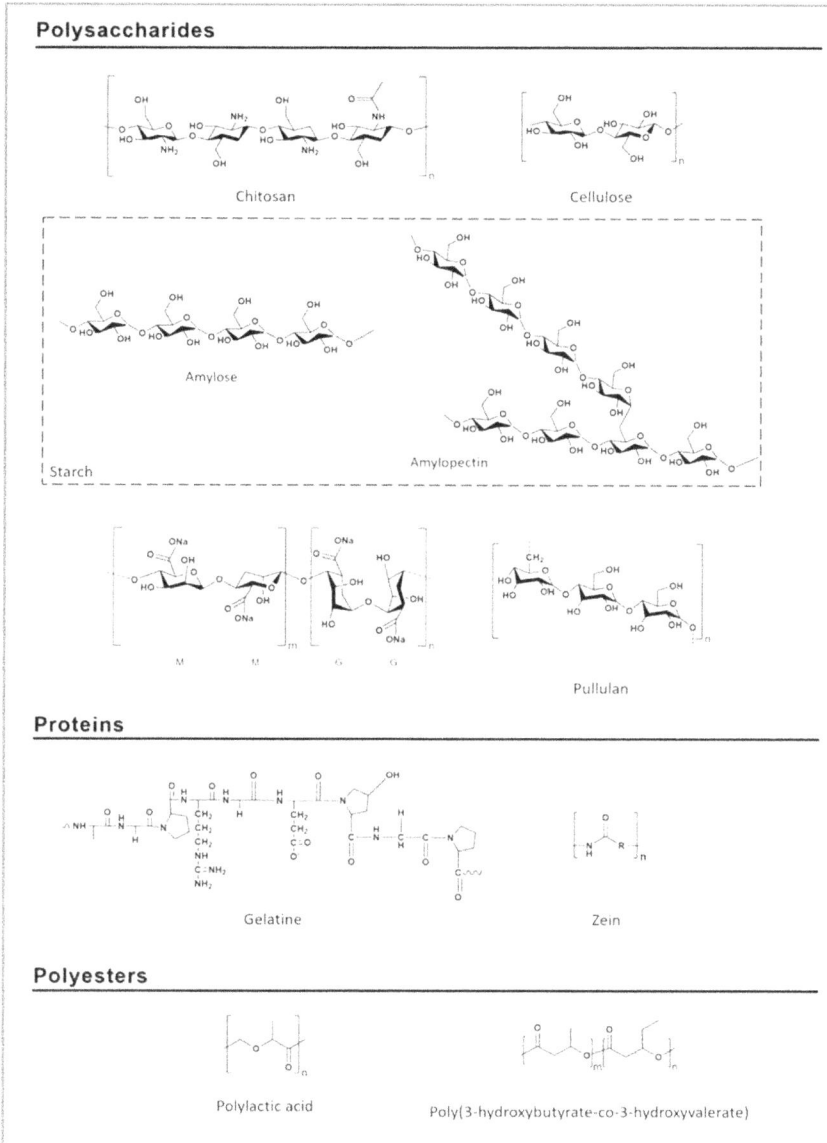

FIGURE 10.3   Chemical structures of various biopolymers (Barra Ana et al., 2020).

Graphene is considered as a suitable filler candidate because it possesses a high electrical conductivity than metals. Similarly, silver nanoparticles have high electrical conductivity and can be added with polymers to make them conductive. It is more important to get a homogenous dispersion of the filler in the matrix. A conductive network is created in the nanocomposites by various methods such as melt blending, solution blending and in situ polymerization (Huang et al., 2019).

A biodegradable polymer becomes a conductor at a particular amount of the filler, which is called percolation threshold. This percolation threshold is attained in polylactic acid with the addition of 0.00094 vol.% of multiwalled carbon nanotubes. Carbon nanotubes are one-dimensional carbon nanomaterials having more applications in nanoelectronics. Polylactic acid is highly porous, which makes embedding three-dimensional multiwalled carbon nanotubes in its surface easy. The obtained polymer composite is more lightweight with a density of $0.045\,g\ cm^{-3}$ (Wang et al., 2018).

The addition of graphene quantum dots to thermoplastic starch increases the conductivity of the thermoplastic starch. Graphene quantum dots are prepared by pyrolysis of biological materials such as citric acid in solution. The carbonization process offers a very good dispersion and high crystallinity to this new advanced material. Starch is subjected to a high temperature and high pressure with glycerin and water to convert it into thermoplastic starch. Thermoplastic starch is found to be more flexible and transparent as synthetic thermoplastics. Thermoplastic starch/graphene quantum dot composites are made by melt extrusion method followed by hot pressing. A maximum conductivity is attained for a filler loading of 10.9 wt.%. The uniform dispersion of graphene quantum dots decreases the resistivity of thermoplastic starch. Thus, the obtained conductive films are very much suitable for optoelectronic packaging applications (Chen et al., 2019).

Graphene-based conductive paper-like films (CPFs) are formed using nanofibrillated cellulose (NFC) as a filler. Graphene oxide/NFC is reduced directly to form graphene-based composites using thermal reduction, hydroiodic acid (HI) reduction and ascorbic acid reduction methods. Figure 10.4a shows the preparation of graphene/NFC films by ultrasonication and filtration processes. Graphene with 16.6% of NFC shows a high electrical conductivity of $168.9\ S\ m^{-1}$ as shown in Figure 10.4b. The electrical conductivity of the HI reduction composite films (10% NFC content) reached $153.8\ S\ m^{-1}$. In the cases of 550°C and 450°C thermal reduction composite films (16.6% NFC content), the electrical conductivity is 168.9 and $86.21\ S\ m^{-1}$, respectively. It is noted that increasing the NFC content by more than 10% and 16.6% in HI reduction and thermal reduction composite films, respectively, reduces the electrical conductivity (Chen et al., 2020).

A conductive paper is designed with a purified paper, which is prepared from ultra-high temperature (UHT) pasteurized milk packaging waste. The UHT milk packaging waste is chemically modified by alkali treatment, which is then immersed into aluminum solution. Thus, a purified cellulose paper coated with aluminum is obtained. The conductive paper can work as an electrode for electrochemical sensor. The presence of aluminum enhances the conductivity of cellulose (Phamonpon et al., 2020). Nanocellulose incorporated with activated carbon films shows electrical conductivity for applying the films in intelligent packaging. These films are demonstrated to be excellent biosensors (Sobhan et al., 2019).

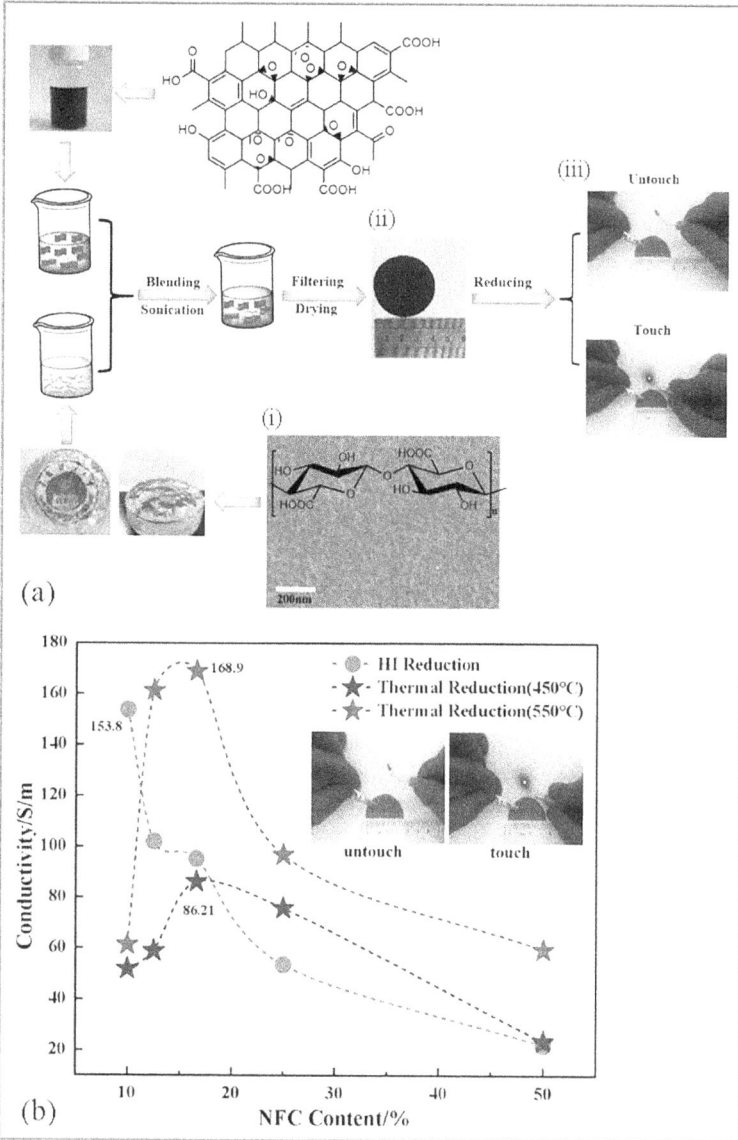

FIGURE 10.4 Fabrication and conductivity measurement of the graphene-based CPFs. (a) Schematic diagram showing graphene-based CPF fabrication. (i) SEM image of the prepared NFC; (ii) GO/NFC composite film; (iii) graphene-based CPFs. (b) Electrical conductivity graph of the graphene-based CPFs prepared (using different NFC contents) in thermal reduction and hydroiodic acid (HI) reduction methods (Chen et al., 2020).

10.4 BIODEGRADABLE/CONDUCTIVE POLYMER BLENDS

Polypyrrole and polyaniline (PANI) are the conducting polymers. Hence, they can be added as fillers in natural polymers to induce electrical conductivity (Guo et al., 2013). Electroactive biomaterials are intrinsically conducting polymers (ICPs). They are considered as "smart" biomaterials of recent generation. PPy, PANI and poly(3,4-ethylenedioxythiophene) have unique electrical and optical properties that lead to potential biomedical applications (Da Silva and Córdoba de Torresi, 2019).

Sago starch is utilized with PANI to prepare a conductive blend. Graphene is added as filler to these hybrid polymers to enhance the electrical conductivity. PANI/sago/graphene is subjected to ultrasonic irradiation using hydrochloric acid and ammonium persulfate to make the nanocomposites. The electrical conductivity of nanocomposites is enhanced from 2.98×10^{-4} to 1.34×10^{-1} S cm^{-1} for less than 1 wt.% of graphene. Sago is a complete natural material, and PANI is a multifunctional and commercial conducting polymer. So, the synergistic effect makes the nanocomposite a suitable candidate for smart packaging purposes (Mohsin et al., 2019). Sago starch is the starch material obtained from the stems of palm plant, *i.e.*, *Metroxylon sagu*.

Hydroxyethyl cellulose and PANI (HEC/PANI) blend is made into film by solution casting process. PANI is incorporated to enhance the electrical conductivity of hydroxyethyl cellulose. An electrical conductivity of up to 2.2×10^5 S cm^{-1} is achieved with this HEC/PANI film. This film is used as a pH sensor by measuring the changes in the electrical properties through its V–I (voltage–current) characteristics (Mustapha et al., 2016).

Poly(sulfobetaine methacrylate)/bacterial nanocellulose (PSBMA/BNC) films are prepared by polymerizing the sulfobetaine methacrylate (SBMA) inside the bacterial nanocellulose nanofibrous network and poly(ethylene glycol) diacrylate as cross-linking agent. The proton conductivity of the composite film is found to be 1.5×10^{-3} S cm^{-1} for 94°C and 98% relative humidity (RH). This property is useful to measure the humidity level in food packaging. Figure 10.5a illustrates the cross-linked PSBMA fabrication. Figure 10.5b shows the dried films of pristine BNC and two samples of PSBMA/BNC nanocomposites. Figure 10.5c shows the Arrhenius plot of the proton conductivity that gives the temperature dependence of the conductivity. The three curves correspond to three different RH levels. The activation energy for proton transport is inversely proportional to the RH. The activation energy for proton transport decreases when the RH increases. Figure 10.5d shows the variation of conductivity with respect to RH at different temperatures (Vilela et al., 2019).

Bionanocomposites prepared by using biodegradable polymers with nanofillers are emerging as an alternative material to synthetic conducting polymers. Carbon nanomaterials are used as fillers in natural polymers to improve electrical conductivity. In such applications, graphene oxide is prepared first by modified Hummer's method, which is further reduced by some toxic reducing agents. An electrical conductivity of 0.7 S m^{-1} in-plane and 2.1×10^{-5} S m^{-1} through-plane

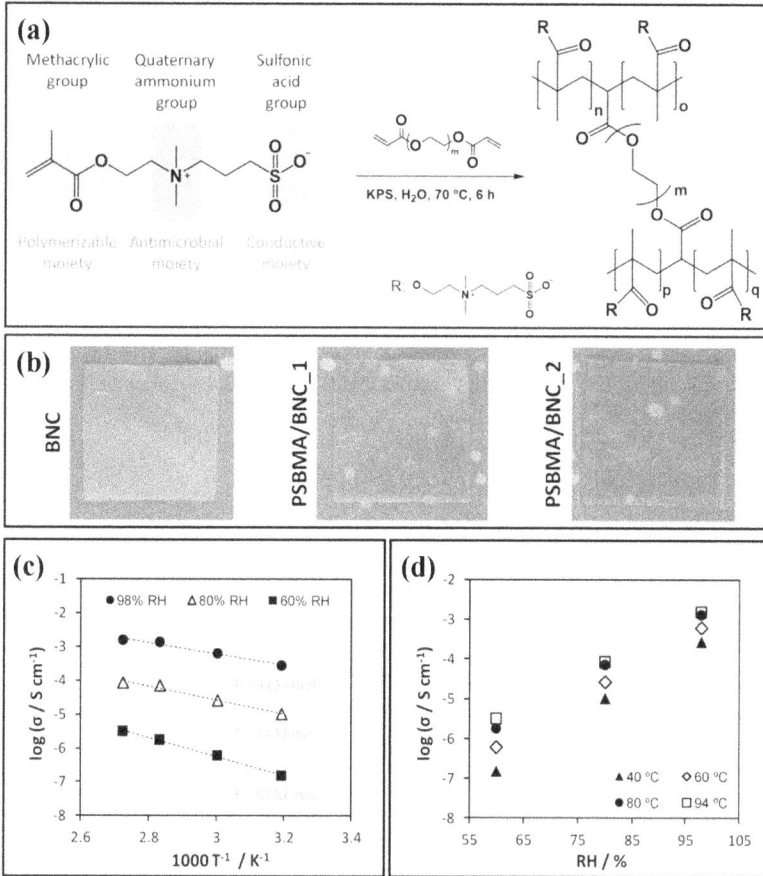

FIGURE 10.5 Fabrication and conductivity of the PSBMA. (a) Cross-linked PSBMA fabrication. Radical polymerization of SBMA is done using cross-linker and radical initiator to obtain the cross-linked PSBMA. Poly(ethylene glycol) diacrylate (PEDGA) and potassium persulfate (KPS) are used as cross-linker and radical initiator, respectively. (b) Dried films of pristine BNC, PSBMA/BNC_1 and PSBMA/BNC_2 nanocomposites. (c) Arrhenius model plot (straight lines are linear fits) shows the in-plane (IP) conductivity of the PSBMA/BNC_2 at various RH and (d) conductivity logarithm as a function of RH at different temperatures (Vilela et al., 2019).

is attained in a chitosan/reduced graphene oxide nanocomposite when graphene oxide is reduced to graphene by caffeic acid, a biocompatible material (Barra et al., 2019).

Smart packaging technology is able to determine a perished food and can support the quality control of foods. Electrically conducting polymers, printed electronics and colorimetric pH sensors using dyes bring out many advances in smart packaging. Bacterial cellulose is modified with conducting polymer polypyrrole and zinc oxide nanoparticles. Polypyrrole has a high electrical conductivity, and

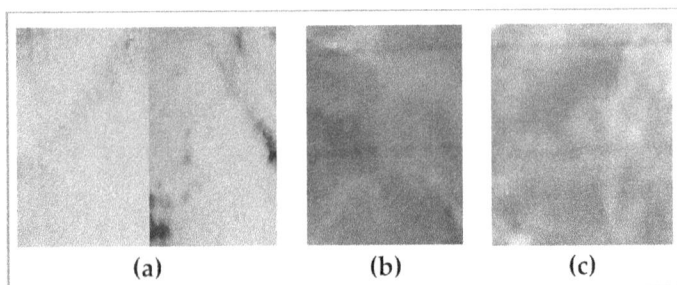

FIGURE 10.6 Polydiacetylene (42.63 mg) and triethyl citrate (17.73%) incorporated (at 71°C temperature after UV light exposure for 18 minutes) cellulose acetate-based films: (a) at air condition; (b) and (c) immersed in pH 4.0 and pH 9.0 solutions, respectively (Ardila-Diaz et al., 2020).

zinc oxide nanoparticles are used in sensors. The electrical resistance of the film changes for various oxidation/reduction reactions and for different volatile compounds (Pirsa et al., 2018).

Polydiacetylene is a conducting polymer, which has been identified as a sensing material. It can detect the biochemical changes in the food packaging environment and can be applied for smart packaging. The polymer exhibits visual changes suitable for colorimetric sensing. Cellulose acetate/polydiacetylene films are prepared by solution casting process using triethyl citrate as plasticizer. Figure 10.6 shows the color transition of the prepared films from blue to red when the films are immersed in solutions of pH from 4.0 to 9.0 (Ardila-Diaz et al., 2020).

10.5   COLORIMETRIC pH SENSORS USING DYES

Several non-toxic nanoparticles such as zinc oxide (ZnO) nanoparticles possess a high sensing ability, which is well demonstrated in many electrochemical sensors. They are impregnated in biodegradable polymers for applications in smart packaging. Cellulose materials such as nanocellulose and bacterial cellulose (BC) are used in intelligent packaging. TEMPO-mediated oxidation process is applied to convert the waste material into value-added products such as nanocellulose. BC is produced from the bacteria *Acetobacter xylinum*. Bacteria can be grown in the medium with glucose, sucrose, ethanol and glycerol. These cellulose materials are fibrous and highly porous biodegradable polymer with high tensile strength.

Cellulose materials support the dyes (pH indicators) utilized in the intelligent packaging. Anthocyanins (pigments of berries) and lycopene (natural dye) are present in plants materials such as fruits and vegetables. As per the report of Kuswandi et al. (2020), chemical dyes are generally used as recognition elements/indicators in intelligent food packaging. However, natural pigments and dyes are emerging as alternative indicators (pH and visual) to chemical dyes in such packaging.

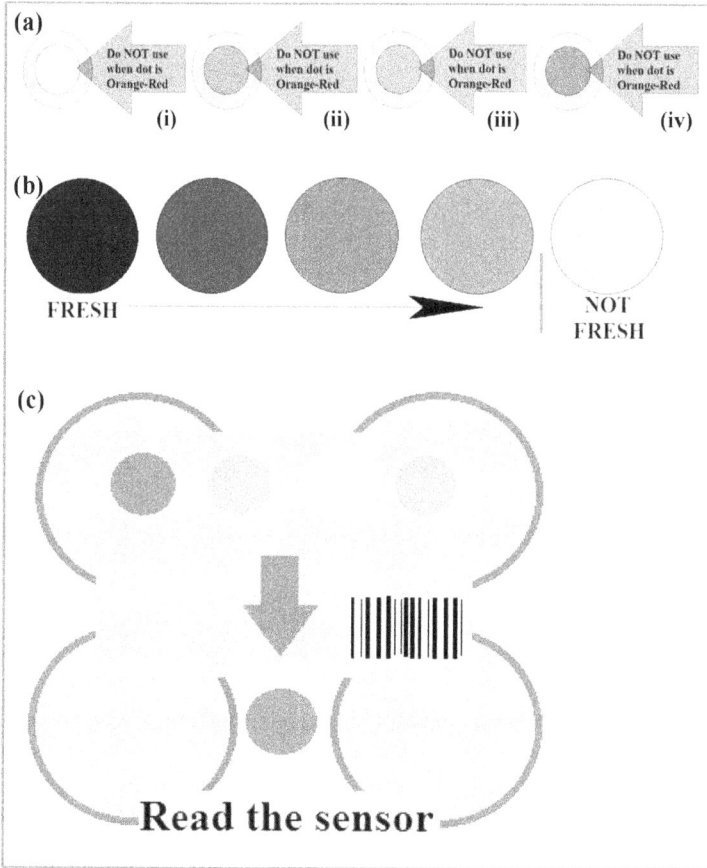

FIGURE 10.7   Schematic diagrams of the different time–temperature indicators (TTIs). (a) Enzyme-based TTI. (i) Inactive state of TTI; (ii) after opening the seal; (iii) 100% expired food; (iv) 120% of the expired condition. (b) UV light-based TTI. The indicator excited by UV light shows the color changes from dark blue to the colorless state. (c) Ripe sensor for fruits; dark gray color—crispy; light gray—firm state; white—juicy fruits; black—expired fruits (Motelica et al., 2020).

Presently, the frozen foods in refrigerators are continuously monitored using TTIs. TTIs are designed as visual indicators. If any chemical reaction is accelerated in the food environment, color change can be seen in TTI. The temperature variation is detected by TTI through mechanisms such as enzymatic, chemical or electrical changes. Monitoring the freshness of food offers effective food safety to the consumer. The TTI sensors are compact and can be integrated on the packaging that facilitates the use.

Figure 10.7 shows the different TTIs. Figure 10.7a shows an enzyme-based TTI. It contains a lipolytic enzyme solution. When the temperature of the environment raises, the pH of the solution changes and produces a color change in the label. Figure 10.7b shows a UV light-based TTI. The label in Figure 10.7b has an organic molecule. The label becomes dark blue color when it is irradiated with UV light. When the temperature of the food packaging environment increases, the color of the label becomes colorless. Figure 10.7c shows a ripe sensor. The TTI of this sensor is used to check the freshness of fruits. The ripening of fruits releases many volatile compounds. These compounds produce visible changes in the labels as demonstrated in Figure 10.7c. The colors dark gray, light gray, white and black indicate crisp, firm, juicy and expired nature of fruits in the packaging, respectively (Motelica et al., 2020).

Cellulose is regenerated by anchoring with acidochromic dye. The modified cellulose is blended with polyvinyl alcohol (PVA) to make a film resulting in the enhancement of mechanical properties. The film exhibits color changes that can be observed using naked eye for the corresponding changes in pH. The film is not subjected to leaching even for very low and high pH changes. The foods such as fish and meat can be monitored in real time by this type of pH sensor (Ding et al., 2020).

BC is modified using anthocyanins from red cabbage. A small amount of PVA is added as a binder to prepare a membrane. The membrane shows color changes for various pH values. This sensor exhibits a good response, and the results are highly reproducible. Food products such as beverages and milk can be monitored for their freshness using this type of colorimetric detection. This pH indicator is particularly suitable for monitoring fruits ripening, because ripening of fruits will not release ethylene gas. Tartaric and maleic acids only participate in fruit ripening and produce changes in pH. The ripeness is indicated by detecting these organic acids. The detection of ripeness is observed in guava, grape and strawberries fruits (Kuswandi et al., 2020).

Chitosan/PVA films are used in smart packaging with anthocyanin as pH indicator. The intensity of the color is high when the film is added with bentonite nanoclays at various concentrations. The addition of black carrot anthocyanins and bentonite improves the tensile strength and reduces the water vapor permeability, respectively. Both anthocyanins and bentonite support the enhancement in the antibacterial effect of the films (Koosha et al., 2019). Hence, chitosan/PVA films can be used to prevent foods from microbial spoilage.

Starch/PVA/ZnO nanoparticles/phytochemicals films are produced by solution casting method. Starch and PVA are dissolved in distilled water at 80°C and dried in hot-air oven to make the film. The starch/PVA film is incorporated with ZnO nanoparticles and the extracts of nutmeg and jamun. The color of jamun extract changes when the pH of packaging environment changes. It is an indication of the spoilage of food. Synthetic dyes are toxic that may cause health issues. Jamun extract is a natural dye, which can be used in food packaging as an alternate to chemical dyes such as bromocresol compounds, chlorophenol red and cresol red (Jayakumar et al., 2019).

Starch from jackfruit seeds is a biodegradable polymer that can be used for food packaging. Starch is highly economical and renewable and can be easily made into films. Starch and PVA are biocompatible. Hence, films prepared using these materials are suitable for food packaging applications. Liu et al. (2017) prepared a starch/PVA film by cross-linking with sodium trimetaphosphate and boric acid for use in intelligent food packaging. Anthocyanins and limonene are added to this composite film as pH indicators. The film shows color changes for solutions with various pH values.

Acetobacter xylinum bacterium is grown in coconut water medium. The minerals needed for bacterial growth are present in this medium. Also, nitrogen source (ammonium sulfate), yeast extract, sucrose and glucose are added in the medium. This method is able to produce high-quality BC. Dyes such as bromothymol blue, phenol red and methyl red are immobilized into the BC as pH indicators. BC membrane is tested by packing with fresh meat for 24 hours in room temperature. Total volatile basic nitrogen is released from meat, resulting in the variation of pH. The changes in the quality of fresh meat can be monitored visually (Dirpan et al., 2019).

Amaranthus leaf extract-incorporated PVA–gelatin films can be used for packaging chilled fish/chicken. Betalains present in the amaranthus leaf extract act as an indicator that produces color change in the film. The film is able to detect pH change, total volatile basic nitrogen amount and the microbial counts inside the packaging. Also, the amaranthus leaf extract improves the characteristics of the prepared film such as water solubility and water vapor permeability (Kanatt, 2020).

Nanocellulose is synthesized from biological waste materials such as sugarcane bagasse using TEMPO-mediated oxidation process. In the next step, the synthesized nanocellulose is converted as a hydrogel by cross-linking with zinc ions. Then hydrogel is attached to bromothymol blue/methyl red dyes. These dyes act as pH indicators. Finally, the hydrogel is applied to check the freshness of chicken breast through colorimetric sensing. Spoiled chicken leads to an increase in the level of both CO_2 and pathogens. The hydrogel response is fast with respect to the level of microorganisms (Lu et al., 2020).

A visual indicator is developed using polylactic acid/titanium dioxide/lycopene nanocomposite films. The prepared film is utilized to analyze the packaging of margarine. It shows excellent color changes whenever variations such as oxidative parameters, storage time and storage temperature occur in the packaging (Pirsa et al., 2021). The visual response of the PVA film is enhanced when it is impregnated with chitosan nanoparticles and mulberry extracts. The film detects the pH changes in the storage of fish (Ma et al., 2018).

10.6   SENSORS IN FOOD PACKAGING

Detection of ammonia gas, volatile compounds, nitrogen, amines, carbon dioxide (CO_2) and pH changes plays a major role in food packaging. Measuring and analyzing the variations of their level supports the determination of freshness of food and spoilage of food as well. Several sensors have been developed to measure their level. Biodegradable composites and nanomaterials such as silver nanoparticles are involved to make such sensors and food packaging materials.

Silver nanoparticles exhibit a resonance peak at 450 nm in UV–Vis spectroscopy that is called surface plasmon resonance peak. Hence, silver nanoparticles are highly useful in developing optical sensors. In this report, BC with silver nanoparticles (in a paper substrate) was prepared as plasmonic sensor. This nanopaper sensor (silver nanoparticles embedded) was exposed to ammonia gas or volatile compounds and analyzed in UV–Vis spectroscopy. The results showed the reduction in the UV–Vis absorbance. When this nanopaper is used for packaging fish and meat, deterioration of food can be seen by color change of the nanopaper. The color change confirms that silver nanoparticles can be employed as food spoilage indicators (Heli et al., 2016).

Carboxymethyl cellulose (CMC)/agar smart films are prepared with cellulose nanocrystals and shikonin. Cellulose nanocrystals are synthesized from onion peel. Shikonin is extracted from the roots of *Lithospermum erythrorhizon*. The multifunctional properties (pH-responsive color change; antioxidant and antibacterial properties) exhibited by the films are due to shikonin (Roy et al., 2021). CMC sodium along with *Artemisia sphaerocephala* Krasch gum and red cabbage extract is made into a film. The film is able to detect ammonia gas based on colorimetric sensing. It can be used as gas and pH sensing label in food packaging applications (Liang et al., 2019).

Ammonia sensor is constructed using the films of cellulose nanofibrils embedded with hydroxyapatite nanoparticles. This sensor shows a very low detection limit of as low as 5 ppm (5 wt.% filler loading). Hydroxyapatite nanoparticles are porous and have a high surface area. The adsorption of ammonia on the film occurs by H-bonding through the water molecules (Narwade et al., 2019). A halochromic nanosensor is designed using cellulose acetate nanofibres and alizarin. The spoilage of fish is analyzed by measuring the parameters such as total volatile basic nitrogen, pH and total viable count. These parameters are found to increase as time passes. A visual color change is exhibited by this sensor, which confirms the sensor as time indicator of fish spoilage (Aghaei et al., 2018).

Sensors to detect water vapor are made with poly(3-hydroxybutyrate-co-3-hydroxyvalerate) (PHBV)/cellulose nanocrystals/iron oxide nanoparticles. Cellulose nanocrystals can increase the hydrophilicity of PHBV. Iron oxide nanoparticles are having more surface area and can increase the sensitivity of the produced films (Abdalkarim et al., 2020). Amines in plants are responsible for fruit development and senescence. Senescence refers to the deterioration of functional characteristics in living organisms. The ripening, microbial contamination and storage conditions of food in a packaging can vary the level of amines (Esti et al., 1998).

CMC films are prepared by attaching with silver nanoparticles/polydiacetylene and glycerol. These films are tested as TTI. The color changes observed on the films are due to the changes in time and temperature of the packaging environment. Silver nanoparticles have a high thermal conductivity. Hence, a small temperature variation on the film can be clearly observed. Polydiacetylene possesses a high surface area that can support the effective sensing ability of the films. Glycerol can offer a symmetrical chemical structure to the films. The films are suitable for packaging fruits and vegetables (Saenjaiban et al., 2020).

TTI is developed by using PVA, chitosan and anthocyanins (derived from red cabbage). The activation test is performed on pasteurized milk. The results exhibited the occurrence of color changes in the film and pH variations. The color change confirms that the chemical composition of the food is modified. Anthocyanins are subjected to structural changes for the changes in pH (Pereira Jr et al., 2015). More carbon dioxide (CO_2) is released from the food during microbial spoilage. Sensors attached on the packaging are able to measure the concentration of CO_2, which in turn indicates the freshness of the food product. A CO_2 sensor is developed using a polypeptide (ε-poly-L-lysine) and anthocyanins. The sensor exhibits color change for various concentrations of CO_2 (Saliu et al., 2018).

10.7 PRINTED ELECTRONICS

Printed electronics is the electronic device that is printed on a substrate (Coatanea et al., 2009; Wikipedia, 2021). It is emerging as an important area of research in smart packaging. The quality of food products can be monitored by printing sensors on the packaging. Conductive inks are the important materials for the development of printed electronics. They are applied to print labels, QR codes and RFID tags on the food packaging. The content of these printed labels, codes and tags can be read or unlocked through an appropriate machine reader.

Conductive inks can be prepared at low cost and can be printed easily on the packaging materials. Advanced materials such as graphene and silver nanoparticles are used to prepare the conductive inks due to their high electrical conductivity. However, conductive inks and films prepared using these nanomaterials have some drawbacks such as high cost and fragile nature. Also, materials such as cellulose nanocrystals are involved in the conductive inks preparation.

The conductive ink should be analyzed for its properties such as resistivity, viscosity, surface tension, adhesion and inkjet printability for use in printed electronics. The curing temperature determines the electrical properties of the inks. Conductive inks are formulated by dispersing metal nanoparticles such as silver, copper and gold nanoparticles or carbon nanomaterials such as graphene and carbon nanotubes in a matrix. The matrix may include water, ethylene glycol, toluene or cyclohexane. The matrix is cured by the sintering process. The conductive ink is then inkjet printed on the packaging material (Fernandes et al., 2020).

10.7.1 CONDUCTIVE PACKAGING FILMS

Cellulose nanocrystals are biodegradable, recyclable and highly porous in nature: these nanocrystals can be used in conductive inks to enhance the rheological properties and stability; their excellent mechanical properties support to improve the strength of the packaging films; also, they are used to prepare flexible and transparent films (Karppinen, 2018). The cost of the cellulose nanocrystals is less and affordable compared to the cost of graphene and silver nanoparticles.

Cellulose nanofibrils are generally extracted from plant materials such as wood and cotton. They have a high surface area, high strength and low density.

Pristine graphene is original, pure, unoxidized form of graphene. A conductive nanocomposite film was prepared using both TEMPO-oxidized cellulose nanofibrils and pristine graphene. The surface of cellulose nanofibrils was modified by TEMPO-mediated oxidation, which leads to the formation of more carboxylate groups on their surface. A maximum electrical conductivity of 568 S cm^{-1} was achieved for 20 wt.% of pristine graphene with TEMPO-oxidized cellulose nanofibrils (Zhan et al., 2019).

Cellulose nanowhiskers are compatible with conductive polymers. This is helpful for film formation. Cellulose nanowhisker/graphene nanoplatelet composite is used to prepare conductive films in solution casting method. The films are highly transparent with improved mechanical, thermal and electrical properties. Also, they have low thermal expansion coefficient, which is right property to apply in smart packaging. The electrical conductivity of the composite films (with 0.5% loading of graphene nanoplatelets) is found to be 4.0×10^{-5} S m^{-1} (Liu et al., 2019).

Cellulose nanofibril films were produced with 40% poly(ethylene glycol) as plasticizer. The plasticizer not only improves the mechanical properties of the films, but also participates in sensor kinetics. The obtained films are demonstrated as humidity sensors that can measure the humidity from 20% to 90%. The interdigitated carbon electrodes printed on the films measure the changes in the impedance of the films (Syrový et al., 2019).

10.7.2   CONDUCTIVE INKS

Conductive inks are made by adding carbon fibers with CMC. The inks are used to print the sensing elements on food packaging applications. The composite (10 wt.% of carbon fibers added without annealing) exhibits a resistance of 300 Ω sq^{-1} at 25°C temperature and 15%RH. The printed sensing elements are able to detect humidity. The sensitivity of the sensing element can be enhanced by heating at 120°C temperature. The sensing elements are stable even after 1000 bending cycles. It confirms their applications in flexible and printed electronics. Also, the conductive ink can be screen printed on cellulose papers (Barras et al., 2017).

The electrode designs prepared using silver nanoparticles are shown in Figure10.8a,b. A close-up view of the electrode design D-6 is shown in Figure10.8c. Figure 10.8d shows the bending of the printed (using I-7 sample ink) electrode. I-7 sample ink is a silver nanoparticles (~16 wt.%)-based conductive ink. It has been formulated using the solvents ethylene glycol, ethanol, PEDOT:PSS dispersion, water, ethanolamine, hyperdispersant (Solsperse 20000) with a volume ratio of solvents of 1.2:2.4:1.4:1:0.04:0.006, respectively. The electrode pattern printed using the I-7 sample ink showed 5.7×10^{-05} Ω cm resistivity (Fernandes et al., 2020).

It is often difficult to print conductive silver ink on cellulose packaging. Hence, conductive ink is prepared by thermal reduction of silver acetate in ionic liquid, namely 1-methylimidazole. The reduction can also be performed by irradiating with UV light. This method can bring out conductivity across the cellulose foil (Samusjew et al., 2018). Silver nanoparticle conductive ink is printed on paper as electrodes. Paper (dielectric material) was used to detect the humidity level by

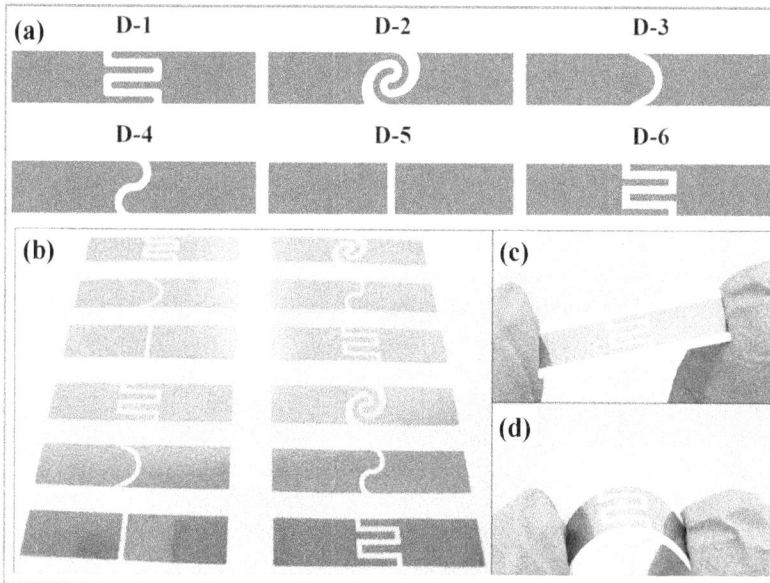

FIGURE 10.8 Designing of the printed electrodes: (a) electrodes designed with the size (27.2 mm width, 5.6 mm height and 0.8 mm gap) for characterization; (b) various designs of printed electrodes; (c) close-up view of the design D-6; (d) bending of the printed electrode with I-7 sample ink formulation (Fernandes et al., 2020).

measuring the capacitance and conductivity. A relative humidity of 35%–80% was measured using this sensor (Mraović et al., 2014).

10.7.3 Sensors and Smart Labels

Smart labels with sensors are attached in food packaging to measure the temperature, humidity and volatile amines. The labels are printed by screen printing and lamination methods on low-cost foils. These smart labels are used to monitor the freshness of food. Smart labels such as NFC chips and QR codes are printed on food packaging that is scanned using devices such as smartphones (Smits et al., 2012). Intelligent packaging is an alternate method to identify the utility periods (manufacturing and expiry dates) of the food packaging. Freshness indicators and TTIs are of this type.

A sensor for oxygen gas was prepared by inkjet printing the ink that consists of ruthenium dye and ethyl cellulose in ethanol on a cellulose paper. The sensing ability can be determined by measuring the fluorescence quenching lifetime for various concentrations of oxygen. The sensitivity was found to be 0.11 μs mg L^{-1} (Maddipatla et al., 2019).

Printed electrical gas sensors are developed to detect the gases such as ammonia, trimethylamine and carbon dioxide. The limit of detection for ammonia gas is demonstrated to be 200 parts per billion. These sensors are implemented into near-field-communication tags to function as wireless and battery-less gas sensors. They are able to interrogate with smartphones. The graphite electrodes printed on cellulose paper measure the electrical impedance of the paper used in the food packaging. Gases present in the water adsorbed by the cellulose fibers can be measured from the electrical properties of the paper. The gases increase the ionic conductivity of paper (Barandun et al., 2019).

The aging and freshness of food can be monitored from a barcode available on the packaging. The sensor barcode can be read from a smartphone camera. The barcode pattern is designed by encapsulating dyes in resin microbeads that are attached in paper substrate. The working of this barcode is tested in the spoilage of chicken at various temperatures. The color of the barcode changes according to the volatiles released from the food (Chen et al., 2017).

The meat freshness was monitored by barcode-based detection technique using deep convolution neural networks. The barcodes were printed on paper by the solution (consists of cellulose acetate, chitosan and dye). These barcodes can be identified by deep convolution neural networks that can be made available in a smartphone. The freshness of meat can be tracked by scanning the barcode (Guo et al., 2020).

Recently, RFID tags have become available at low cost, which leads them to various applications. Chipless RFID sensors are prepared using chitosan hydrogel. They were designed to detect pH in smart food packaging. The electromagnetic characterization of the chitosan hydrogel was performed in ultra-wideband frequencies. The physical properties of the hydrogel change according to the variations in pH (Athauda et al., 2020).

10.8 CONCLUSIONS

Biodegradable polymers such as cellulose, starch and their blends with other polymers are emerging as food packaging materials. These materials will become alternative materials to synthetic food packaging materials in future. They possess high sensing ability and can be used to monitor the conditions present inside the food packaging. TTIs, freshness indicators, chemical sensors, biosensors and gas sensors are the various types of sensors in smart packaging that are used to identify the freshness of food and spoilage of food as well. Sensor labels such as bar codes and RFID tags have also been developed using the technology of printed electronics. New renewable and economically viable biodegradable polymers are available to make this intelligent packaging to bring out with better commercial packaging materials. The information about the quality of food provided by the smart packaging will give a promising support to food safety.

ACKNOWLEDGMENTS

The author expresses immense thanks to her husband Mr. G. Sankar for his assistance in this work. Also, she acknowledges the assistance of International Research Center, Kalasalingam Academy of Research and Education (Deemed University), Krishnankoil, 626 126, India, for providing necessary support and facilities.

REFERENCES

Abdalkarim, Somia Yassin Hussain, Yanyan Wang, Hou-Yong Yu, Zhaofeng Ouyang, Rabie A.M. Asad, Mengya Mu, Yujun Lu, Juming Yao, and Lianyang Zhang. "Supermagnetic cellulose nanocrystal hybrids reinforced PHBV nanocomposites with high sensitivity to intelligently detect water vapor." *Industrial Crops and Products* 154 (2020): 112704.

Aghaei, Zahra, Bahareh Emadzadeh, Behrouz Ghorani, and Rassoul Kadkhodaee. "Cellulose acetate nanofibres containing alizarin as a halochromic sensor for the qualitative assessment of rainbow trout fish spoilage." *Food and Bioprocess Technology* 11, no. 5 (2018): 1087–1095.

Alam, Arif U., Pranali Rathi, Heba Beshai, Gursimran K. Sarabha, and M. Jamal Deen. "Fruit quality monitoring with smart packaging." *Sensors* 21, no. 4 (2021): 1509. https://doi.org/10.3390/s21041509

Ardila-Diaz, Lina D., Taíla V. de Oliveira, and Nilda de F.F. Soares. "Development and evaluation of the chromatic behavior of an intelligent packaging material based on cellulose acetate incorporated with polydiacetylene for an efficient packaging." *Biosensors* 10, no. 6 (2020): 59. https://www.mdpi.com/2079-6374/10/6/59

Athauda, Tharindu, Parama Chakraborty Banerjee, and Nemai Chandra Karmakar. "Microwave characterization of chitosan hydrogel and its use as a wireless pH sensor in smart packaging applications." *IEEE Sensors Journal* 20, no. 16 (2020): 8990–8996.

Barandun, Giandrin, Matteo Soprani, Sina Naficy, Max Grell, Michael Kasimatis, Kwan Lun Chiu, Andrea Ponzoni, and Firat Güder. "Cellulose fibers enable near-zero-cost electrical sensing of water-soluble gases." *ACS Sensors* 4, no. 6 (2019): 1662–1669.

Barra, Ana, Jéssica D.C. Santos, Mariana R.F. Silva, Cláudia Nunes, Eduardo Ruiz-Hitzky, Idalina Gonçalves, Selçuk Yildirim, Paula Ferreira, and Paula AAP Marques. "Graphene derivatives in biopolymer-based composites for food packaging applications." *Nanomaterials* 10, no. 10 (2020): 2077. https://doi.org/10.3390/nano10102077

Barra, Ana, Nuno M. Ferreira, Manuel A. Martins, Oana Lazar, Aida Pantazi, Alin Alexandru Jderu, Sabine M. Neumayer et al. "Eco-friendly preparation of electrically conductive chitosan-reduced graphene oxide flexible bionanocomposites for food packaging and biological applications." *Composites Science and Technology* 173 (2019): 53–60.

Barras, Raquel, Inês Cunha, Diana Gaspar, Elvira Fortunato, Rodrigo Martins, and Luis Pereira. "Printable cellulose-based electroconductive composites for sensing elements in paper electronics." *Flexible and Printed Electronics* 2, no. 1 (2017): 014006.

Chen, Jie, Zhu Long, Shuangfei Wang, Yahui Meng, Guoliang Zhang, and Shuangxi Nie. "Biodegradable blends of graphene quantum dots and thermoplastic starch with solid-state photoluminescent and conductive properties." *International Journal of Biological Macromolecules* 139 (2019): 367–376.

Chen, Junjun, Hailong Li, Lihui Zhang, Chao Du, Tao Fang, and Jian Hu. "Direct reduction of graphene oxide/nanofibrillated cellulose composite film and its electrical conductivity research." *Scientific Reports* 10, no. 1 (2020): 1–10. https://www.nature.com/articles/s41598-020-59918-z

Chen, Yu, Guoqing Fu, Yael Zilberman, Weitong Ruan, Shideh Kabiri Ameri, Yu Shrike Zhang, Eric Miller, and Sameer R. Sonkusale. "Low-cost smart phone diagnostics for food using paper-based colorimetric sensor arrays." *Food Control* 82 (2017): 227–232.

Coatanea, E., V. Kantola, J. Kulovesi, L. Lahti, R. Lin, and M. Zavodchikova. Printed electronics, now and future. In Neuvo, Y., & Ylönen, S. (eds.), *Bit Bang – Rays to the Future*. Helsinki University Print, Helsinki, Finland, 2009, 63–102. ISBN 978-952-248-078-1, http://lib.tkk.fi/Reports/2009/isbn9789522480781.pdf

Da Silva, A.C., and S.I. Córdoba de Torresi. "Advances in conducting, biodegradable and biocompatible copolymers for biomedical applications." *Frontiers in Materials* 6 (2019): 98. https://doi.org/10.3389/fmats.2019.00098

Ding, Lei, Xiang Li, Lecheng Hu, Yunchong Zhang, Yang Jiang, Zhiping Mao, Hong Xu, Bijia Wang, Xueling Feng, and Xiaofeng Sui. "A naked-eye detection polyvinyl alcohol/cellulose-based pH sensor for intelligent packaging." *Carbohydrate Polymers* 233 (2020): 115859.

Dirpan, A., I. Kamaruddin, A. Syarifuddin, A. N. F. Rahman, R. Latief, and K. I. Prahesti. "Characteristics of bacterial cellulose derived from two nitrogen sources: Ammonium sulphate and yeast extract as an indicator of smart packaging on fresh meat." In *IOP Conference Series: Earth and Environmental Science*, vol. 355, no. 1, p. 012040. IOP Publishing, Bristol, United Kingdom, 2019. doi:10.1088/1755-1315/355/1/012040

Esti, M., G. Volpe, L. Massignan, Dario Compagnone, E. La Notte, and G. Palleschi. "Determination of amines in fresh and modified atmosphere packaged fruits using electrochemical biosensors." *Journal of Agricultural and Food Chemistry* 46, no. 10 (1998): 4233–4237.

Fang, Zhongxiang, Yanyun Zhao, Robyn D. Warner, and Stuart K. Johnson. "Active and intelligent packaging in meat industry." *Trends in Food Science & Technology* 61 (2017): 60–71.

Fernandes, Iara J., Angélica F. Aroche, Ariadna Schuck, Paola Lamberty, Celso R. Peter, Willyan Hasenkamp, and Tatiana LAC Rocha. "Silver nanoparticle conductive inks: Synthesis, characterization, and fabrication of inkjet-printed flexible electrodes." *Scientific Reports* 10, no. 1 (2020): 1–11. https://www.nature.com/articles/s41598-020-65698-3

Guo, Baolin, Lidija Glavas, and Ann-Christine Albertsson. "Biodegradable and electrically conducting polymers for biomedical applications." *Progress in Polymer Science* 38, no. 9 (2013): 1263–1286.

Guo, Lingling, Ting Wang, Zhonghua Wu, Jianwu Wang, Ming Wang, Zequn Cui, Shaobo Ji, Jianfei Cai, Chuanlai Xu, and Xiaodong Chen. "Portable food-freshness prediction platform based on colorimetric barcode combinatorics and deep convolutional neural networks." *Advanced Materials* 32, no. 45 (2020): 2004805.

Halonen, Niina, Petra S. Pálvölgyi, Andrea Bassani, Cecilia Fiorentini, Rakesh Nair, Giorgia Spigno, and Krisztian Kordas. "Bio-based smart materials for food packaging and sensors–a review." *Frontiers in Materials* 7 (2020): 82.

Heli, Bentolhoda, Eden Morales-Narváez, Hamed Golmohammadi, A. Ajji, and Arben Merkoçi. "Modulation of population density and size of silver nanoparticles embedded in bacterial cellulose via ammonia exposure: Visual detection of volatile compounds in a piece of plasmonic nanopaper." *Nanoscale* 8, no. 15 (2016): 7984–7991.

Huang, Yao, Semen Kormakov, Xiaoxiang He, Xiaolong Gao, Xiuting Zheng, Ying Liu, Jingyao Sun, and Daming Wu. "Conductive polymer composites from renewable resources: An overview of preparation, properties, and applications." *Polymers* 11, no. 2 (2019): 187.

Jayakumar, Aswathy, K. V. Heera, T.S. Sumi, Meritta Joseph, Shiji Mathew, G. Praveen, Indu C. Nair, and E.K. Radhakrishnan. "Starch-PVA composite films with zinc-oxide nanoparticles and phytochemicals as intelligent pH sensing wraps for food packaging application." *International Journal of Biological Macromolecules* 136 (2019): 395–403.

Kanatt, Sweetie R. "Development of active/intelligent food packaging film containing Amaranthus leaf extract for shelf-life extension of chicken/fish during chilled storage." *Food Packaging and Shelf Life* 24 (2020): 100506.

Karppinen, A. 2018. Cellulose fibrils for printed electronics. [online] Exilva.com. Available at: <https://www.exilva.com/blog/cellulose-fibrils-for-printed-electronics> [Accessed 26 August 2021].

Koosha, Mojtaba, and Sepideh Hamedi. "Intelligent Chitosan/PVA nanocomposite films containing black carrot anthocyanin and bentonite nanoclays with improved mechanical, thermal and antibacterial properties." *Progress in Organic Coatings* 127 (2019): 338–347.

Kuswandi, Bambang, Ni PN Asih, Dwi K. Pratoko, Nia Kristiningrum, and Mehran Moradi. "Edible pH sensor based on immobilized red cabbage anthocyanins into bacterial cellulose membrane for intelligent food packaging." *Packaging Technology and Science* 33, no. 8 (2020): 321–332.

Liang, Tieqiang, Guohou Sun, Lele Cao, Jian Li, and Lijuan Wang. "A pH and NH_3 sensing intelligent film based on Artemisia sphaerocephala Krasch. gum and red cabbage anthocyanins anchored by carboxymethyl cellulose sodium added as a host complex." *Food Hydrocolloids* 87 (2019): 858–868.

Liu, Bin, Han Xu, Huiying Zhao, Wei Liu, Liyun Zhao, and Yuan Li. "Preparation and characterization of intelligent starch/PVA films for simultaneous colorimetric indication and antimicrobial activity for food packaging applications." *Carbohydrate Polymers* 157 (2017): 842–849.

Liu, Dongyan, Yu Dong, Yueyue Liu, Na Ma, and Guoxin Sui. "Cellulose nanowhisker (cnw)/graphene nanoplatelet (gn) composite films with simultaneously enhanced thermal, electrical and mechanical properties." *Frontiers in Materials* 6 (2019): 235.

Lu, Peng, Yang Yang, Ren Liu, Xin Liu, Jinxia Ma, Min Wu, and Shuangfei Wang. "Preparation of sugarcane bagasse nanocellulose hydrogel as a colourimetric freshness indicator for intelligent food packaging." *Carbohydrate Polymers* 249 (2020): 116831.

Ma, Qianyun, Tieqiang Liang, Lele Cao, and Lijuan Wang. "Intelligent poly (vinyl alcohol)-chitosan nanoparticles-mulberry extracts films capable of monitoring pH variations." *International Journal of Biological Macromolecules* 108 (2018): 576–584.

Maddipatla, Dinesh, Binu B. Narakathu, Manuel Ochoa, Rahim Rahimi, Jiawei Zhou, Chang K. Yoon, Hongjie Jiang et al. "Rapid prototyping of a novel and flexible paper-based oxygen sensing patch via additive inkjet printing process." *RSC Advances* 9, no. 39 (2019): 22695–22704.

Mohsin, M.E. Ali, A. Arsad, A Hassan, N.K. Shrivastava, and M.A. Ahmad Zaini. "Preparation and characterization of conductive Polyaniline/Sago/Graphene nanocomposites via ultrasonic irradiation." In *IOP Conference Series: Materials Science and Engineering*, vol. 522, no. 1, p. 012002. IOP Publishing, 2019.

Motelica, Ludmila, Denisa Ficai, Ovidiu Cristian Oprea, Anton Ficai, and Ecaterina Andronescu. "Smart food packaging designed by nanotechnological and drug delivery approaches." *Coatings* 10, no. 9 (2020): 806. https://www.mdpi.com/2079-6412/10/9/806

Mraović, Matija, Tadeja Muck, Matej Pivar, Janez Trontelj, and Anton Pleteršek. "Humidity sensors printed on recycled paper and cardboard." *Sensors* 14, no. 8 (2014): 13628–13643.

Mustafa, Fatima, and Silvana Andreescu. "Chemical and biological sensors for food-quality monitoring and smart packaging." *Foods* 7, no. 10 (2018): 168. https://doi.org/10.3390/foods7100168

Mustapha, Nooranis, Nozieana Khairuddin, Ida Idayu Muhamad, Shahrir Hashim, and Md Bazlul Mobin Siddique. "Characterization of HEC/PANI film as a potential electro-active packaging with pH sensor." *Sains Malaysiana* 45, no. 7 (2016): 1169–1176.

Narwade, Vijaykiran N., Shaikh R. Anjum, Vanja Kokol, and Rajendra S. Khairnar. "Ammonia-sensing ability of differently structured hydroxyapatite blended cellulose nanofibril composite films." *Cellulose* 26, no. 5 (2019): 3325–3337.

Pereira Jr, Valdir Aniceto, Iza Natália Queiroz de Arruda, and Ricardo Stefani. "Active chitosan/PVA films with anthocyanins from Brassica oleraceae (Red Cabbage) as time–temperature indicators for application in intelligent food packaging." *Food Hydrocolloids* 43 (2015): 180–188.

Phamonpon, W., N. Ruecha, N. Rodthongkum, and S. Ummartyotin. "Development of electrochemical paper-based analytical sensor from UHT milk packaging waste." *Journal of Materials Science: Materials in Electronics* 31 (2020): 10855–10864.

Pirsa, Sajad, and Sima Asadi. "Innovative smart and biodegradable packaging for margarine based on a nano composite polylactic acid/lycopene film." *Food Additives & Contaminants: Part A* 38, no. 5 (2021): 856–869.

Pirsa, Sajad, Tohid Shamusi, and Ehsan Moghaddas Kia. "Smart films based on bacterial cellulose nanofibers modified by conductive polypyrrole and zinc oxide nanoparticles." *Journal of Applied Polymer Science* 135, no. 34 (2018): 46617.

Roy, Swarup, Hyun-Ji Kim, and Jong-Whan Rhim. "Synthesis of carboxymethyl cellulose and agar-based multifunctional films reinforced with cellulose nanocrystals and shikonin." *ACS Applied Polymer Materials* 3, no. 2 (2021): 1060–1069.

Saenjaiban, Aphisit, Teeranuch Singtisan, Panuwat Suppakul, Kittisak Jantanasakulwong, Winita Punyodom, and Pornchai Rachtanapun. "Novel color change film as a time–temperature indicator using polydiacetylene/silver nanoparticles embedded in carboxymethyl cellulose." *Polymers* 12, no. 10 (2020): 2306.

Saliu, Francesco, and Roberto Della Pergola. "Carbon dioxide colorimetric indicators for food packaging application: Applicability of anthocyanin and poly-lysine mixtures." *Sensors and Actuators B: Chemical* 258 (2018): 1117–1124.

Samusjew, Aleksandra, Alice Lassnig, Megan J. Cordill, Krzysztof K. Krawczyk, and Thomas Griesser. "Inkjet printed wiring boards with vertical interconnect access on flexible, fully compostable cellulose substrates." *Advanced Materials Technologies* 3, no. 4 (2018): 1700250.

Sand, C., and Z. Boz. 2018. Packaging that communicates freshness. *Food Technology*, [online] (72 (11). Available at: <https://www.ift.org/news-and-publications/food-technology-magazine/issues/2018/november/columns/packaging-that-detects-when-food-products-are-degrading> [Accessed 23 August 2021].

Smits, Edsger, Jeroen Schram, Matthijs Nagelkerke, Roel Kusters, Gert van Heck, Victor van Acht, Marc Koetse, Jeroen van den Brand, and G. Gerlinck. "Development of printed RFID sensor tags for smart food packaging." In *Proceedings of the 14th International Meeting on Chemical Sensors*, Nuremberg, Germany, pp. 20–23, 2012.

Sobhan, Abdus, Kasiviswanathan Muthukumarappan, Zhisheng Cen, and Lin Wei. "Characterization of nanocellulose and activated carbon nanocomposite films' biosensing properties for smart packaging." *Carbohydrate Polymers* 225 (2019): 115189.

Syrový, Tomáš, Stanislava Maronová, Petr Kuberský, Nanci V. Ehman, María E. Vallejos, Silvan Pretl, Fernando E. Felissia, María C. Area, and Gary Chinga-Carrasco. "Wide range humidity sensors printed on biocomposite films of cellulose nanofibril and poly (ethylene glycol)." *Journal of Applied Polymer Science* 136, no. 36 (2019): 47920.

Vilela, Carla, Catarina Moreirinha, Eddy M. Domingues, Filipe M.L. Figueiredo, Adelaide Almeida, and Carmen SR Freire. "Antimicrobial and conductive nanocellulose-based films for active and intelligent food packaging." *Nanomaterials* 9, no. 7 (2019): 980. https://www.mdpi.com/2079-4991/9/7/980

Wang, Guilong, Long Wang, Lun Howe Mark, Vahid Shaayegan, Guizhen Wang, Huiping Li, Guoqun Zhao, and Chul B. Park. "Ultralow-threshold and lightweight biodegradable porous PLA/MWCNT with segregated conductive networks for high-performance thermal insulation and electromagnetic interference shielding applications." *ACS Applied Materials & Interfaces* 10, no. 1 (2018): 1195–1203.

Wikipedia. 2021. *Printed electronics*. [online] Available at: <https://en.wikipedia.org/wiki/Printed_electronics> [Accessed 26 August 2021].

Zhan, Yang, Chuanxi Xiong, Junwei Yang, Zhuqun Shi, and Quanling Yang. "Flexible cellulose nanofibril/pristine graphene nanocomposite films with high electrical conductivity." *Composites Part A: Applied Science and Manufacturing* 119 (2019): 119–126.

11 Migration Studies of Biodegradable Composites

Atanu Kumar Paul
Indian Institute of Technology Guwahati

CONTENTS

11.1 INTRODUCTION

Food safety, in general, and packaging material safety, in particular, are based on three pillars: a substance's toxicity, the amount of migration of the substance/ biodegradable composites into food, and the level of exposure to that food. Toxicity is often identified by standard procedures that determine an acceptable or tolerated amount of daily exposure. However, determining the degree of exposure is difficult and fraught with uncertainty. We must know the migration of

DOI: 10.1201/9781003227908-11

biodegradable composites into food and the consumption of that food. However, food consumption is influenced by individuals and local dietary habits, whereas migration is not. As a result, exposure can only be estimated as a population average based on certain exposure models. Most regulatory regimes assess exposure and maximum migration levels cautiously. Countries restrict packaging systems, while others allow individual substance movement. Authorities will always have a way of enforcing any system. Industry spends a lot of time and money proving that their products meet the standards. This chapter also discusses latex migration as a packaging material.

Food packaging is ubiquitous nowadays. Almost all supermarket foods are pre-packaged in multifunctional packaging designed to protect foods and extend shelf life (Scarsella et al., 2019). Food packaging frequently uses inks and varnishes to attract potential buyers and convey product information. Food packaging is made from a variety of materials such as metals, glass, plastics, and paper to improve the functional qualities of certain foods. These materials are commonly used in laminate constructions, food container sealing, and labelling. Environmental issues have rapidly escalated to become a serious concern in our world (Techochaingam et al., 2013). Plastic, cloth, and other non-biodegradable wastes are major causes for concern. Many people try to avoid this by using biodegradable materials. Packaging is one of the most common uses of biodegradable materials.

Petroleum-based plastics have become commonplace, demonstrating that different conventional materials can be replaced (Aaliya et al., 2021). The usage of bio-based materials is therefore an alternate means of developing ecologically secure products (Aaliya et al., 2021). Bio-based materials are made up of compounds derived from living stuff as their building blocks.

11.2 BIODEGRADABLE GREEN COMPOSITES

Conventional polymers are often derived from petroleum resources such as polyolefins and are suitable for a wide range of applications such as packaging, construction materials, commodities, and consumer products. Polyolefin-based polymers, which are inexpensive in cost, robust, solvent resistant, waterproof, and resistant to physical ageing, have become a cornerstone of contemporary society. Polyolefin materials' resistance to breakdown by microbes is both a benefit and a disadvantage in the long run. In 2002, packaging industries were expected to consume 41% w/w of total worldwide plastic output, with 47% of that production used to package foodstuffs (Ray and Bousmina, 2005). Most oil packaging is non-recyclable or ineffective for recycling or is quickly converted into a trash disposal, which is a significant quantity of non-degradable waste (Figure 11.1). In sites of deposition, microorganisms cannot destroy traditional plastics (Mueller, 2006) and are hence extremely persistent in the environment (Shimao, 2001; Kalia, 2016).

Biodegradable green polymers are an emerging group of polymers that can replace synthetic polymers. The quest for innovative goods and methods that

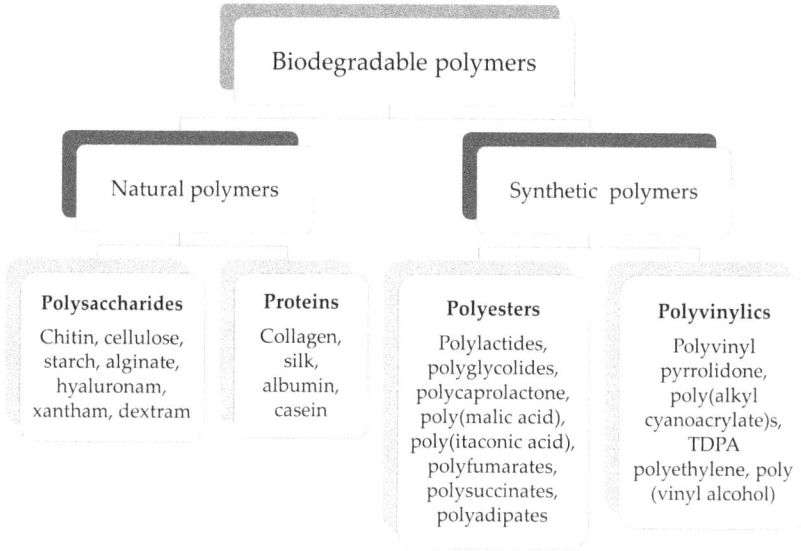

FIGURE 11.1 Classification of biodegradable polymers in four families.

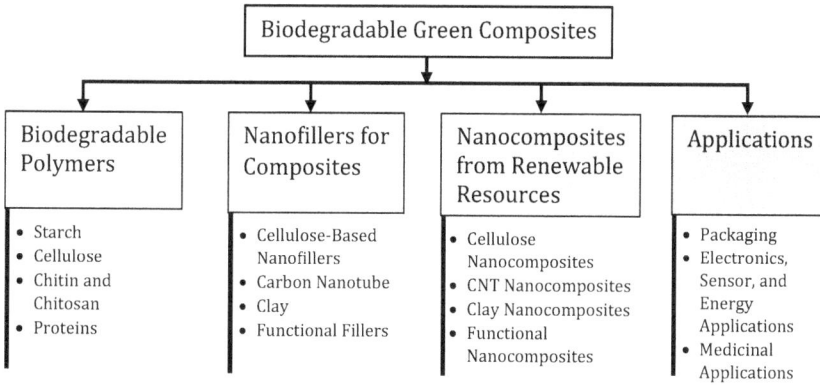

FIGURE 11.2 Classification of biodegradable green composites.

comply with the environment has been spurred by increasing environmental legislation. Green composites now emerge as the workable alternative to glass fibre-reinforced composites (Figure 11.2).

11.3 NATURAL RUBBER LATEX

Natural rubber latex is a white sap from the *Hevea brasiliensis* tree. This sap is further purified and compounded to make it easier to process and to maximize its physical characteristics. Natural latex products tend to be highly pure and have increased physical qualities recognized for natural rubber latex – excellent elongation, tear property, and recovery.

Natural milky white thick colloidal solution containing hydrocarbon polymer is latex rubber (Curtis and Klykken, 2008). Rubber latex is the name given to the product manufactured from it. The commercial latex rubber is made from the sap of the Para rubber tree *(Hevea brasiliensis)*, which is named for the Brazilian state where the species was found. With large molecular weight molecules, it possesses viscoelastic characteristics. Rubber latex is cis-1,4-polyisoprene in structure. Isoprene is a diene, and the addition of 1, 4 results in a double bond in each isoprene unit in the polymer, as illustrated in Figure 11.3.

Once in the Amazon in large numbers, natural latex was manufactured commercially. In recent years, natural latex manufacturing has shifted to Malaysia, Indonesia, and other countries of the Far East. More than 90% of the world's natural rubber output is currently produced in Asia and well over half of the total originates from these nations. Thailand, India, and Sri Lanka are also major Asian producers. The rubber output in China and the Philippines has risen significantly ("Latex," n.d.). The typical composition of natural rubber latex is given in Table 11.1 (Beaudouin et al., 1994).

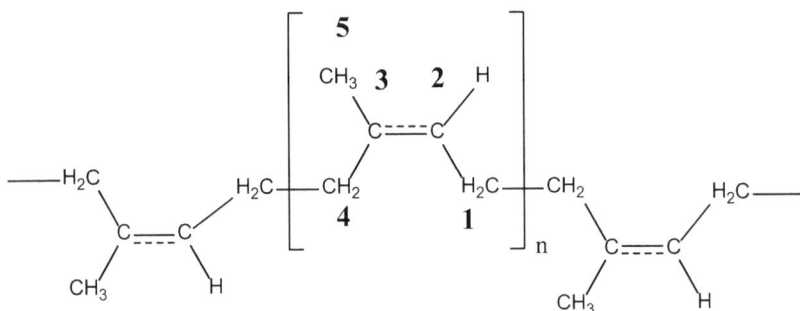

FIGURE 11.3   Structure of cis-1,4-polyisoprene unit.

TABLE 11.1
Chemical Composition of Latex

Percent	Component
55–65	Water
30–40	Cis-1,4-polyisoprene particles
2–3	Plant proteins
1.5–3.5	Resin
1.0–2.0	Sugars
0.5–1.0	Ash
0.1–0.5	Sterol glycosides

FIGURE 11.4 Milky latex extracted from rubber tree *Hevea brasiliensis* as a source of natural rubber.

The use of latex gloves grew significantly in the 1980s because of the HIV pandemic. Since then, the incidence of latex allergy has been estimated at between 8% and 17% among medical workers (Figure 11.4). The establishment of a latex-free operating room will benefit both patients and caregivers ("Glove Selection Guide | Office of Environment, Health & Safety," n.d., "NIOSH alert: preventing allergic reactions to natural rubber latex in the workplace," 1997; Brown et al., 1998).

The extensive use of medical gloves and other latex medical equipment (natural rubber latex), in particular among health professionals, led to a substantial increase in acute type I illnesses of hypersensitivity. In this working group, the prevalence of latex sensitization is between 5% and 17% (Lagier et al., 1992; Turjanmaa et al., 1996). An itchy skin rash, urticaria, angioedema, rhinoconjunctivitis, asthma, and anaphylactic shock are all symptoms of latex allergy (Table 11.2). There have also been reports of fatalities (Kaczmarek et al., 1996).

11.3.1 Latex Manufacturing Process

A complex combination of cis-1,4-polyisoprene particles in a phospholipoprotein envelope and a serum comprising sugars, nucleic acids, lipids, minerals, and proteins makes up latex, a natural rubber. Type I hypersensitivity is caused by these proteins, which cause severe, rapid anaphylactic responses (Jaeger et al., 1992). The sap of the commercial rubber tree, *Hevea brasiliensis*, is used to make latex. Rubber trees are mostly grown in Malaysia, Indonesia, and Thailand, where they are tapped by cutting spiral or diagonal grooves in the bark and placing collecting cups to catch the flowing sap. Ammonia is applied as a preservative after the sap is collected to avoid bacterial contamination and autocoagulation. Ammonia

TABLE 11.2
Allergic and Irritant Glove-Related Diseases

Disease	Pathogenesis	Accelerators
Allergic contact dermatitis	Cell-mediated allergy	Vulcanizers
		Antioxidants
		Organic pigments
		Lubricant powder crystals
Irritant contact dermatitis	Mechanical effect	Lubricant powder crystals
	Occlusion	Alkaline soaps
	Associated use of disinfectants	Bacterial toxins
		Ethylene oxide
Irritant or pseudo-allergic urticaria	Non-immunological mechanism	Lubricant powder
		Heat
		Pressure
Contact urticaria, protein contact dermatitis	IgE-mediated allergy	Latex proteins
Rhinitis, asthma	IgE-mediated allergy	Latex proteins
Angioedema	IgE-mediated allergy	Latex proteins
Anaphylactic shock	IgE-mediated allergy	Latex proteins

disturbs the collected sap, resulting in a two-phase product that is centrifuged to concentrate. Although ammonia efficiently stabilizes latex, excessive amounts might cause skin discomfort. To address this issue, most producers employ a low ammonia level in conjunction with secondary preservatives. Accelerators and antioxidants are added to the concentrated material to increase the latex product's strength, stretch, and durability. These accelerators and antioxidants have been linked to contact dermatitis, which is a type IV hypersensitivity (van Ketel, 1984). Vulcanization, a technique that includes curing the latex with heat and sulphur to facilitate cross-linking of the polymer chains, makes the latex material heat-stable and stretchy.

11.3.2 Global Legislation for Rubber Materials in Contact with Food

The Community legislation comprises general rules applicable to all materials and articles laid down in the Framework Regulation (EC) No. 1935/2004 and specific rules only applying to certain materials or certain substances (Parliament, 2004). The two general principles on which legislation on food contact materials is based are the principles of inertness and safety. A general overview is presented in Figures 11.5 and 11.6.

Rubber is an elastomer-based product which is supplemented by one or more processing aids. In a wide variety of applications, rubber contacts food and/or beverages. Three separate types of rubber materials are developed, each with its own needs and positive effects list (Council of Europe, 2014).

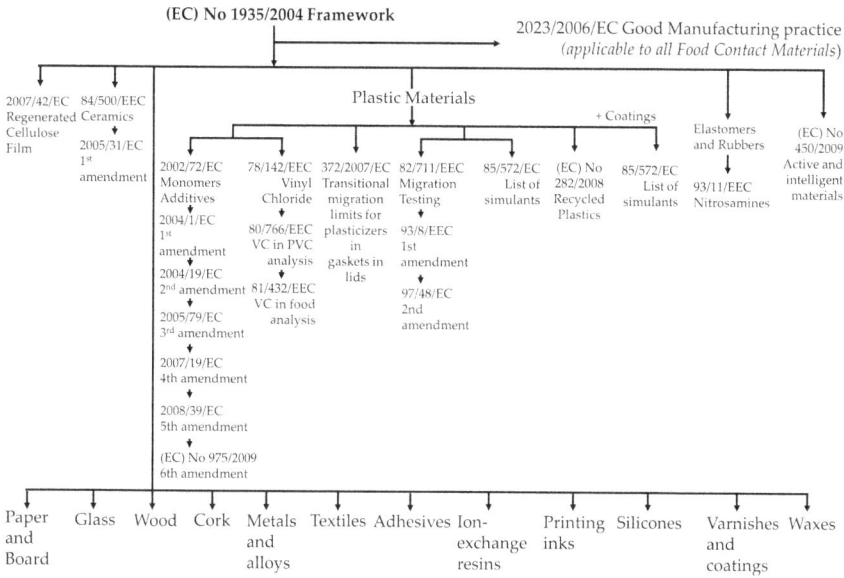

FIGURE 11.5 Overview of Community legislation (last update October 20, 2009).

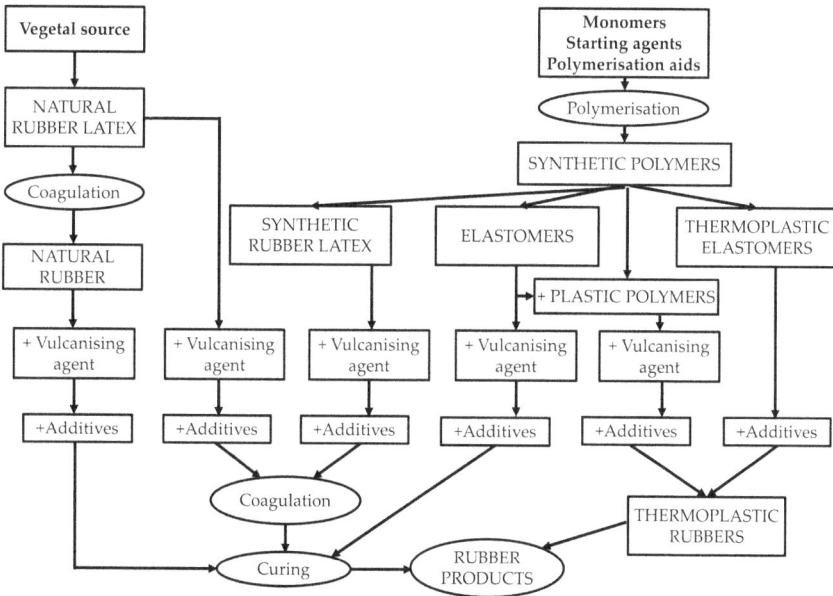

FIGURE 11.6 Scheme of manufacturing rubber products.

TABLE 11.3
Rubber Components Used in Food Contact Applications

Location	Component
Food transportation	Conveyor belts, rubber skirting, and rubber paddle lips
Pipe work components	Seals, gaskets, flexible connectors, and butterfly valves
Pumps	Progressive cavity pump stators, diaphragm pumps
Plate heat exchangers	Gaskets
Machinery/storage vessels	General seals and gaskets
Cans/bottles	Can sealants and bottle seals
Food handling/preparation	Gloves and feather pluckers
Food manufacturing	Silicone sweet moulds
Food wrapping	Meat and poultry nets

The Title 21 part of the Code of Federal Regulations covers all devices that come into contact with humans and are made of or include natural rubber, as well as packaging or components made of or containing natural rubber. Natural rubber latex, dry natural rubber, and synthetic latex or synthetic rubber containing natural rubber are all included in the phrase "natural rubber." It also states that devices that come into contact with humans and include natural rubber latex must have the following statement in bold font on the device labelling: "Caution: This Product May Cause Allergic Reactions." This statement must appear on all device labels and other labelling, as well as on the device packaging's main display panel, the outer package, container, or wrapper, and the device package, container, or wrapper immediately adjacent to it (Code of Federal Regulations, 2020).

11.3.3   RUBBER AND CHEMICAL MIGRATION INTO FOOD

Rubbers, in contrast to plastics, are rarely utilized in the packaging of food. The use of rubber in flip top seals on beer bottles and the seal on food cans is an exception (Table 11.3). This is owing to rubber's unique characteristics, which allow it to be employed in a wide range of goods (Barnes et al., 2007).

11.4   BIOPOLYMER COMPOSITES

Biopolymer composites are made by reinforcing various natural fibres from animal and plant sources to natural and/or synthetic biopolymers (Christian, 2016). Natural fibre (reinforcement agent) is introduced to the biopolymer matrix at a discontinuous phase in a continuous phase to increase the rigidity and strength of the produced composite (Noor Azammi et al., 2020).

The maximal stiffness and tensile strength of biocomposites typically vary from 1 to 4 GPa and 20 to 200 MPa (Christian, 2016). Thermal properties, electrical conductivity, morphological qualities, crystallinity, degradability, and production cost of biopolymer composites are all affected by natural fibre reinforcements (Behnam Hosseini, 2020) (Table 11.4).

TABLE 11.4

Improved Properties of Various Biopolymer Composites Prepared by Different Processing Techniques

Matrix System	Reinforcement	Processing Technique	Remarkable Characteristics of Biopolymer Composites	References
PLA	Silane-treated flax fibre	Manual hot plate compression moulding	The mechanical properties improved	González-López et al. (2019)
Corn starch	Hemp fibre	Dry incorporation method	The mechanical properties improved. The composite boards can be used in building industries	Kremensas et al. (2019)
Tapioca starch	PALF	Compression moulding and machine processing	The mechanical properties improved	Jaafar et al. (2018)
PVA	Acylated and chitosan-grafted silk fibre	Film casting technique	Thermal stability increased. The composite films exhibit good anti-bacterial and anti-fungal properties, which is suitable for wound healing applications	Sheik et al. (2018)
Sugar palm starch	Seaweed blended sugar palm fibre	Melt blending	The thermal stability and tensile and flexural properties improved. Composites exhibited good water and biodegradation resistance	Jumaidin et al. (2017)
Chitosan	Emu feather fibre	Mixture was dropped into an alkaline coagulant solution	The composite can be used as an absorbent for treating industrial waste water containing lead	Anantha and Kota (2016)
PHBV	Silk fibre	Matrix-assisted pulsed laser evaporation technique	Resistance improved and degradation rate decreased. The composites can be used for drug release schemes and other biomedical applications	Thakur et al. (2016)
PLA/ PCL blend	Silanized oil palm fibre	Melt blending	The mechanical properties and thermal stability improved	Eng et al. (2014)
PLA	Alkali, silane, and maleinized polybutadiene rubber-treated chicken feather fibre	Twin-screw extrusion	The mechanical properties improved	Huda et al. (2013)
PBS	Kenaf fibre	Melt mixing	The crystallization rate, tensile modulus, and storage modulus improved	Liang et al. (2010)

There are two techniques for producing fibres and matrices: bulk and laminate composites. In bulk composites, the fibres are randomly oriented in three dimensions, virtually showing isotropic behaviour. Laminate composites, on the other hand, are orthotropic, with fibres aligned in several layers and bonded together in the matrix. Each fibre layer in this composite has a two-dimensional orientation (Christian, 2016). Inter-diffusion, adsorption, chemical bonding, reaction bonding, electrostatic attraction, and mechanical bonding can all be used to adhere natural fibre to polymer matrix. Many factors influence adhesion processes, including chemical composition and molecular conformation, morphological properties of natural fibre, diffusivity of element materials, and the atomic arrangement of fibre and matrix (Noor Azammi et al., 2020). The type of fibre, percentage of fibre content, moisture absorption of fibre, surface modification method of fibre, structure and design of composite, interfacial adhesion between fibre and matrix, the presence of voids, and incorporation of additives such as plasticizers, compatibilizers, nanofillers, and binding agents all have an impact on the properties of biopolymer composites (Kabir et al., 2012; Sudamrao Getme and Patel, 2020). The density, water sensitivity, gas permeability, degradability, and shelf life of biopolymer composites are all affected by the reinforcing materials and plasticizers used. As a result of the hydrophilicity of natural fibre and the hydrophobicity of biopolymer, researchers alter the surface of fibres to increase their adherence in a composite. Chemical alteration improves the performance of biopolymer composites based on the processing type, processing needs, and environmental circumstances (Vinod et al., 2020). Biocompatibility and durability of biopolymer composites are important concerns since no appropriate solution has been developed to entirely regulate these two aspects.

11.5 MIGRATION BEHAVIOUR OF LUBRICANTS IN POLYPROPYLENE COMPOSITES

Lubricant surface migration affects the quality of thermoplastic polymer composites. Bak et al. (2021) investigated the surface migration of lubricants in polypropylene composites in order to improve the composites' quality. Polypropylene (PP)/lubricant composites were made using a co-rotating twin-screw extruder and injection moulding, and the lubricant migration phenomena in the PP/lubricant composites were studied for 72 hours under accelerated ageing conditions with temperatures ranging from 20°C to 90°C and humidity of 100%. Thermoplastic polymers are of significant interest to industry and academics. Polypropylene (PP) is a well-known and frequently used thermoplastic. The majority of these lubricants are low molecular weight molecules that are used in trace amounts (Holec et al., 2020).

Surface migration of these lubricants has been observed to cause quality degradation, particularly in initial quality assessments and vehicle dependability studies. Migration is described by the International Union of Pure and Applied

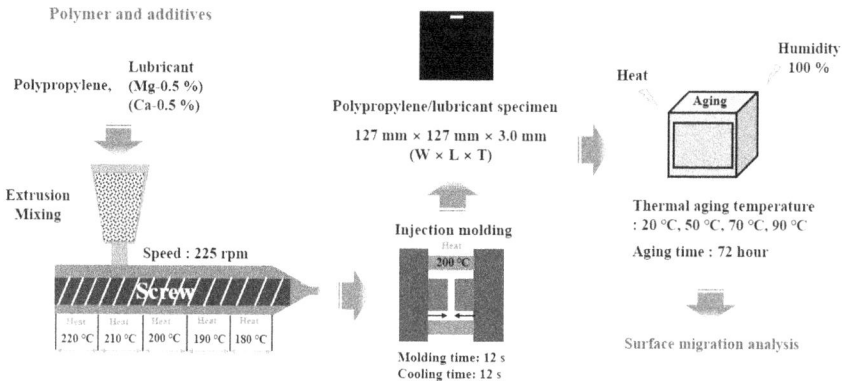

FIGURE 11.7 Schematic of procedure followed for the surface migration analysis of polypropylene/lubricant composites by thermal and humidity ageing.

Chemistry as a process in which one component of a polymer mixture, generally not a polymer, undergoes phase separation and migration to the mixture's exterior surface (Chougule et al., 2013; Nouman et al., 2017). This migratory phenomenon has the potential to change the surface characteristics necessary for specific applications such as vehicle interiors and food packaging materials. The migration of lubricants and antistatic chemicals, in particular, not only alters the look of the finished product, but also lowers the antistatic capabilities of PP. These adverse variations in the PP surface composition cause significant issues in mechanical failure, quality control rejection owing to surface contamination, and surface inhomogeneity (Li and Yang, 2006; Médard et al., 2002). As a result, additive characterization at the PP surface is required to enhance production processes in order to improve product quality and competitiveness (Figure 11.7). As a result, research is being done to develop techniques for confirming lubricant surface migration in PP/lubricant composites (Bang et al., 2012; Vitrac and Hayert, 2005).

In addition, some lubricants are well recognized to be cancerous to humans and to produce endocrine disrupting substances. It is therefore highly necessary to analyse the exposure of consumers to lubricants from polymer materials (Wang et al., 2018).

11.6 FOOD SIMULANTS

The cause of food contamination may be relevant in migration. In order to verify compliance, the migration depends on the four pillars shown in Figure 11.8. When planning migration experiments, each of the criteria must be addressed.

The movement depends on the type of food. For example, components that are hydrophobic will better move to foods on the surfaces that have a large fat content. The EU has identified several simulants that can be employed in order to avoid this testing for all (possible) kinds of foodstuffs (Figure 11.8).

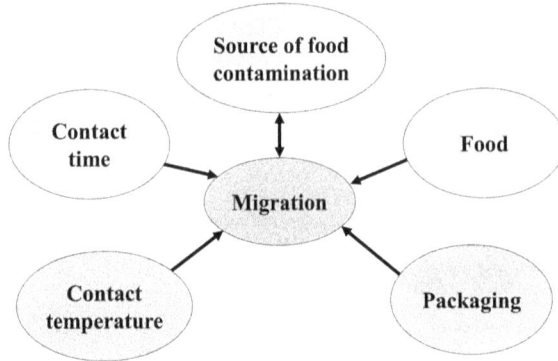

FIGURE 11.8   Migration-related parameters.

These food simulants are as follows:

- Simulant A: water, mimicking aqueous foodstuffs.
- Simulant B: 3% acetic acid, mimicking acidic foodstuffs with pH <4.5.
- Simulant C: 10% ethanol, mimicking alcoholic foodstuffs.
- Simulant D: olive oil, or other approved oils, mimicking fatty foodstuffs.
- 50% ethanol: mimicking high-alcohol-containing products and milk and some dairy products.

Food contact materials (FCMs) are often believed to represent all foods in 3% acetic acid, 10% ethanol, and olive oil. However, not all simulants are essential for specific meals or dietary categories. The EU has developed a table for the selection of the necessary food simulators. This table was published in Directive 85/572/EEC ("Council Directive 85/572/EEC of 19 December 1985 laying down the list of simulants to be used for testing migration of constituents of plastic materials and articles intended to come into contact with foodstuffs - Publications Office of the EU," n.d.) and was amended in Directive 2007/19/EC ("Commission Directive 2007/19/EC of 2 April 2007 amending Directive 2002/72/EC relating to plastic materials and articles intended to come into contact with food and Council Directive 85/572/EEC laying down the list of simulants to be used for testing migration of constituents of plastic materials and articles intended to come into contact with foodstuffs (Text with EEA relevance) - Publications Office of the EU," n.d.). In Table 11.5, an excerpt of Directive 85/572/EEC is given. It shows that for fried potatoes, only simulant D is needed for testing. In addition, a simulant D correction factor (DRF), which is explained below, of 5 is assigned to the fried potatoes.

Simulants B and C both simulate migration to aqueous meals, so no testing with simulant A is required. If the food isn't on the table, the user must assess it himself. To determine whether a food is fatty, CEN 14481 can be used to test for fatty content in all foods. Testing some foods may necessitate a technique that

TABLE 11.5

Excerpt from the Directive 85/572/EEC That Needs To Be Used for the Selection of the Simulants

Reference Number	Description of Foodstuffs	Simulants To Be Used			
		A	B	C	D
7.05	Rennet:				
	A. In liquid or viscous form	X(a)	X(b)		
	B. Powdered or dried				
08	Miscellaneous products				
08.1	Vinegar		X		
08.02	Fried or roasted foods:				
	A. Fried potatoes, fritters, and the like				X/5
	B. Of animal origin				X/4
08.03	Preparations for soups, broches, in liquid, solid, or powder form (extracts, concentrates); homogenized composite food preparation, prepared dishes:				
	A. Powdered or dried:				
	I. With fancy substances on the surface				X/5
	II. Other				

works between $-20°C$ and $100°C$. If the food contains water, use simulant A. If the meal's pH is 4.5 or lower, simulant B must be used instead of simulant A ("EN 14233:2002 - Materials and articles in contact with foodstuffs - Plastics" n.d.; Rijk and Veraart, 2010).

11.7   NO MIGRATION

FDCA permits firms to evaluate whether or not a certain ingredient employed in a food contact material is a food additive. This is done by the Federal Food, Drug, and Cosmetic Act (FDCA). When a producer finds that when food contact material is utilized as intended, the substance is not reasonably expected to go into food, the substance is a non-food additive under the definition of food additive (Code and States, 2021). The properly conceived and carried out extraction study can support this "no migration" viewpoint. Calculations might also demonstrate that the degree of migratory potential is insignificant (Heckman, n.d.).

These techniques nevertheless raise the issue of whether migration should, literally, not exist or whether a little amount of movement is acceptable. However, business relies on different sources for advice, for example the "Ramsey Proposal" and the judgment Monsanto V. Kennedy. The FDA (Food and Drug Administration of the USA) never established definite criteria in order to determine whether a chemical may properly be considered as a food element (Monsanto vs. Kennedy, 1979).

The Monsanto case featured an appeal from the FDA Commissioner's determination that the chemical acrylonitrile copolymer, which is used to make unbreakable beverage containers, was a hazardous food additive. The FDA contended that the migration of any amount of a chemical is sufficient to classify it as a food additive, allowing the FDA to realize the need for a food additive clearance for any food contact material, even if there is no proof that any component of it migrates to food. The United States Court of Appeals rejected FDA's argument, however, stating:

> Congress did not intend that the component requirement of "food additive" would be satisfied by . . . a mere finding of any contact whatever with food. . . .
>   For the component element of the definition to be satisfied, Congress must have intended the Commissioner to determine with a fair degree of confidence that a substance migrates into food in more than insignificant amounts ("*Ibid.* at 955," n.d.).

According to the Ramsey Proposal and the Monsanto case, chemicals that move to food in minute amounts are not food additives. This perspective is supported by a 1974 statement from the FDA's Office of General Counsel, which states that the legal decision that a chemical is a food additive must be based on more than proof of a trace quantity of migration (Peter Barton Hutt, 1974).

The amount of migrating substance should be proportionally less for substances of relatively high toxicity, such as heavy metals, or those that will have widespread use in food contact applications. Carcinogenic impurities in substances must be examined on a case-by-case basis utilizing risk assessment techniques.

11.8   PRINCIPAL ISSUES IN GLOBAL FOOD CONTACT: INDIAN PERSPECTIVE

11.8.1   THE INDIAN SUBCONTINENT: A STUDY IN CONTRAST

India has one of the hottest and driest climates in the world on the climatic front. In this location, high temperatures and humidity contrast with low temperatures and moisture levels. The Indian subcontinent has the biggest proportion of students and analphabets in the globe.

11.8.2   FOOD CONTACT LEGISLATION

The food contact legislation was defined for the first time in India by the "Prevention of Food Adulteration Act" of 1954 and Rules of 1955 (Reference: Akalank's BARE ACT). Maximum emphasis and stress is given to packaging labelling.

11.8.3   INDIAN STANDARDS FOR DIRECT FOOD CONTACT

This standard lists the permissible pigments and colours for using plastics on the basis of BIS (Bureau of Indian Standards) Standard 9833:1981, which are acceptable to use in the interaction with food, pharmaceuticals, and potable water.

11.8.4 Methods of Analysis and Determination of Specific and Overall Migration Limits

The BIS Standard 9845:1998 prescribes the method of analysis for the determination of overall migration of constituents of single- or multi-layered heat-sealable films, single homogeneous non-sealable films, and finished containers and closures for sealing as lids, in the finished form, preform, or converted form.

BIS Standard 10171:1986 – gives the guidelines on the suitability of plastics for food.

BIS Standard 10146:1982 – gives the following plastic resins: polyethylenes, polypropylenes, ionomers, acid copolymers, nylon, polystyrene, polyester, EVA, and so on.

11.8.5 Acceptability Criteria

As the FDA (Food and Drug Administration) lacks skilled workers, there is insufficient knowledge of such authorities. Consequently, owing to the lack of packaging expertise and the requisite labelling characteristics, few audits are undertaken. In comparison with the package labelling, which is substantially low, the focus is on food components and labelling. For this food contact law, there are no strict norms and requirements for packaging to be applied. This is extremely well known to the Indian company. A food contact certification is required of all packaging providers. Test laboratories for migration testing are also available, and PVC is permitted for use in direct food contact.

11.8.6 Future

The whole system is still in its infancy, but awareness is rising at the same time. There are extremely powerful media in India, which can ensure compliance with the law on food contacts. Much emphasis is placed on compliance with HACCP and BRC. Another important component of food contact laws is batch traceability. The paper by Sameer Mehendale contains a detailed explanation of the "Conditions for Sale and License," "Package of Food to Carry a Label," such as "Oil, Milk, and Infant Food Substitutes," and other legislation and regulations for migration and food interaction from an Indian viewpoint (Mehendale, 2010).

11.9 CONCLUSIONS

Biodegradable green polymers are an emerging group of polymers that can replace synthetic polymers in many different potential uses. New bio-based goods are a huge potential to develop, but to design sustainable bio-based products is the true difficulty. High costs are the main constraints of the existing biodegradable polymers.

The quest for innovative goods and methods that comply with the environment has been spurred by increasing environmental legislation and society concerns. Their particular property balance would offer new prospects for the commercial growth of biocomposites in the world of green materials of the 21st century. It may be stated that biodegradable green composites are vital to the promotion and implementation of environmentally friendly materials and production processes in agricultural, car manufacturing, composites and materials scientists as well as in cultivation. The biodegradable green composites may be used successfully as structural, medicinal, vehicle, and electrical materials. Green composites now emerge, particularly in automotive and construction sectors, as the workable alternative to glass fibre-reinforced composites. It also may have some issues with the special characteristics of rubber that make it so good for food contacts. In packaging, rubber may be an essential material. Due to the minimally cross-linked polymers, it is regarded safe for the use of rubber material for food packaging. The loose connection between latex and the food allows for the flow of chemicals. This exchange of latex rubber with food-grade materials therefore made it unsuitable for food packing. Some additives are present, and this quality has been improved by using plasticizers. In general, rubber latex could not concern food contact material owing to its restrictions. A food contact substance should be used without reactivity and dangerous release in food. The regulation on food contact materials, as laid forth in the European Union (EU), includes items in contact with water for human consumption. The adjustment of the food contact material also influences circumstances such as temperature, time, and field. To understand the quality of rubber store the Rubber Resolution of the Council of Europe has recorded and justified the area of application of the rubber material. In the case of diene rubbers (e.g., NR, NTBR, and SBR), the highly reactive character of rubber is caused either by insaturations or by a significant number of aliphatic hydrogen atoms. The use of certain anti-degradants to safeguard the polymer for the maintenance of material characteristics would thus need rubber latex to be biodegradable in nature. Increasing the usage of certain kinds of rubbers by some migrants might be limited. Unlike materials such as thermoplastics, rubber technology has been developed for a very long time, most of which has been before the time when health and safety issues have profoundly influenced production practices and research and development operations. Thus, it became more suited to commercialization in modern civilization in such climate circumstances because of the presence of different compounds in rubber latex. A greater awareness of issues became apparent with the increased usage of packaging materials. Inventories of food contamination risk have been initiated by the responsible authorities. For whatever action taken by the government, consumer protection is the driving force. Thus, the migration of materials from packing materials in one manner or another is limited. Many countries have created their own regulatory structure, which means that there are diverse methods throughout the world. Some nations acknowledge other countries' standards and approve the safety of a material for packaging in conformity with these regulations.

REFERENCES

Aaliya, B., Sunooj, K.V., Lackner, M., 2021. Biopolymer composites: a review. *Int. J. Biobased Plast.* 3, 40–84. https://doi.org/10.1080/24759651.2021.1881214

Anantha, R.K., Kota, S., 2016. Removal of lead by adsorption with the renewable biopolymer composite of feather (Dromaius novaehollandiae) and chitosan (Agaricus bisporus). *Environ. Technol. Innov.* 6, 11–26. https://doi.org/10.1016/j.eti.2016.04.004

Bak, M.G., Won, J.S., Koo, S.W., Oh, A., Lee, H.K., Kim, D.S., Lee, S.G., 2021. Migration behavior of lubricants in polypropylene composites under accelerated thermal aging. *Polymers (Basel).* 13. https://doi.org/10.3390/polym13111723

Bang, D.Y., Kyung, M., Kim, M.J., Jung, B.Y., Cho, M.C., Choi, S.M., Kim, Y.W., Lim, S.K., Lim, D.S., Won, A.J., Kwack, S.J., Lee, Y., Kim, H.S., Lee, B.M., 2012. Human risk assessment of endocrine-disrupting chemicals derived from plastic food containers. *Compr. Rev. Food Sci. Food Saf.* 11, 453–470. https://doi.org/10.1111/j.1541-4337.2012.00197.x

Barnes, K.A., Sinclair, C.R., Watson, D.H., 2007. *Chemical Migration and Food Contact Materials*, Woodhead Publishing Limited. CRC Press. https://doi.org/10.1201/9781439824474

Beaudouin, E., Prestat, F., Schmitt, M., Kanny, G., Laxenaire, M., Moneret-Vautrin, D., 1994. High risk of sensitization to latex in children with spina bifida. *Eur. J. Pediatr. Surg.* 4, 90–93. https://doi.org/10.1055/s-2008-1066075

Behnam Hosseini, S., 2020. Natural fiber polymer nanocomposites. *Fiber-Reinforced Nanocomposites Fundam. Appl.* 279–299. https://doi.org/10.1016/B978-0-12-819904-6.00013-X

Brown, R.H., Schauble, J.F., Hamilton, R.G., 1998. Prevalence of latex allergy among anesthesiologists: Identification of sensitized but asymptomatic individuals. *Anesthesiology* 89, 292–299. https://doi.org/10.1097/00000542-199808000-00004

Chougule, R., Khare, V.R., Pattada, K., 2013. A fuzzy logic based approach for modeling quality and reliability related customer satisfaction in the automotive domain. *Expert Syst. Appl.* 40, 800–810. https://doi.org/10.1016/j.eswa.2012.08.032

Christian, S.J., 2016. Natural fibre-reinforced noncementitious composites (biocomposites). *Nonconv. Vernac. Constr. Mater.* 111–126. https://doi.org/10.1016/B978-0-08-100038-0.00005-6

Code of Federal Regulations, 2020. Title 21--Food and Drugs Chapter I--Food and Drug Administration Department of Health and Human Services Subchapter H - Medical Devices. https://doi.org/21CFR801.437

Code, U.S., States, U., 2021. 21 U.S. Code § 321- Definitions; generally.

Commission Directive 2007/19/EC of 2 April 2007 amending Directive 2002/72/EC relating to plastic materials and articles intended to come into contact with food and Council Directive 85/572/EEC laying down the list of simulants to be used for testing migration of constituents of plastic materials and articles intended to come into contact with foodstuffs (Text with EEA relevance) - Publications Office of the EU [WWW Document], n.d. URL https://op.europa.eu/en/publication-detail/-/publication/40822ea6-da11-47da-987c-f6ac6fcff3b6/language-en (accessed 8.18.21).

Council Directive 85/572/EEC of 19 December 1985 laying down the list of simulants to be used for testing migration of constituents of plastic materials and articles intended to come into contact with foodstuffs - Publications Office of the EU [WWW Document], n.d. URL https://op.europa.eu/en/publication-detail/-/publication/bd664554-cb4a-4d7e-8e95-e36da6c900bf/language-en (accessed 8.18.21).

Council of Europe, 2014. Policy statement on rubber intended to come into contact with food of the Council of Europe.

Curtis, J., Klykken, P., 2008. A Comparative Assessment of Three Common Catheter Materials. Dowcorningcom 1–8.

EN 14233:2002- Materials and articles in contact with foodstuffs - Plastics [WWW Document], n.d. URL https://standards.iteh.ai/catalog/standards/cen/a9c70753-218a-4a3e-9e26-f14dbb999fd3/en-14233-2002 (accessed 8.18.21).

Eng, C.C., Ibrahim, N.A., Zainuddin, N., Ariffin, H., Yunus, W.M.Z.W., 2014. Impact strength and flexural properties enhancement of methacrylate silane treated oil palm mesocarp fiber reinforced biodegradable hybrid composites. *Sci. World J.* 2014. https://doi.org/10.1155/2014/213180

Glove Selection Guide | Office of Environment, Health & Safety [WWW Document], n.d. URL https://ehs.berkeley.edu/glove-selection-guide (accessed 8.18.21).

González-López, M.E., Pérez-Fonseca, A.A., Cisneros-López, E.O., Manríquez-González, R., Ramírez-Arreola, D.E., Rodrigue, D., Robledo-Ortíz, J.R., 2019. Effect of maleated PLA on the properties of rotomolded PLA-agave fiber biocomposites. *J. Polym. Environ.* 27, 61–73. https://doi.org/10.1007/s10924-018-1308-2

Heckman, n.d. Heckman, supra note 2, at 102.

Holec, D., Dumitraschkewitz, P., Vollath, D., Fischer, F.D., 2020. Surface energy of au nanoparticles depending on their size and shape. *Nanomaterials* 10. https://doi.org/10.3390/nano10030484

Huda, M.S., Schmidt, W.F., Misra, M., Drzal, L.T., 2013. Effect of fiber surface treatment of poultry feather fibers on the properties of their polymer matrix composites. *J. Appl. Polym. Sci.* 128, 1117–1124. https://doi.org/10.1002/app.38306

Ibid. at 955, n.d.

Jaafar, J., Siregar, J.P., Piah, M.B.M., Cionita, T., Adnan, S., Rihayat, T., 2018. Influence of selected treatment on tensile properties of short pineapple leaf fiber reinforced tapioca resin biopolymer composites. *J. Polym. Environ.* 26, 4271–4281. https://doi.org/10.1007/s10924-018-1296-2

Jaeger, D., Kleinhans, D., Czuppon, A.B., Baur, X., 1992. Latex-specific proteins causing immediate-type cutaneous, nasal, bronchial, and systemic reactions. *J. Allergy Clin. Immunol.* 89, 759–768. https://doi.org/10.1016/0091-6749(92)90385-F

Jumaidin, R., Sapuan, S.M., Jawaid, M., Ishak, M.R., Sahari, J., 2017. Thermal, mechanical, and physical properties of seaweed/sugar palm fibre reinforced thermoplastic sugar palm Starch/Agar hybrid composites. *Int. J. Biol. Macromol.* 97, 606–615. https://doi.org/10.1016/j.ijbiomac.2017.01.079

Kabir, M.M., Wang, H., Lau, K.T., Cardona, F., 2012. Chemical treatments on plant-based natural fibre reinforced polymer composites: An overview. *Compos. Part B Eng.* 43, 2883–2892. https://doi.org/10.1016/j.compositesb.2012.04.053

Kaczmarek, R.G., Silverman, B.G., Gross, T.P., Hamilton, R.G., Kessler, E., Arrowsmith-Lowe, J.T., Moore, R.M., 1996. Prevalence of latex-specific IgE antibodies in hospital personnel. *Ann. Allergy, Asthma Immunol.* 76, 51–56. https://doi.org/10.1016/S1081-1206(10)63406-0

Kalia, S., 2016. Biodegradable green composites. In: *Biodegradable Green Composites.* pp. 1–345. https://doi.org/10.1002/9781118911068

Kremensas, A., Kairyte, A., Vaitkus, S., Vejelis, S., Balčiunas, G., 2019. Mechanical performance of biodegradable thermoplastic polymer-based biocomposite boards from hemp shivs and corn starch for the building industry. *Materials (Basel).* 16. https://doi.org/10.3390/ma12060845

Lagier, F., Vervloet, D., Lhermet, I., Poyen, D., Charpin, D., 1992. Prevalence of latex allergy in operating room nurses. *J. Allergy Clin. Immunol.* 90, 319–322. https://doi.org/10.1016/S0091-6749(05)80009-0

Latex [WWW Document], n.d. How Prod. are Made. URL http://www.madehow.com/Volume-3/Latex.html#ixzz4MErKMRRm (accessed 8.17.21).

Li, X.M., Yang, R.J., 2006. Study on blooming of tetrabromobisphenol A bis(2,3-dibromopropyl ether) in blends with polypropylene. *J. Appl. Polym. Sci.* 101, 20–24. https://doi.org/10.1002/app.23089

Liang, Z., Pan, P., Zhu, B., Dong, T., Inoue, Y., 2010. Mechanical and thermal properties of poly(butylene succinate)/plant fiber biodegradable composite. *J. Appl. Polym. Sci.* 115, 3559–3567. https://doi.org/10.1002/app.29848

Médard, N., Poleunis, C., Vanden Eynde, X., Bertrand, P., 2002. Characterization of additives at polymer surfaces by ToF-SIMS. *Surf. Interface Anal.* 34, 565–569. https://doi.org/10.1002/sia.1361

Mehendale, S., 2010. Principal issues in global food contact: Indian perspective. In: *Global Legislation for Food Packaging Materials.* pp. 337–344. https://doi.org/10.1002/9783527630059.ch19

Monsanto vs. Kennedy, 613, 1979. Monsanto vs. Kennedy, 613 F.2d 947 (D.C. Cir. 1979.

Mueller, R.-J., 2006. Biological degradation of synthetic polyesters—Enzymes as potential catalysts for polyester recycling. *Process Biochem.* 41, 2124–2128. https://doi.org/10.1016/j.procbio.2006.05.018

NIOSH alert: Preventing allergic reactions to natural rubber latex in the workplace. [WWW Document], 1997. Hosp. Technol. Ser. URL https://webcache.googleusercontent.com/search?q=cache:7aIG4ZpFiO0J:https://www.cdc.gov/niosh/docs/97-135/pdfs/97-135.pdf+&cd=2&hl=en&ct=clnk&gl=in (accessed 8.18.21).

Noor Azammi, A.M., Ilyas, R.A., Sapuan, S.M., Ibrahim, R., Atikah, M.S.N., Asrofi, M., Atiqah, A., 2020. Characterization studies of biopolymeric matrix and cellulose fibres based composites related to functionalized fibre-matrix interface. *Interfaces Part. Fibre Reinf. Compos. Curr. Perspect. Polym. Ceram. Met. Extracell. Matrices,* 29–93. https://doi.org/10.1016/B978-0-08-102665-6.00003-0

Nouman, M., Saunier, J., Jubeli, E., Yagoubi, N., 2017. Additive blooming in polymer materials: Consequences in the pharmaceutical and medical field. *Polym. Degrad. Stab.* 143, 239–252. https://doi.org/10.1016/j.polymdegradstab.2017.07.021

Parliament, E., 2004. Regulation (EC) No 1935/2004 of the European Parliament and of the Council of 27 October 2004 on materials and articles intended to come into contact with food and repealing Directives 80/590/EEC and 89/109/EEC, Official Journal of the European Union.

Peter Barton Hutt, 1974. Memorandum from Peter Barton Hutt, General Counsel, FDA, to Sam D. Fine (October 31, 1974) (on file with author).

Ray, S.S., Bousmina, M., 2005. Biodegradable polymers and their layered silicate nanocomposites: In greening the 21st century materials world. *Prog. Mater. Sci.* 50, 962–1079. https://doi.org/10.1016/j.pmatsci.2005.05.002

Rijk, R., Veraart, R., 2010. *Global Legislation for Food Packaging Materials, Global Legislation for Food Packaging Materials.* Wiley-VCH Verlag GmbH & Co. KGaA, Weinheim, Germany. https://doi.org/10.1002/9783527630059

Scarsella, J.B., Zhang, N., Hartman, T.G., 2019. Identification and migration studies of photolytic decomposition products of UV-photoinitiators in food packaging. *Molecules* 24. https://doi.org/10.3390/molecules24193592

Sheik, Sareen, Sheik, Sana, Nagaraja, G.K., Chandrashekar, K.R., 2018. Thermal, morphological and antibacterial properties of chitosan grafted silk fibre reinforced PVA films. *Mater. Today Proc.* 5, 21011–21017. https://doi.org/10.1016/j.matpr.2018.06.493

Shimao, M., 2001. Biodegradation of plastics. *Curr. Opin. Biotechnol.* 12, 242–247. https://doi.org/10.1016/S0958-1669(00)00206-8

Sudamrao Getme, A., Patel, B., 2020. A review: Bio-fiber's as reinforcement in composites of polylactic acid (PLA). *Mater. Today Proc.* 26, 2116–2122. https://doi.org/10.1016/j.matpr.2020.02.457

Techochaingam, T., Netpradit, S., Tanprasert, K., Boochathum, P., 2013. Application of para rubber latex mixed with organic dyes as printing ink for biodegradable plastic film. *J. Met. Mater. Miner.* 23, 25–28.

Thakur, K., Kalia, S., Kaith, B.S., Pathania, D., Kumar, A., Thakur, P., Knittel, C.E., Schauer, C.L., Totaro, G., 2016. The development of antibacterial and hydrophobic functionalities in natural fibers for fiber-reinforced composite materials. *J. Environ. Chem. Eng.* 4, 1743–1752. https://doi.org/10.1016/j.jece.2016.02.032

Turjanmaa, K., Alenius, H., Mäkinen-Kiljunen, S., Reunala, T., Palosuo, T., 1996. Natural rubber latex allergy. *Allergy Eur. J. Allergy Clin. Immunol.* 51, 593–602. https://doi.org/10.1111/j.1398-9995.1996.tb04678.x

van Ketel, W.G., 1984. Contact urticaria from rubber gloves after dermatitis from thiurams. *Contact Dermatitis* 11, 323–324. https://doi.org/10.1111/j.1600-0536.1984.tb01026.x

Vinod, A., Sanjay, M.R., Suchart, S., Jyotishkumar, P., 2020. Renewable and sustainable biobased materials: An assessment on biofibers, biofilms, biopolymers and biocomposites. *J. Clean. Prod.* 258. https://doi.org/10.1016/j.jclepro.2020.120978

Vitrac, O., Hayert, M., 2005. Risk assessment of migration from packaging materials into foodstuffs. *AIChE J.* 51, 1080–1095. https://doi.org/10.1002/aic.10462

Wang, Z.W., Li, B., Lin, Q.B., Hu, C.Y., 2018. Two-phase molecular dynamics model to simulate the migration of additives from polypropylene material to food. *Int. J. Heat Mass Transf.* 122, 694–706. https://doi.org/10.1016/j.ijheatmasstransfer.2018.02.004

12 Degradation Studies of Biodegradable Composites

Francis Luther King
Swarnandhra College of Engineering and Technology (A)

CONTENTS

DOI: 10.1201/9781003227908-12

12.1 INTRODUCTION

Degradation is a change in the chemical body of a polymer chain, which leads to a decrease in the relative molecular mass of the polymer over time by breaking large molecules into fine fragments of different sizes and structures. It is a well-known fact that our current existence would be impossible to maintain without the use of man-made polymers. PET and polyurethane (PU) have grown in popularity. PET is a polymer composed of polyethylene terephthalate (PET) that is commonly used to make flasks and storage tanks, drain channels, water cans and bottles, and beverage containers. As the world's population has expanded, so has the use of plastic. As a result, plastics have become indispensable in a wide range of products, and they have the ability to enable previously unthinkable feats. Plastics are reshaping the planet. To fully benefit from the merits of polymers, however, their components must be correctly recovered and managed at the moment of termination (Scott G, 1999).

More research into the subject of packaging has been spurred by the continuing increase in polymer-based materials for logistics packaging applications, as well as their resilience to degradation and exposure throughout the environment once discarded. According to estimates, 2% of all polymers wind up in the environment, contributing considerably to the current ecological disaster (Psomiadou et al., 1997). Several materials claim to be biodegradable to sell various polymers, using terms such as degradable, biological, compostable, green materials, Oxo-biodegradable, and decomposable. The usage and accumulation of petrochemical-based goods and plastics in the environment after the service period creates trash. Academics and engineers have expressed concern regarding economic and environmental waste management as a result of this. It's crucial to develop perishable materials and composites that can be disposed of safely and efficiently. Figure 12.1 shows the life cycle of green materials.

12.1.1 BIOPOLYMERS

Biopolymers derived from natural sources were plentiful, filaments of (spider) silk, which deteriorate in aqueous medium (Zhao et al., 2010; Arai et al., 2004), and the scaffold material chitin, which degrades slowly in soil (Sato et al., 2010; Krsek and Wellington, 2001; Nakashima et al., 2005). Chitosan is

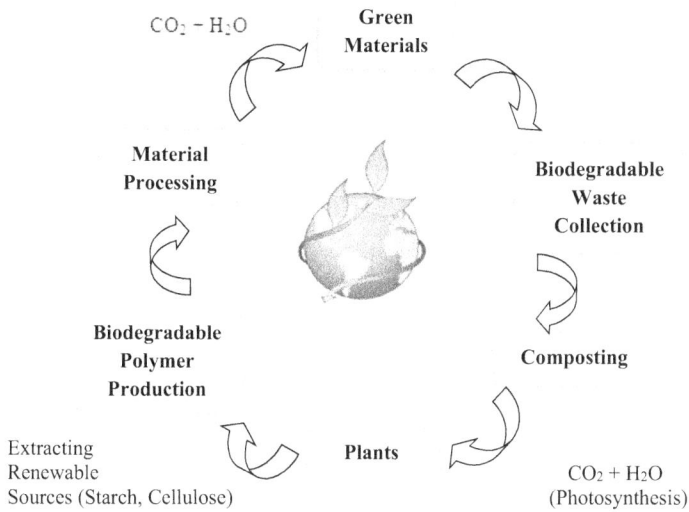

$CO_2 - H_2O$

Green Materials

Material Processing

Biodegradable Waste Collection

Biodegradable Polymer Production

Composting

Extracting Renewable Sources (Starch, Cellulose)

Plants

$CO_2 + H_2O$ (Photosynthesis)

FIGURE 12.1   Life cycle of green materials. (Reproduced from King et al., 2020.)

extremely similar to chitin and takes many months in the soil to break down (Sawaguchi et al., 2015). Three types of biodegradable polymers may be identified. The first type of biopolymer is synthetic polymers. Polyesters, polyanhydrides, polyamides, polycarbonates, polyurethane, polyureas, and polyacetals are all affected by microbial hydrolysis (Jang et al., 2001; Chandra and colleagues, 1997). Microbial polymers such as poly (3-hydroxybutyrate) (PHB) and poly (3-hydroxybutyrate-covalerate) (PHBV) constitute the second category of classification (Saad and Seliger, 2004). Finally, the combinations of polyethylene with starch are the best example of polymeric blends and bio-based composites are commonly called the third class of polymers that microbes rapidly digest. (Sharma et al., 2001; Albertson et al., 1994). Generally, bio-based biodegradable polymers are categorized into three major classifications: (i) synthesized from bio-derived monomers, (ii) produced by microorganisms, and (iii) directly extracted from biomass with partial modification. Typically, starch is utilized as a thermoplastic. Thermoplastic starch (TPS), on the other hand, has poor mechanical characteristics because of its sensitivity to humidity and water content. Despite this, starch has been used in a variety of goods due to its inexpensive nature and broad availability.

12.1.2   BIOPOLYMERS FROM PETRO-SOURCES

Biopolymers which are not made from renewable basic sources may potentially degrade. One such polymer is polycaprolactone (PCL), which is synthesized via ring-opening polymerization of ε-caprolactone. In seawater, PCL degrades entirely in just a few weeks (Tsuji and Suzuyoshi, 2002; Rutkowska et al., 2002;

Heimowska et al., 2011). The conditions in anaerobic sewage sludge are likewise adequate (Rutkowska et al., 2002); however, the delayed degradation rate of 3-mm-thick tensile test samples is noticeable once again, with less than 5% mass loss after 120 days (Bastioli et al., 1995). PBS (polybutylene succinate) itself is derived from petroleum crude oil, but it will be accessible in the future in a (partially) bio-based form. It degrades in soil, and yet only 11% after 180 days. Degradation occurs at a rate of 28% in the case of films. After 180 days, there has been a 9% reduction in bulk, but the biological cause has yet to be determined. Deterioration via emitted CO_2 resulted in a 65% decay (Silvia et al., 2020).

12.1.2.1 Cellulose Based

Cellulose is the most common organic substance in our surroundings, and it is frequently employed as a positive primary source in biodegradability tests, such as ASTM D5209-92, due to its easy biodegradability in all environmental circumstances. Cotton has 95%–97% cellulose and is used extensively in textile manufacturing. Cotton fabric cellulose has a high degree of crystallinity (up to 73%), making biological breakdown more difficult. The molecular chains are less firmly linked to one other in amorphous areas, making them more susceptible to enzyme assault. A cotton T-shirt takes 25 months to deteriorate fully in seawater, according to Nancy Wallace (2018).

12.1.2.2 Protein Based

Like starch and cellulose, protein-based polymers are naturally derived amino acids. Casein is the commonly identified protein-based polymer from milk. When subjected to isotonic saline solution, after 30 days casein exhibits a loss of 45% total mass. Meanwhile, during the degradation of casein in the saline solution, ammonia gets released and inhibits the activity of the microorganisms during degradation (Austin and Austin, 2007). Gluten is a water-soluble protein found in wheat flour that is frequently employed as a protein source. Its permeability, like that of thermoplastic starch, is reliant on the application of plasticizers. Gluten dissolves entirely in soil in 30–50 days, with results ranging from 15% total mass loss in 70 days (Lim et al., 1999; John et al., 1998) to 100% biodegradation in 12 days for breakdown in compost settings.

12.1.2.3 Starch Based

Corn, potatoes, and wheat may all be used to make starch, making it a renewable resource. There are several phases in the processing that must be completed before pure starch can be extracted. The abundance of polar groups in native starch facilitates the development of hydrogen bonds and, as a result, significant water absorption. Although it is thought that starch is totally biodegradable in all environments, the researchers are only aware of experiments on native starch breakdown in compost. Researchers demonstrate that starch films disintegrate entirely in a short period of time (30–84 days), but cellulose, as a control material, takes just 10 days under same circumstances (Vaverková et al., 2012; Torres et al., 2011).

12.1.2.4 CO_2 Based

The conversion of CO_2 into polymeric polymers provides a potential greenhouse gas recycling alternative. Polypropylene carbonate (PPC) is made by copolymerizing carbon dioxide with propylene oxide. The biological breakdown of PPC is aided by high temperatures. After 6 months, a total mass loss of just 3.2% was detected in garden soil (Du, L.C et al., 2004). A mass loss of 8% was recorded at 40°C in a composting setting, followed by a significant rise in disparity (Bahramian et al., 2016). After only 3 months in an industrial composting operation (at 60°C), a full absolute mass loss may be recorded (Luinstra, 2008).

12.1.3   POLYMERS AND SURROUNDINGS

Polymer materials are currently utilized in a variety of applications, including consumables, building materials, architectural interiors, chemicals and petroleum products, food processing sectors, and medical uses. Around 14% of used polymeric packaging is presently reused across the globe. This figure represents the financial difficulties that have arisen as a result of collecting the post-consumer handling of various packaging materials and forms, frequently adopting underdeveloped post-consumer utilization systems (Francis, 2016).

12.1.4   POLYMER IMPACT ON ENVIRONMENT

Both water and land may introduce polymer-based compounds into the environment. Littering, disposal of unwanted plastic goods, movements from landfills, and rejection during garbage collection were all common methods in which polymer-based pollutants were discharged into nature (Gregory, 2009; Tharpes, 1989). According to the US National Institutes of Health, 44% of seabird species have consumed synthetic polymers mistaken for food, resulting in millions of deaths each year. As seabirds perform an essential ecological function in regulating the population numbers of fish and crustaceans, this widespread loss of shoreline birds poses a severe environmental hazard. The US National Institutes of Health experts have described it as "a fast rising, long-term danger." Before taking any initiative to eradicate this type of pollution, it's crucial to understand how synthetic polymers damage ecosystems (James Ducker, 2021).

12.1.4.1   By Ocean

Ocean-based sources include items lost or abandoned by fishing operations, offshore oil or gas rigs, and rubbish dumped by recreational boats. In the past, prefabricated polyethylene and polypropylene pellets were purportedly used on ship decks to reduce resistance when loading and unloading cargo. As a result, waves and currents wash away and disperse the majority of these particles (Tharpes, 1989). Seventy per cent of this garbage is anticipated to wind up on the bottom, 15% on beaches, and 15% floating in the water. Plastics account for 70% of the trash, with metal and glass accounting for the majority of the remaining 30%.

More than 1200 aquatic species have been found to ingest trash, live in or on it, or become entangled in it, according to studies. This litter affects a wide range of animals, including mammals, fish, and crabs (James Ducker, 2021). Garbage dumping in the ocean has been prohibited under international law in 1973 (MARPOL 73/78 Annex V), which took effect in 1988 and restricted ship functional discharges (do Sul and Costa, 2007).

12.1.4.2 By Land

Man-made polymers are a huge concern on ground since they are frequently discarded, in which they will stay for generations, gradually leaching toxins into ground as time progresses. Dumping, both deliberate and inadvertent, is a major source of polymer-based materials waste discharged into the environment from land-based sources (Gregory, 2009). Littering is an issue at event sites, especially when there aren't enough trash disposal options (Cierjacks et al., 2012). Wind-blown garbage from containers or processing and disposal facilities, on the other hand, results in unintentional littering (Tharpes, 1989). As per the Clean Air Council, an American is using average 102.1 billion plastic bags (a synthetic polymer) every year, with much less than 1% of these bags being reused. Not only do such polymeric materials gradually infiltrate toxic contaminants, but its durability and non-biodegradability mean that as man-made polymer use continues to expand, new landfills will be required on a regular basis (James Ducker, 2021).

12.1.4.3 By Industries

Commercial manufacturing processes have changed civilized life, from the simplest electronic gadgets to the greatest automobiles. Unfortunately, contamination is a significant side consequence of rising manufacturing output (James Ducker, 2021). Breathing, the freshwater we consume, the soil we travel on, as well as the lighting and sounds we receive may all be affected by industrial pollution.

Pharmacological medicines can enter the environment in two ways:

- Chemicals and compounds are expelled by people and animals, primarily through urine, and travel into the environment either knowingly or unknowingly through open drains.
- Production and processing plants which produce the active ingredient can discard of medicines into the ecosystem, through either domestic water or via municipal solid waste management.

12.1.4.4 By Landfills

When polymer-based products reach the end of their useful life, they are frequently discarded in landfills (Barnes et al., 2009). In most industrialized regions of the globe, garbage is regularly carried to landfills and absorbed into the soil (Rayne, 2008). Persistent organic pollutants, namely the insecticides DDT and toxaphene, are proven poisons that can stay in the surroundings for several years. Researchers from the University of the Pacific tested man-made polymers identified at coastal locations in the north Pacific Ocean in 2007 and discovered the existence of hazardous poisons in every test. If consumed, such polymeric materials may leak poisonous compounds into fishes and animals, posing a hazard to the community of ocean fisheries that people consume (James Ducker, 2021).

12.1.4.5 Sewage Debris

Sewage-derived trash and debris also provide a route for polymer-based products to pollute the environment. Personal hygiene goods connected to polymer-based materials including condoms, cotton buds, and small PE beads observed in some hand cleansers and face washes (Ashley et al., 2005; Williams and Simmons, 1999) are the inputs to a sewage system that is largely uncontrolled in many countries. Large objects are typically eliminated via monitoring; however, they may get up in the environment during sewage outbursts following periods of heavy rain. The wastewater treatment plants' capacity to manage small particles and filaments has been questioned (Browne et al., 2011).

12.1.5 ENVIRONMENTAL CIRCUMSTANCES

Polymer-based materials have now been identified as the most common source of coastal, marine, and terrestrial pollution in a number of studies conducted across the world, with regional differences in polymer prevalence noted. When polymer-based products are released into the atmosphere, they are transported and spread over different ecological areas. The distance that rubbish and debris will travel is determined by their size, shape, and weight. Winds or rainfall may readily transport lighter objects over large distances, eventually collecting in the oceans. During heavy rainstorms, highway debris and rubbish can be carried into ditches and gullies and delivered to the sea when geology permits. Macro-plastics are large bits of polymer trash that have been labelled as such and categorized as items with a diameter of more than 5 mm, since this size enables for an evaluation to determine the item's origin (Browne et al., 2011; Thompson et al., 2009).

12.1.6 FACTORS AFFECTING POLYMER DEGRADATION CHARACTERISTICS

The degradation rate of polymeric materials is influenced by polymer properties. The complexity of a polymer's structure that forms highly structured networks and composition (co-polymers) can have an impact on the overall degradability by affecting enzyme accessibility (Artham and Doble, 2008). PBMs (e.g. PE, PP, and PET) with short and regular repeating units, high symmetry, and strong inter-chain hydrogen bonding limit accessibility and are less vulnerable to enzyme assault (Artham and Doble, 2008).

The following are some of the factors that might affect polymer biodegradation (Sharma, 2011).

1. Chemical structure.
2. Chemical composition.
3. Presence of ionic groups.

    4. Presence of unexpected units or chain defects.
    5. Molecular weight distribution.
    6. Morphology (amorphous/semi-crystalline, microstructures, and residual stresses).
    7. Presence of low-molecular-weight compounds.
    8. Processing conditions.
    9. Annealing.
   10. Sterilization process.
   11. Configuration structure.
   12. Storage history.
   13. Shape.
   14. Adhesion of atoms, ions, or molecules and water, lipids, ions, etc.
   15. Physicochemical factors (ion exchange, ionic strength, and pH).
   16. Mechanisms of hydrolysis (enzymes versus water).

12.2 DEGRADATION OF POLYMERS

It is typically categorized as listed based on the variables.

    1. Biological degradation
    2. Chemical degradation
    3. Thermal degradation
    4. Weather degradation
    5. Mechanical degradation

12.2.1 BIOLOGICAL DEGRADATION

The growing usage of polymeric products in recent decades has prompted the development of biopolymers and polymer biodegradation. The science of producing biodegradable, perishable materials is known as biodegradable technology. It incorporates today's processes with science-based plant genetics systems. Wetness, temperature, pH, salinity, the presence or absence of oxygen, and the availability of various nutrients all have a role in the microbial breakdown of plastics. Polymers are frequently mixed with additives to improve processability or offer additional functions. The vast majority of plastics are not intended to degrade naturally. The stability and resistance of plastics are what distinguishes them. Microbial degradation is the biological breakdown or "digestion" of carbon dioxide, water, methane, biomass, or inorganic substances by bacteria or microorganisms (Gautam et al., 2007). The six primary kinds of microorganisms/microbes are bacteria, archaea, fungus, protozoa, algae, and viruses. Microbes' proclivity for breaking down polymers is directly proportional to their capacity to attach to the polymer's surfaces. Polymers are broken down by microorganisms in two ways: metabolically and enzymatically. The degradation process begins with breakdown and progresses to mineralization (Krzan et al., 2006).

12.2.1.1 Polymer Biodegradation Mechanism

Biochemical transformation and microbial digestion are used by microorganisms to break down composite components into simpler forms. Changes in polymer characteristics such as microbial enzyme digestion, molecular weight loss, mechanical strength loss, and surface quality loss are all examples of polymer biodegradation (Hadad et al., 2005). The following depicts microorganisms involved in various phases of polymer breakdown.

a. Bio deterioration: Biodegradable materials are broken down into tiny parts by microbial and other decomposer species, as well as abiotic forces.

b. Depolymerisation: Microorganisms create catalytic agents (enzymes and free radicals) capable of cleaving polymer structures and molecules lowering their molecular mass. This process produces tiny units such as oligomers, dimers, and monomers that can pass through the semi-permeable outer bacterial membranes.

c. Assimilation: Microbes use transferred molecules as a source of carbon and energy in the cytoplasm to make storage, new biomass, and a variety of metabolites, which aid in the maintenance of cellular activity, structure, and reproduction. As a result, microorganisms reproduce and consume nutritional substrates from the surrounding.

d. Mineralization: Metabolites can be excreted and reach the extracellular environment through mineralization. CO_2, N_2, CH_4, and H_2O are some of the molecules that are released into the environment. Aerobic microbes create microbial biomass, CO_2, and H_2O when oxygen is present. Anaerobic bacteria generate microbial products in anoxic conditions. As indicated in equation (12.1), aerobic biodegradation occurs when oxygen is present and carbon dioxide is produced. Anaerobic biodegradation occurs when oxygen is not available, and methane is generated instead of carbon dioxide, as indicated in equation (12.2) (Grima et al., 2002).

12.2.1.2 Aerobic Biodegradation

Anaerobic biodegradation happens in the presence of oxygen where oxygen is the electron acceptor. The organic components will be broken down by microbes in the presence of oxygen. Aerobic bacteria, often known as aerobes, are responsible for breaking down organic molecules quickly. These bacteria's metabolism is oxygen-dependent. In a process known as cellular respiration, aerobes use oxygen to oxidize substrates (sugars and fats) to obtain energy. Before cellular respiration, glucose molecules are broken down into two smaller molecules in the cytoplasm of the aerobes. The smaller molecules are subsequently carried to the mitochondrion, where aerobic respiration takes place. Oxygen is used in chemical reactions as shown in equation (12.1) that break down small molecules into water and carbon dioxide and release energy. Methane gas is not produced under anaerobic environments (Fritsche and Hofrichter, 2008).

$$\text{Polymer} + O_2 \rightarrow CO_2 + H_2O + \text{Biomass} + \text{Residues} \qquad (12.1)$$

12.2.1.3   Anaerobic Biodegradation

Biodegradation occurs in anaerobic circumstances when there is no oxygen present and the electron acceptor is nitrate, sulphate, or another molecule. When anaerobic bacteria dominate aerobic microbes under anaerobic circumstances, biodegradation and assimilation occur. The process begins with bacterial hydrolysis of the input materials to break down insoluble organic polymers, such as carbohydrates, and make them accessible to other microorganisms as indicated in equation (12.2). Bacteria, such as acetic acid-forming bacteria and methane-forming bacteria, are engaged in the digestive process in anaerobic circumstances. These bacteria consume the initial feedstock, which is transformed into immediate compounds such as sugars, hydrogen, and acetic acid before being turned to biogas through a series of steps. Because it reduces the volume and bulk of the input material, anaerobic biodegradation is commonly employed to treat wastewater sludge and biodegradable trash. It also decreases landfill gas emissions into the atmosphere. Anaerobic digestion is a sustainable energy source since it generates biogas rich in methane and carbon dioxide, which may be used to generate electricity and replace fossil fuels. After digestion, nutrient-rich solid wastes can be utilized as a fertilizer.

$$\text{Polymer} \rightarrow CH_4 + CO_2 + H_2O + \text{Biomass} + \text{Residues} \qquad (12.2)$$

12.2.1.4   Principles in Testing Biodegradable Polymers

For the biodegradability test of plastics, it is critical to select the appropriate type of testing method. Specimens are buried in the soil according to the criteria in an optimum setting for a full-scale composting process in field testing. A few drawbacks of field testing include the inability to regulate surrounding variables such as temperature range, pH of the soil, and humidity. The weight loss and apparent changes in the SEM pictures are used to assess the disintegration of the specimens. Even though the environment is much closer to the field test, the parameters may be controlled, monitored, and modified since disintegration can take place in soil, compost, or sea water. In laboratory experiments for reproducible disintegration of plastics, defined media and parameters are employed. Polymers disintegrate at a considerably faster rate than they would under normal conditions. This is the most significant benefit when investigating the fundamental principles of polymer biodegradation, but only limited conclusions can be drawn from laboratory measurements on the absolute disintegration rate of plastics in natural environments.

12.2.1.5   Soil Burial Degradation Mechanism

Soil burial is a classic and typical approach for the deterioration of polymers. According to a study by Francis et al. (2020), experiments on soil degradation were carried out in the open air in Puducherry, India. As per ASTM standards, standard samples are dumbbell-shaped and square-shaped. ASTM G160-12 standard is used for the biodegradation testing of composites used in landfill tests

FIGURE 12.2   Samples buried in natural soil.

with minor changes (King et al., 2020). The samples were buried in natural soil at a depth of 15 cm, as shown in Figure 12.2, with a relative humidity of 65% and a temperature of 29°C–31°C. The test was carried out in this work by burying various weight fractions of PLA hybrid composites to a depth of 15 cm below ground level.

For the biodegradability test, ASTM standard samples weighing 13 gm and 11 gm were chosen and buried in the soil for 150 days. To keep the microorganisms active, the soil was kept at a moisture level of around 20%. Throughout the study, the buried samples were dug up at 30-, 60-, 90-, 120-, and 180-day interval, cleaned in distilled water, and dried in a vacuum oven at 50°C for 24 hours. After that, the samples were weighed to assess the weight loss, and the mass of each sample was determined using an electric balance before and after. Every month for 150 days, the rate of biodegradability and weight loss of the hybrid composites were recorded. Every 4 weeks, the buried samples were dug up and washed to remove dirt residues from the surface, then dried at 35°C to achieve a consistent dry weight for each sample, which was then weighed and assessed for degradability analysis. Studies on soil degradation of PLA basalt/bagasse composites results showed that the inclusion of PLA and basalt weight % within the mixture resulted in a reduction in water consumption. Since the sugarcane fibres expanded and deteriorated, the weight increased. This makes sense because PLA and basalt are hydrophobic, but bagasse is hydrophilic.

The polarity nature of bagasse fibre increases as the bulk % of sugarcane fibre for various formulations rises, and thus water penetration and absorption rises. The samples represented with 7 wt. % of sugarcane fibres exposed to soil burial with a water content of 25% showed the most significant deterioration. These circumstances encourage the development of organisms (bacteria and/ or fungi) in the soil, which accelerates the decomposition of composite surfaces and thus begins the degradation. Figure 12.3 shows the complete degradation of the polymer composites. The amorphous portion of the materials were vulnerable to microbial attack during the soil degradation process sample's total degree of crystallinity increased. Figure 12.4 shows the optical image of soil-degraded samples: (a) 30 days; (b) 60 days; (c) 90 days; and (d) 150 days of polymer composite.

Thomas Bayerl et al. (2014) experimented to see how natural fibres, namely flax fibres, affected the rate of PLA breakdown during home composting. It was

FIGURE 12.3   Complete degradation of polymer composites (King et al., 2020).

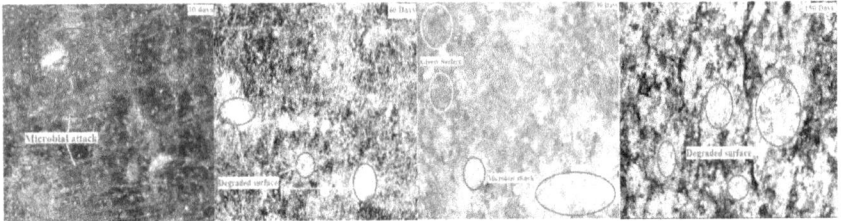

FIGURE 12.4   Optical image of soil-degraded samples: (a) 30 days; (b) 60 days; (c) 90 days; and (d) 150 days of polymer composite. (Reproduced with permission from King et al., 2020.)

discovered that the inclusion of flax fibres aided PLA breakdown by increasing the polymer surface area available for a possible hydrolysis reaction and microbial destruction. The primary weight loss throughout the trial period, however, was due to the breakdown of flax fibres, according to the research.

However, specimens where the fibres come in contact with a wet environment and the surface area that may be influenced by water reported a rise in brittleness. This was witnessed for unidirectional fibres with high weight fraction. For a preliminary examination, the deteriorated samples after cleaning and drying are shown in Figure 12.5 and were evaluated based on morphological alterations. The matrix in all of the samples changed from clear to opaque as the deterioration time increased. After only 2 weeks, UD-30 samples became opaque, but plain PLA samples took more than 4 weeks to become opaque. The colour of the fibres changed as they faded out in the first several weeks. However, as the test period went on, a darkening of the fibres was noticed, although this was due to the usage of coffee powder to keep the compost at the desired temperature. The filaments were observed to function as conduits for water and microbes, promoting the creation of fractures and crazes around the filaments. SEM examination revealed that the cracks, fads were more prevalent in the micropores of deteriorated fibres. The inclusion of microbes in the composting aided PLA breakdown because the deteriorating activity resulted in higher temperatures, which aided the enzymatic hydrolysis. Furthermore, the presence of microbes was primarily responsible for the degradation of the fibres.

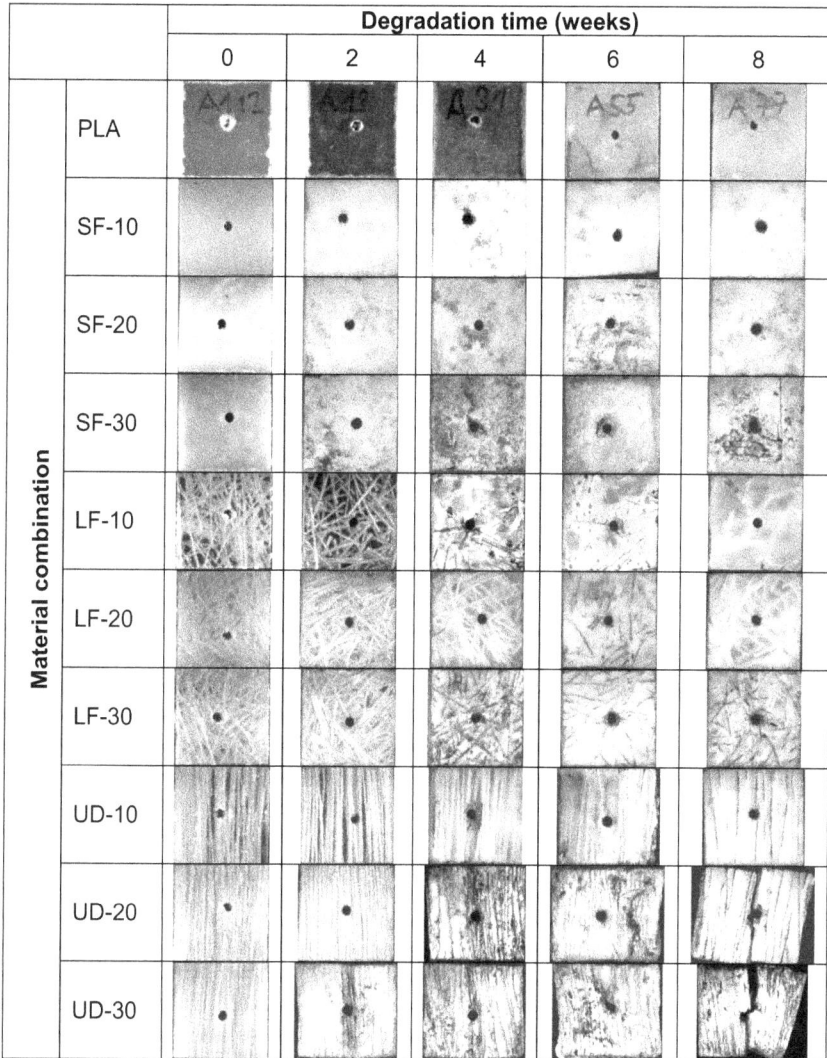

		Degradation time (weeks)				
		0	2	4	6	8
Material combination	PLA					
	SF-10					
	SF-20					
	SF-30					
	LF-10					
	LF-20					
	LF-30					
	UD-10					
	UD-20					
	UD-30					

FIGURE 12.5 Morphological changes of degraded samples. (Reproduced with permission from Bayerl et al., 2014.)

Figure 12.6 shows an example of the skeletal structure of PLA following flax fibre degradation. The presence of fibres and the geometry of the fibre structure in a PLA composite can have a significant impact on the composite's degradation behaviour and thus should be considered when designing a part that must meet biodegradability specifications and is used in applications such as panels in the automotive or construction industries. Structural defects and flaws were noticed in large number in scanning electron microscope images, which attributes to fibre deterioration during composting.

FIGURE 12.6 Hole over the surface of a UD-30 specimen originating from degraded flax fibre. (Reproduced with permission from Bayerl et al., 2014.)

12.2.1.6 Water Degradation

Water degradation mostly affects water-sensitive polymers, particularly those that absorb a lot of moisture. Polymers absorb water to varying degrees depending on their molecular structure, fillers, and additives, which fractures the polymeric chain. Plastic must be exposed to moisture and heated to initiate the hydrolytic breakdown process. Hydrolytic breakdown is also advantageous in the case of polymeric drug delivery systems. Passive hydrolysis is the most significant kind of breakdown for many biodegradable/compostable polymers.

The PLAGA polymer's deteriorating process has been studied extensively. Degradation is frequently seen as a hydrolytic process. When an ester bond is broken, a carboxyl end group and a hydroxyl end group are formed. The carboxyl end groups produced in this method can be used to catalyse the hydrolysis of more ester bonds, a process known as autocatalysis. Polymer composite matrix material is homogeneous at first, meaning that the average molecular weight is uniform throughout the matrix. After placing the samples in an aqueous solution, water penetrates the samples, triggering hydrolytic breakage of ester bonds, which speeds up the hydrolysis of the remaining ester bonds. Finally, when the inner material has been entirely transformed into soluble oligomers, void geometry is created when it dissolves in watery environments.

12.2.2 Chemical Degradation

The depolymerization reaction in polymers and other chemical compounds, which occurs wholly or partially to create monomers and oligomers, is known as chemical degradation. Polymers may be readily broken down by reagents such as acids, alcohols, water, amines, and glycols since they are made up of a range of organic and inorganic chemicals. The non-biodegradability of man-made polymers is a major problem. As a remedy, many techniques of polymer reprocessing and recycling have been proposed. Chemical breakdown is one of the most efficient recycling techniques in this situation. A multitude of approaches can be used to cause chemical degradation of polymeric materials. They are as follows:

- Base monomer regeneration.
- Oligomer transformation.
- Transformation of glycolysis output into valuable goods, industrial chemicals, and usable coating intermediates.

Chemical degradation is categorized as follows depending on the type of reagents employed:

- Acidolysis
- Hydrolysis
- Glycolysis
- Aminolysis
- Alcoholysis.

12.2.3 THERMAL DEGRADATION

Thermal degradation studies on a variety of polymers and polymer-based materials, such as polyethylene (PE), polypropylene (PP), polystyrene (PS), polymethyl methacrylate (PMMA), ethylene vinyl acetate copolymer, acrylonitrile butadiene styrene (ABS) terpolymers, and rubber, have been investigated by different researchers (Blanco and Bottino, 2016; Crompton, 2013). There are several strategies such as thermogravimetric analysis (TGA), spectroscopic calorimetry (SC), differential scanning calorimetry (DSC), Fourier transform infrared analysis (FTIRA), nuclear magnetic resonance, and hyphenated instruments such a liquid chromatography/mass spectrometry and evolved gas analysis (EGA) that are used to investigate the deterioration of polymers.

Polymers when exposed to high temperature will become soft and degrade, which limits the usage of polymer-based materials in several engineering applications. If heated to a sufficiently high temperature, polymers will continue to degrade even in an inert environment. The heat degradation of polymers is connected to chain scission, a series of chemical reactive reactions. Chain scission is commonly used to start multistep free radical processes (initiation, propagation, and termination) (Ray and Cooney, 2018). The degradation of biopolymers is accelerated when oxygen concentration is high (Ray and Cooney, 2018). Lignocellulosic material may be broken down by heat (Taherzadeh and Karimi, 2008). Many of these biopolymers begin to break down early and continue in a multistep process. In the presence of air, the glass transition temperature (T_g)

is the critical threshold that separates between glassy and rubbery behaviour. Amorphous materials, notably natural polymers and biopolymers, have glass transition temperatures. With the exception that early degradation occurs at a lower temperature, these polymers have properties that are remarkably similar to those seen in an inert environment.

When it comes to the mechanism of heat degradation, several types of polymers must be identified.

1. Scissions in polymers occur largely at the centre of the chain. Such polymers are likely to evaporate completely at exceptionally high temperatures.
2. In polymers, scissions typically occur between the backbone carbon atoms and the functional groups. Such scissions result in the formation of double bonds in the chain, as well as the possibility of cross-connections across chains. After prolonged heating, such polymers become rather stable in the form of a partially carbonized residue. Polymers with a high cross-linking percentage the polymers are converted into a honeycomb pattern of carbonized residues when heated (Madorsky and Straus, 1959).

12.2.3.1 Steps of Thermal Degradation

Over the last six decades, several studies have been done in an attempt to better understand how polyolefin degradation occurs. Thermal degradation of polymers is a complex process that involves the simultaneous synthesis and breakdown of hydroperoxides. Degradation is accelerated by oxygen, dampness, and straining, resulting in brittleness, cracking, and fading. Random degradation of polyethylene occurs when a hydrogen atom migrates from one carbon to the next, resulting in additional fragments. The dynamics of PP and PE degradation are crucial because they show how a complicated radical chain reaction works, as well as how waste incineration and other recycling processes work. A sequence of reactions involving chain scission are the effects associated with the degradation of polymers when subjected to elevated temperatures. Based on the fact that PP and PE are mostly utilized for packaging and represent for the bulk of plastic trash in household waste. As previously stated, the chemistry behind the reaction process is influenced by a number of elements. The majority of chain scission reactions follow a multistep free radical pathway that includes (i) initiation, (ii) propagation, and (iii) termination stages.

12.2.3.2 Pyrolysis

The primary processes that occur during the non-oxidative thermal breakdown of polymers are the following:

• Depolymerization
• Random chain scission

- Side group removal
- Cross-linking
- Substitution
- Within side group reaction.

The thermal properties of polystyrene at relatively high temperatures are well recognized. When heated in a vacuum or in a neutral atmosphere, it progressively vaporizes at temperatures between 250°C and 400°C. Boonstra and van Amerongen pyrolysed polyisoprene and other polymers in nitrogen at various pressures up to atmospheric and temperatures up to 775°C. Polyisobutylene yields 46% monomer at 775°C and 15 mm pressure. At 775°C and 5 mm pressure, it was 14% polybutadiene. At 675°C and 15 mm pressure, the monomer yield for polystyrene was 50%. The same monomer yields 5% for polyisoprene, 20% for polyisobutylene, 5% for polybutadiene, and 40% for polystyrene at temperatures as high as 400°C and in a vacuum arc. This polymer's pyrolysis produced the following results: At 500°C, stabilization occurs, and the further loss due to volatilization from 500°C to 800°C is just around 10%. The specimen was slowly heated to 800°C for 135 minutes and then maintained at that temperature for 5 minutes. Volatilization losses were reduced due to the slow pace of heating. Higher pyrolysis temperatures, as with polystyrene, result in greater fragmentation of the volatile chemicals, as seen by the relative yields of CH_4 and H_2 (Madorsky and Straus, 1959).

12.2.4 WEATHER DEGRADATION

When polymeric materials are subjected to high temperatures and UV radiation, their properties deteriorate, making them unsuitable for their intended applications. High temperatures can cause both short-term physical and long-term chemical changes in polymer materials. The systematic degradation of polymers and polymer coatings in the laboratory is known as weathering of polymeric materials. If polymers are exposed to UV radiation for a lengthy period of time, they will discolour and lose their mechanical properties.

12.2.4.1 Weathering Mechanism

When polymer-based materials, such as composite materials, are subjected to extremely high temperatures, UV light changes the structure and behaviour of the polymer. Photon energy from sunshine is a critical component that promotes photodeterioration in polymers. Ultraviolet rays and visible light-induced degradation affect the majority of polymers. The sun's UV light (290–400 nm) has a significant impact on the durability of polymers in outdoor applications (Singh, 2008). The ether portion of the amorphous phase deteriorates as a result of light irradiation, which creates ester, aldehyde, formate, and propyl end groups. Degradation occurs at different wavelengths depending on the kind of bonds in the polymer. To destroy polyetylene, for example, a wavelength of light of approximately 300 nm is required, whereas polypropylene requires a wavelength of around 370 nm.

The sequence of happenings of photodegradation is highlighted below (modified from Bhuvaneswari. G, Harini (2018)):

Initiation

$$Polymer \rightarrow R* \; (Free \; radical) \tag{12.3}$$

Propagation

$$R + O_2 \rightarrow RO_2 \tag{12.4}$$

$$RO_2 + PH \rightarrow ROOH + R \tag{12.5}$$

$$ROOH \rightarrow \underset{\downarrow \beta\text{-transmission}}{RO} + OH \downarrow \tag{12.6}$$

Termination

$$2\,RO_2 \rightarrow nonradical \; products \tag{12.7}$$

Photocatalytic breakdown cannot commence without the presence of a chromophore component in the polymers. Ketones, quinones, and peroxides function as main degradation targets via absorption of photons in the 380 nm wavelength region, producing radical stimulation or splitting.

12.2.5 MECHANICAL DEGRADATION

Mechanical degradation is the breakdown of polymers under physical stress. The materials' molecular bonds remain intact, which distinguishes this process from deterioration. When hydrodynamic forces are imparted over polymers during the flow, they fail withstand and undergo mechanical degradation, which is the first reason of a deterioration, and the efficiency diminishes slowly. Coiled elastic polymers must be fully stretched by pure extension flow before it shatters (Edson, 2020). During pure extension of the polymers, macromolecules are stretched and the applied mechanical forces will be very high at its junction. Polymers are subjected to a variety of mechanical degradations in the field, including ageing and breakage caused by air weathering, water turbulence, freeze–thaw cycles, pressure caused by burial beneath soil or snow, and damage caused by animals or birds. A polymer's mechanical scission is determined by a variety of factors (Edson, 2020):

1. The polymer's concentration
2. Molecular weight
3. Chemical structure
4. Arrangement of a linear polymer
5. The weather conditions
6. Intensity of turbulence (typically Reynolds number)

7. The interaction of the polymer with the solvent (the solvent's quality)
8. The residence time (the amount of time the molecules are subjected to turbulent flow)
9. Polydispersity
10. Geometry and local occurrences, such as contraction and expansion.

There are four major routes that determine mechanical degradation. Some of these include regrinding, adhesive pressing, compression, and injection moulding.

12.2.5.1 Regrinding

Regrinding is also known as making powders. Crushing polymer-based material waste into particles and reusing those particles are examples of mechanical regrinding techniques. Two-roll milling and ball milling are the most efficient methods for recycling polymers. In contrast, a ball mill can grind stiff polymers and polymeric-based materials. The powdered particles collected can be used as a filler in the production of foam or elastomers.

12.2.5.2 Adhesive Pressing

Adhesive pressing is a simple and rapid method of recycling. The outside portions of polyurethane particles are sprayed with a sticky binder and fused in a hot press in this process. It is a traditional method for mechanically recycling flexible polyurethane foam. This is the easiest method for recycling flexible polyurethane, and it may be used to produce mats, carpet underlay, gym flooring components, and automotive soundproofing.

12.2.5.3 Compression Moulding

Compression moulding is a technique for processing polyurethane fine particles at a high temperature and pressure of 180°C under 350 bar without the need of additional binders. Reaction injection moulding has been utilized to efficiently recycle polyurethane waste into car components. Coloured polyurethane, on the other hand, is difficult to recycle. Fine aggregates are commonly co-processed with mudguards and sporting grounds. Reaction injection is about 6% re-ground. Composites, which include 15% polymer, can be utilized to manufacture car doors and display panels.

12.2.5.4 Injection Moulding

Polymer degradation may occur during the fabrication of composites using injection moulding due to poor raw material processing prior to forming, also during plasticizing and first phase injection. If the polymer has more wetness, a chemical change inside the injection mixing chamber can occur, causing the plastics to deteriorate. Even when polymers are properly dried, extreme melt temperatures, extended residence times, and high shear rates can cause material degradation inside the machine chamber and during injection. Barrel screw transfers most of the thermal energy necessary to bring the polymer to its operating temperatures via compression and shearing. If the material left in the barrel is for too long, then the energy absorbed may be enough to disrupt the covalent bonds within the chains, allowing the system to degrade (David Rose, 2019).

12.3   CONCLUSIONS

Over the last few generations, the use of polymers and polymeric materials in everyday life has steadily risen. In industrialized nations, the amount of polymer used per person has risen, while the production of synthetic polymers has also increased. The wide range of characteristics and chemical composition of recovered polymers has limited their use. The kind of polymer and many other factors influence the manner of degradation. The reality is that the majority of polymer-based wastes are produced in municipal solid trash, combined with many other types of residues, reducing their chances of being recycled. As a result, landfilling is the sole option for storing polymeric waste that will not degrade for decades. The degradation mechanisms for several polymeric materials have been described in this chapter. To summarize, the proper degrading methods and procedures will be effective in the recycling of polymers for the conversion of polymeric trash into useful chemicals that may be utilized as fuels or raw materials in the chemical industry. Given the importance of stability, recyclability, and environmental concerns, this topic deserves further research.

REFERENCES

Albertsson AC, Griffin GJL, Karlsson S, Nishimoto K, Watanabe, Y (1994) Spectroscopic and mechanical changes in irradiated starch-filled LDPE. *Polym Degrad Stabil* 45(2): 173–178.

Arai T, Freddi G, Innocenti R, Tsukada M (2004) Biodegradation of Bombyx mori silk fibroin fibers and films. *J Appl Polym Sci* 91: 2383–2390.

Artham T, Doble M (2008) Biodegradation of aliphatic and aromatic polycarbonates. *Macromol Biosci* 8(1): 14–24.

Ashley R, Blackwood D, Souter N, Hendry S, Moir J, Dunkerley J, Davies J, Butler D, Cook A, Conlin J, Squibbs M, Britton A, Goldie P (2005) Sustainable disposal of domestic sanitary waste. *J Environ Engineering-Asce* 131(2): 206–215.

Austin B, Austin D (2007) *Bacterial Fish Pathogens: Disease of Farmed and Wild Fish*. Dordrecht, The Netherlands: Springer, ISBN 9781402060687.

Bahramian B, Fathi A, Dehghani F (2016) A renewable and compostable polymer for reducing consumption of non-degradable plastics. *Polym Degrad Stab* 133: 174–181.

Barnes DKA, Galgani F, Thompson RC, Barlaz M (2009) Accumulation and fragmentation of plastic debris in global environments. *Philos Trans R Soc B: Biol Sci* 364(1526): 1985–1998.

Bastioli C, Cerutti A, Guanella I, Romano GC, Tosin M (1995) Physical state and biodegradation behavior of starch-polycaprolactone systems. *J Environ Polym Degrad* 3: 81–95.

Bayerl T, Geith M, Somashekar AA, Bhattacharyya D (2014) Influence of fibre architecture on the biodegradability of FLAX/PLA composites. *International Biodeterioration & Biodegradation* 96(2014): 18–25.

Bhuvaneswari. G, Harini (2018) *Recycling of Polyurethane Foams || Degradability of Polymer*, 29–44. Chennai: Elsevier Inc., CIPET.

Blanco I, Bottino FA (2016) Kinetics of degradation and thermal behaviour of branched hepta phenyl POSS/PS nanocomposites. *Polym Degrad Stabil* 129: 3749.

Browne MA, Crump P, Niven SJ, Teuten E, Tonkin A, Galloway T, Thompson R (2011) Accumulation of microplastic on shorelines worldwide: Sources and sinks. *Environ Sci Technol* 45(21): 9175–9179.

Chandra R, Rustgi R (1997) Biodegradation of maleated linear low density polyethylene and starch blends. *Polym Degrad Stabil* 56(2): 185–202.

Cierjacks A, Behr F, Kowarik (2012) Operational performance indicators for litter management at festivals in semi-natural landscapes. *Ecol Indic* 13: 328–337.

Crompton TR (2013) *Thermal Methods of Polymer Analysis*. Shropshire, UK: Smithers Rapra.

do Sul JAI, Costa MF (2007) Marine debris review for Latin America and the Wider Caribbean Region: From the 1970s until now, and where do we go from here? *Mar Pollut Bull* 54(8): 1087–1104.

Du LC, Meng YZ, Wang SJ, Tjong SC (2004) Synthesis and degradation behavior of poly(propylene carbonate) derived from carbon dioxide and propylene oxide. *J Appl Polym Sci* 92: 1840–1846.

Ducker J (2021) Environmental Problems Caused by Synthetic Polymers. sciencing. com, https://sciencing.com/environmental-problems-caused-by-synthetic-polymers 12732046.html

Francis R (2016) *Recycling of Polymers: Methods, Characterization and Applications*. Hoboken, NJ: John Wiley & Sons, ISBN 978-3-527-33848-1.

Fritsche W, Hofrichter M (2008) Aerobic degradation by microorganisms. *Biotechnology: Environmental Processes II*, Vol. 11, WILLEY-VCH Verlag GmbH, D-69469 Weinheim (Federal Republic of Germany), 2000.

Gautam R, Bassi AS, Yanful EK (2007) A review of biodegradation of synthetic plastic and foams. *Appl Biochem Biotechnol* 141(1): 85–108.

Gregory MR (2009) Environmental implications of plastic debris in marine settings entanglement, ingestion, smothering, hangers-on, hitch-hiking and alien invasions. *Philos Trans R Soc B: Biol Sci* 364(1526): 2013–2025.

Grima S, Bellon-Maurel V, Feuilloley P, Silvestre F (2002) Aerobic biodegradation of polymers in solid-state conditions: A review of environmental and physicochemical parameter settings in laboratory simulation. *J Polymer Environ* 8: 4.

Hadad D, Geresh S, Sivan A (2005) Biodegradation of polyethene by the thermophilic bacterium Brevibacillus borstelensis. *J Appl Microbiol* 98(5): 1093–1100.

Heimowska A, Krasowska K, Rutkowska M (2011) Degradability of different packaging polymeric materials in sea water. *Int Polym Sci Technol* 1: 262–268.

Jang BC, Huh SY, Jang JG, Bae YC (2001) Mechanical properties and morphology of the modified HDPE/starch reactive blend. *J Appl Polym Sci* 82(13): 3313–3320.

John J, Tang J, Bhattacharya M. (1998) Processing of biodegradable blends of wheat gluten and modified polycaprolactone. *Polymer* 39: 2883–2895.

King M FL, Robert Singh G, Baruch LJ, Srinivasan V (2020) Moisture absorption and biodegradation of poly lactic acid hybrid composites. *Waffen-Und Kostumkunde J* 11(6): 61–71.

Kliem S, Kreutzbruck M, Bonten C (2020) Review on the biological degradation of polymers in various environments. *Materials* 13: 4586.

Krsek M, Wellington EM (2001) Assessment of chitin decomposer diversity within an upland grassland. *Antonie van Leeuwenhoek* 79: 261–267.

Krzan A, Hemjinda S, Miertus S, Corti A, Chiellini E (2006) Standardization and certification in the area of environmentally degradable plastics. *Polym Degrad Stabil* 91(12): 281–933.

Lim SW, Jung IK, Lee KH, Jin BS (1999) Structure and properties of biodegradable gluten/ aliphatic polyester blends. *Eur Polym J* 35: 1875–1881.

Luinstra G (2008) Poly(propylene carbonate), old copolymers of propylene oxide and carbon dioxide with new interests: Catalysis and material properties. *Polym Rev* 48: 192–219.

Madorsky SL, Straus S (1959) Thermal degradation of polymers at high temperatures. *Journal of Research of the National Bureau of Standords-A. Physics and Chemistry* 63A(3): 261–268.

Nakashima T, Nakano Y, Bin Y, Matsuo M (2005) Biodegradation characteristics of chitin and chitosan films. *J Home Econ Jpn* 56: 889–897.

National Oceanic and Atmospheric Administration Marine Debris Program. Director: Nancy Wallace. In Clean Guide; NOAA 101: Washington, DC, USA, 2018.

Psomiadou E, Arvanitoyannis I, Biliaderis CG, Ogawa H, Kawasaki N (1997) Biodegradable films made from low density polyethylene (LDPE), wheat starch and soluble starch for food packaging applications. Part 2. *Carbohydr Polym* 33(4): 227–242.

Ray S, Cooney RP (2018) Thermal degradation of polymer and polymer composites. *Handbook of Environmental Degradation of Materials*, 185–206. Auckland, New Zealand: Elsevier Inc., The University of Auckland.

Rayne S (2008) The need for reducing plastic shopping bay use and disposal in Africa. *African J Environ Sci Technol* 3(3), eISSN: 1996-0786.

Rose D (2019) Plastic Degradation during Injection Molding, March 26th, 2019. https:// aim.institute/plastic-degradation-during-injection-molding

Rutkowska M, Krasowska K, Heimowska A, Steinka I (2002) Effect of modification of poly(ε-Caprolactone) on its biodegradation in natural environments. *Int Polym Sci Technol* 29: 77–84.

Saad GR, Seliger H (2004) Biodegradable copolymers based on bacterial poly((R)-3-hydroxybutyrate): Thermal and mechanical properties and biodegradation behaviour. *Polym Degrad Stabil* 83(1): 101–110.

Sato K, Azama Y, Nogawa M, Taguchi G, Shimosaka M (2010) Analysis of a change in bacterial community in different environments with addition of chitin or chitosan. *J Biosci Bioeng* 109: 472–478.

Sawaguchi A, Ono S, Oomura M, Inami K, Kumeta Y, Honda K, Sameshima-Saito R, Sakamoto K, Ando A, Saito A (2015) Chitosan degradation and associated changes in bacterial community structures in two contrasting soils. *Soil Sci Plant Nutr* 61: 471–480.

Scott G (1999) *Polymers and the Environment*. Cambridge, UK: Royal Society of Chemistry.

Sharma K, Singh V, Arora A (2011) Natural biodegradable polymers as matrices in transdermal drug delivery. *Int J Drug Dev Res* 3(2): 85–103.

Sharma N, Chang LP, Chu YL, Ismail H, Ishiaku US, Ishak ZAM (2001) A study on the effect of pro-oxidant on the thermooxidative degradation behaviour of sago starch filled polyethylene. *Polym Degrad Stabil* 71(3): 381–393.

Singh B, Sharma N (2008) Mechanistic implications of plastic degradation. *Polym Degrad Stab* 93: 561–584.

Soares EJ (2020) Review of mechanical degradation and de-aggregation of drag reducing polymers in turbulent flows. *Journal of Non-Newtonian Fluid Mechanics* 276: 104225.

Taherzadeh M, Karimi K (2008) Pretreatment of lignocellulosic wastes to improve ethanol and biogas production: A review. *Int J Mol Sci* 9: 1621–1651.

Tharpes YL (1989) International environmental law: Turning the tide on marine pollution. *U Miami Inter-Am L Rev* 20(3): 579–614.

Thompson RC, Moore CJ, vom Saal FS, Swan SH (2009) Plastic, the environment & human health: Current consensus & future trends. *Philos Trans Royal Soc* 364(21): 53–66.

Torres FG, Troncoso OP, Torres C, Díaz DA, Amaya E (2011) Biodegradability and mechanical properties of starch films from Andean crops. *Int J Biol Macromol* 48: 603–606.

Tsuji H, Suzuyoshi K (2002) Environmental degradation of biodegradable polyesters 2. Poly(ε-caprolactone), poly[(R)-3-hydroxybutyrate], and poly(L-lactide) films in natural dynamic seawater. *Polym Degrad Stab* 75: 357–365.

Vaverková M, Toman F, Adamcová D, Kotovicová J (2012) Study of the biodegrability of degradable/biodegradable plastic material in a controlled composting environment. *Ecol Chem Eng S* 19: 347–358.

Williams AT, Simmons SL (1999) Sources of riverine litter: The River Taff, South Wales, UK. *Water Air Soil Pollut* 112(1–2): 197–216.

Zhao YQ, Cheung HY, Lau KT, Xu CL, Zhao DD, Li HL (2010) Silkworm silk/poly(lactic acid) biocomposites: Dynamic mechanical, thermal and biodegradable properties. *Polym Degrad Stab* 95: 1978–1987.

13 Rheological Studies of Biodegradable Composites

Gourhari Chakraborty
National Institute of Technology Andhra Pradesh

Sayan Kumar Bhattacharjee and Vimal Katiyar
Indian Institute of Technology Guwahati

CONTENTS

DOI: 10.1201/9781003227908-13

13.1   INTRODUCTION

Polymer is a major component for most of the conventional packaging materials in application for short-term and long-term preservation of goods. Application domain covers almost all the fields including packaging, biomedical, structural, paint and electrical [1–3]. Commodity plastics from petrochemical feedstock are non-degradable in nature, resulting in solid waste management problem and thus causing life-threats for living organisms. It takes several decades for the decomposition of polymeric wastes because of non-degradability under microbial attack. In view of environmental sustainability and depletion of fossil fuel-based feedstock, the present trend of polymer technology is shifting towards bio-based and biodegradable polymeric systems [4].

Biodegradable polymers have become obvious replacement of plastic goods; however, poor processability, low melt strength, less thermal stability and lower mechanical strength limit their applicability in real fields. In recent years, developments in the field of polymer technology have been taking place addressing these drawbacks either through development of novel nanomaterial-reinforced composites or by introducing new technology for processing polymeric substances. Biodegradable materials pose additional advantages such as less greenhouse gas emissions, bio-base, less toxicity, comparable strength and modulus and biocompatibility. Bio-based composites thus are application-specific, sustainable and eco-friendly materials and possess a wide variety of applications, which makes the product economically viable [5].

Biodegradable composites are mostly synthesized in three different routes, i.e. solution processing, melt processing and in situ polymerization. Solution casting is mostly carried out for thermally less stable polymers and hydrophilic polymers. Targeted applications are drug delivery, implants, tissue culture, suture, etc., whereas in order to do melt processing, different techniques such as extrusion, moulding, thermoforming and melt mixing of composites are carried out for different composites based on filler, matrix nature and application. Reinforcement filler can be of different types based on properties and application perspectives such as particulate and fibre (aspect ratio), inorganic and organic (chemical structure), and conductive and non-conductive (based on conductivity). Biocomposite properties are dependent on the nature of matrix, nature of filler, interfacial interaction between matrix and filler and nature of the processing. In order to prepare perfect biocomposites, the interfacial compatibility between matrix and filler should be good [6–9]. Rheology investigation is a tool by which the nature of composite, dispersion of filler within matrix, processability, melt strength and ability of moulding can be studied. Molecular weight deterioration is a limitation for biodegradable polymers during melt processing. In order to track the nature of the degradation, mechanism of degradation, extent of degradation and rectification of molecular weight deterioration can be examined using melt rheology. Compatibility between matrix and reinforcement determines different properties such as mechanical strength and thermal stability of composite materials, and the compatibility can be improved by grafting, wrapping, cross-linking and using compatibilizer.

This processing technique sometimes modifies chain relaxation behaviour; sometimes, it destroys the homopolymeric nature of matrix that can be properly inferred by melt rheology behaviour study. Similarly, the impact on melt strength due to the addition of nanomaterial in the biodegradable polymers can be studied by characterizing flow behaviour of the composite. In recent years, master batch dilution technique has come out as a facile technique for uniform dispersion of filler in polymer matrix. In this case of successful fabrication, composite can be studied and different processing parameters can be optimized through rheology investigation. In case of hydrophilic polymers such as cellulose and chitosan, gel-forming ability can be seen using rotational flow investigation. In case of polymeric blends, paints and coating application impact of rheology is enormous [10,11].

Through this chapter, a discussion about the impact of rheology investigation on biodegradable composites is carried out. Instrumentation used for different flow behaviour investigations including rheometer and viscometer are detailed. A comprehensive review of the different rheological investigations is carried out for different biodegradable composites, including solution viscosity, melt rheology investigation and use of model fitting for biodegradable composites. This chapter is concluded by highlighting summary and outlook at the end of the chapter.

13.2   RHEOLOGY AND BIODEGRADABLE COMPOSITES

Rheology is a powerful tool and is attaining much attention because of the ability to analyse various properties related to filler–matrix interaction, and it may provide information on the deformation of polymer due to the impact of fillers, impact of filler orientation, nature of dispersion of nanomaterial within polymer matrix and structure–property correspondence in biodegradable composites [12]. In addition to this, rheology investigation also provides information related to the processability of any polymer matrix and filler combination. Different processing characteristics through rheological investigation of these composites can lead to information about fillers loading, temperature and screw speed for the fabrication of final composite materials [13].

Different aspects of the matrix–reinforcement interactions of composites can be understand using rheological data. Oscillatory mode experiments including frequency sweep and strain sweep using fix plate geometry provide information on the complex viscosity, storage and loss moduli, phase angle and polymer chain relaxation. The nature of complex viscosity in varying frequency plots indicates shear thinning behaviour of the biocomposites along with the extent of adherence to power law. The nature of interaction of polymer filler, i.e. miscibility and dispersion, can be visualized from the Cole–Cole plots (a plot between imaginary and real viscosity) and the Han plot (a plot between storage and loss moduli). In addition to this, molecular weight and impact of processing on distribution can be known from the data of the crossover point, which is the value of frequency at which the loss modulus (G'') is the same as the storage modulus (G'). Nonlinear flow behaviour of storage modulus in low-frequency domain indicate the possible network formation within the composite in the melt state. The impact of temperature on the flow behaviour, i.e. on G', G'' and η, can be determined, and flow activation energy of polymer matrix and nanocomposites can be investigated. Through melt rheology investigation and analysing rheological data processability and operation condition optimization of different melt processes can be done for a particular biodegradable polymer matrix and filler [14].

Along with normal homopolymeric composite system processing different composite systems with chain modifiers such as cross-linker, plasticizer and compatibilizer can be studied using rheology tool. In case of composite containing polymer blend as matrix, rheological behaviour gives variation in characteristic parameters, which are proportional to the nature of the blend as well as filler–polymer interaction. The extent of cross-linking and change in chain relaxation behaviour can be examined from rheology data for different cross-linking systems (peroxide linker for reactive extrusion; chemical linker for gel) [15].

Rheology investigation for polymeric solution using rotational field gives information about the nature of flow of polymeric solution, i.e. shear stress (τ) with respect to shear rate ($\gamma°$) and viscosity (η) with shear rate. These investigations are significant for cellulose, chitosan, protein and other hydrophilic polymers, which are capable of gel formation. Rheology investigation generates information about the flowability and gel point of these polymer, which helps in the optimization of polymer loading and gel preparation parameters [16]. Rheological investigation of magnetorheological polymeric solutions gives an indication of applicability of these polymers in the arena of ferrofluid [17].

Rheology investigation also provides information about the molecular weight alteration of matrix due to the incorporation of different nanofillers, chain modifiers and processing conditions. Different deferent rheology parameters give information about the deformability, melt strength, nature of flow under shear, yield stress and percolation point of a composite system. Rheology information reveals the effectivity of a composite and suitability of the processing path (Figure 13.1).

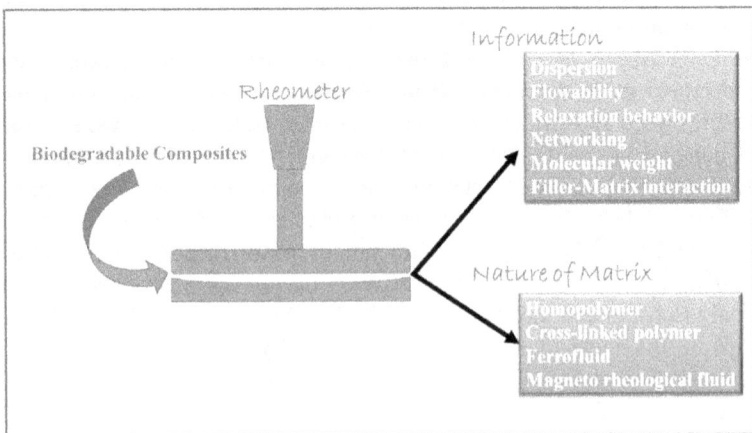

FIGURE 13.1   Impact of rheology study on biodegradable composites.

13.3 INSTRUMENTATION OF DIFFERENT FLOWABILITY INVESTIGATIONS

13.3.1 Rheometer

Flowability and flow nature of a liquid, suspension or slurry under shear force are investigated or measured using laboratory device rheometer. In case of some materials, it is not possible to define the actual flow behaviour using normal viscosity calculation. This is also used to study the behaviour of polymer in molten state (melt). Rheometers can be of two types based on the application of force on the fluid: rotational rheometer, in which measurements are carried out under controlled applied shear stress or shear strain, and extensional rheometer, which operates under extensional stress or extensional strain. Depending on the nature of the study, oscillatory rheology investigation can be of two types: (i) strain sweep (where angular frequency ω is kept constant and investigation is carried out in variation of strain%) and (ii) frequency sweep (where strain% is kept constant and measurements are carried out by varying angular frequency ω). In case of rotational rheology operation, either measurement is carried out in varying torque or at a constant torque within a time interval [18].

13.3.1.1 Shearing Geometries

a. **Cone/plate (CP) or cone–plate measuring systems:**
CP systems are suitable for all types of fluids; however, particle size determines their applicability to dispersions investigations. According to the ISO standard, a cone angle of $\alpha = 1°$ is used; cones with an angle of more than $4°$ are considered to be sub-standard. For precise measurements, CP systems with truncated cone tips are the systems of choice.

b. **Concentric cylinder (CC) measuring systems:**
CC systems are commonly used for tests on low-viscosity liquids. There is a restriction on the radius ratio of cup and bob, and it must not exceed 1.0487.

c. **Parallel plate (PP) measuring systems:**
PP systems are mostly used for different polymeric systems including polymeric melts, solution and gels. The gap widths between the plates are kept in the range of 0.5–1.0 mm. Depending on the type of investigation and the nature of fluid, other parameters are set (Figure 13.2).

13.3.2 Viscometer

The flow property of biodegradable polymers is determined from the measurement of viscosity using an instrument called a viscometer. Viscosity is a measure of resistance towards flow of fluid due the application of surface drag force. The measurement is carried out in the laminar flow condition. Depending on the solvent, composition and concentration, viscosity is different for different polymeric systems [19].

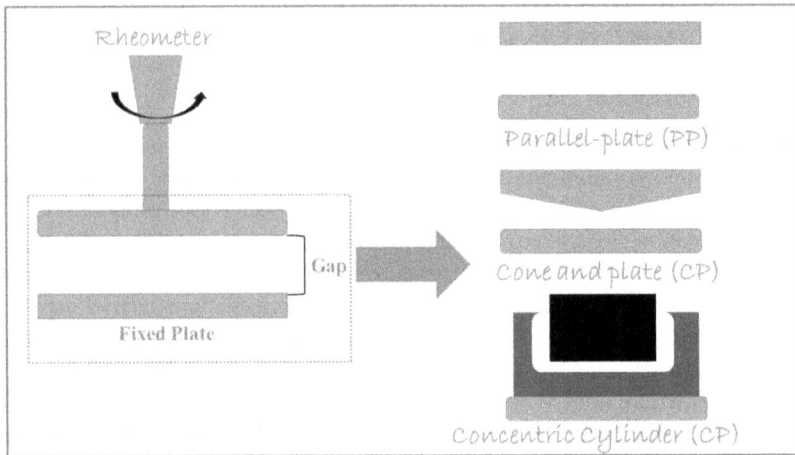

FIGURE 13.2   Different types of rheometer [18].

In some cases, for industrial production of composites, melt flow index (MFI) is measured. MFI is measured on the principle that the amount of polymer passed through the capillary in 10 minutes under the application of a predefined load at a particular temperature. In case of maintenance of polymer quality and composite consistency, MFI is measured. MFI is an indicator of flowability, and it depends on the type of polymer, loading of filler and temperature.

13.4   RHEOLOGY STUDIES OF BIODEGRADABLE COMPOSITES

13.4.1   Solution Viscosity

Hydrogels and organogels have high commercial application in the areas such as drug delivery, adsorption, tissue culture, food and cosmetics. Cellulose, chitosan, gelatine and different derived cellulose materials were used as a matrix for the preparation of composite gels for different applications. In order to determine the effectivity of gel and consistency of the prepared gel solution viscosity with concentration, pH and shear rate are measured. Cellulose nanoparticle-based hydrogels were synthesized by different groups, and the gelation behaviour was studied. Different cross-linking methods are methods used for the preparation of gels. Gel stability depends on the strength of cross-linking. Physical cross-linking, the use of non-solvent and the use of surfactant can create weak cross-linking bonds, whereas the use of chemical cross-linker and cross-linking through irradiation can provide strong cross-linked gels. Rheology investigation can give information about the nature of binding as well as the stability of nanocomposite gels made of biodegradable polymers. Structure rheology investigation of cellulose nanocrystals (CNCs) suspension through viscosity measurement was carried out using water as media. The viscosity of CNC suspensions was observed to gradually increase with rise in the concentration, which is possibly due to the growth in the collision of CNCs. The viscosity of suspension was observed to be dependent on CNC concentration, shear rate and aspect ratio of CNC [20]. Switch grass and cotton fibres-based CNC for gel preparation was investigated by [21]. It is reported that an increase in the CNC content increased the storage and loss moduli and the viscosity of the suspension. Aspect ratio and loading are found to be the controlling parameters for the gel point, and the presence of two critical points was noticed. Similar investigations were carried out for cellulosic gel of carboxymethyl cellulose in aqueous media in the presence of HCl and NaOH by Lopez et al. (2021). The stability of the gel was inspected taking different samples after 2 weeks and after 8 months. It was observed from the oscillatory rheology investigation that the prepared gel of concentration over 5 wt.% loading was stable within the duration. However, in case of 4 wt.% gel loss in gel strength was noticed around 30% within same time interval. Large-amplitude oscillatory shear response was observed to be positive for the gels. The addition of NaCl promoted the cross-linking with the increase in salt concentration; storage and loss moduli were found to be increasing in nature. The addition of NaOH decreased the gel formation, whereas HCl was found to have positive impact on gel formation [22]. Rheology investigation of gel-based inks was carried out in order to check the product consistency by several groups. A cellulose–aloe vera-based biogel was synthesized to develop 3D print of products by direct ink writing technique. In this case, product flow properties were studied using parallel plate (PP) and cone and plate (CP) rheometer. In case of ink application, high viscosity of gels is required for extrusion. From this investigation, it was found that ink possessed viscosity around 2800–4400 mPa s^{-1} at a shear rate of 0.01 s^{-1} depending upon the cellulose loading and the values were comparable to bacterial cellulose/alginate gel-based inks [23]. Similarly, oscillatory rheology investigation was carried out by Markstedt et al. (2017) for cellulose and cross-linkable xylan-based biomimetic hydrogels. Storage and loss moduli were found to be increasing with an increase in cellulose loading between 2 and 3 wt.% [24]. In some cases, rotational rheology investigation was carried out to derive information about product consistency. Alginate/methyl cellulose-based pastes' consistency upon sterilization was investigated by using viscosity measurement within a shear rate 0–100 s^{-1}. It was observed that for all sterilization techniques there was no impact on the product shear thinning profile. However, in case of γ-radiation viscosity was found to be lower and almost identical within shear rate range [25]. Skin penetration properties of cellulose–ether hydrogel and the impact of viscosity were studied by Binder et al. (2019). In this particular report, the pseudo-plastic nature was investigated within 1–100 s^{-1} shear rate [26]. Other studies such as cation-induced gelation of cellulose were carried out by Ju et al. [27]. Rheology investigation was also carried out in case of chitosan-based hydrogels. Adhikari et al. (2021) developed a chitosan-based nanocomposite hydrogel introducing hydroxyapatite and alginate along with chitosan for biomaterial ink. Rheology investigation was carried out within 0–1000 s^{-1} shear rate. The highest viscosity was noticed for 5% alginate with 2% chitosan and 0.4% hydroxyapatite composition. Phase angle was found to be decreased for the composite gels [28]. Chitosan-based porous hydrogels were synthesized dissolving in different organic solvents such as acetic acid, formic acid and lactic acid. Rheological investigations of the gels were carried out in oscillatory mode using 1% stain amplitude. It was observed that viscosity is dependent on acid concentration, type of acid and chitosan concentration. Dynamic viscosity was highest in case of acetic acid-based gel, i.e. 3.72 Pa.s. Porosity and pore volume were noticed to be higher for acetic acid-based gels [29]. Similar investigations were carried out in a chitosan–gelatine-based hydrogel with the incorporation of ZnO by Karakus et al. [16] and in a thermosensitive drug-loaded chitosan-based gel by Gholizadeh et al. [30].

Polymeric coating nowadays is one of major applications of biopolymers and biopolymeric composites. Spreadability, thickness and coating uniformity of polymeric coating are dependent on various rheological parameters. Different coating conditions and unit operations are based on pumping, filling and spraying, which are dependent on the flowability of the composite solution. High apparent viscosity is desirable for better film formation through dip coating at low shear rates [31]. In another case of edible coating was carried out using biodegradable polymer chitosan and gelatine and pomegranate peel extract (PPE) was incorporated into it to valorize agro-waste [32]. Rheology investigation indicated increments of viscous behaviour due to the incorporation of PPE, which is favourable for fruit coating. PPE interaction with polymer mixture was observed to be weak through frequency sweep measurements. The introduction of the chitin nanofibrils with high anisotropy was observed to alter the rheological nature of chitosan solutions significantly [33]. Solid-like behaviour of chitosan/chitin slurries and the flow instabilities were noticed during rheology investigation of the same. It showed rubber-like properties as dynamic modulus turned out to be independent of frequency. The self-assembly processes of the chitosan-based solutions were elucidated by measuring the time-dependent nature of dynamic moduli up to reaching the gel point. The rheological behaviour of the polymeric solutions is described by physical networks, which were tough and elastic in nature.

Solution viscosity was also used in the determination of viscosity average molecular weight of biodegradable polymers during new synthesis and sometimes in order to study the impact of processing. In some of the applications where the solution of polymer is used viscosity average molecular weight appears to be more relevant to the properties of polymer. Pan et al. (2018) correlated stereo-complexation of melt-drawn PLA fibres with viscosity average molecular weight [34]. The molecular weight of PLA was measured through viscosity calculation by Liu et al. [35]. The impact of recycling of PLA/silk nanocomposite was measured by the calculation of intrinsic viscosity using chloroform solvent by Tesfaya et al. (2017). It was noticed that during each recycling time intrinsic viscosity dropped significantly, which indicates the possible degradation of PLA chains during extrusion in each recycling [36]. Thus, different solution-based rheology parameters, particularly viscosity, can give information about the consistency of production, structural nature, flow nature, optimum synthesis condition and effectivity in applications of biodegradable polymer-based composite solutions, gels, inks and pastes (Figure 13.3).

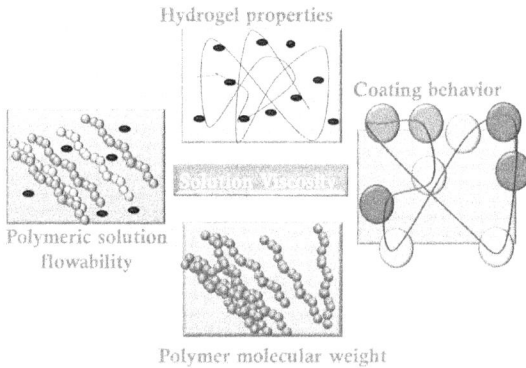

FIGURE 13.3 Applications of solution rheology in the field of biodegradable polymeric composites.

13.4.2 MELT RHEOLOGY

Industrial applications of polymers need different melt processing operations during the synthesis of composite as well as during the application-oriented object casting. Biodegradable polymeric composite production efficiency also depends on proper selection of processing condition. As melt strength, dispersion of filler and deteriorations of molecular weight are the prime factors that determine the process cost-efficiency and composite quality.

13.4.2.1 Melt Flow Properties of Composites

Rheological properties of polymeric melts such as storage modulus (G′), loss modulus (G″) and complex viscosity (η) give information on the flowability and flow nature of the composites. Composite loading was also determined from the optimum values of the rheology parameters. Shojaeiarani et al. (2019) investigated different flow properties of melt PLA/CNC nanocomposites with different loadings of CNC. Rheology study was conducted in oscillatory mode using parallel plate geometry at 170°C. The incorporation of CNC increased the viscosity of the nanocomposite for both the fabrication pathways (film casting and spin coating). G′ and G″ were found to be increasing with respect to the CNC loading; it indicates the possible increase in the melt strength and chain entanglements due to the addition of CNC within PLA matrix [37]. Recycling of polymer is a major domain that can reduce the cost of production and will reduce the disposal problem. It is a challenge to keep intact the original properties of polymer even after reprocessing. The recycling ability of biodegradable polymer can also be examined by measuring the rheological parameters of different recycle products. In case of silk-incorporated PLA nanocomposites, melt extrusion at 200°C was carried out and recycled for four times. The effectivity of recycling was investigated using different characterization techniques, and changes in rheology properties of the composites were also inspected. Lowering of crossover was noticed for NPLA and SNC/NPLA composites with the increase in recycle number possibly due to the degradation of molecular weight and variation in polydispersity. Storage modulus was found to decrease due to recycling for NPLA (710 kPa) to 3 times recycled NPLA (670 kPa). However, in case of SNC reinforced storage modulus almost similar for all the recycled composites. Thus, in this study rheology investigation implies suitability of SNC as a filler for PLA products where the product can be recycled [36]. Thus, for recasting of polymer composites adequate melt strength is required for the maintenance of actual shape for respective application. High melt strength also increases the cost of operation. So, in case of melt casting, proper information of rheology parameters is needed for the synthesis of cost-effective composite. Similar investigation was conducted by Lee et al. (2019) for clay-reinforced composites of PLA-based biodegradable polymeric systems. In this study, PLA-based blends were used as matrix with 20% incorporation of poly(ε-caprolactone) (PCL), poly(butylene adipate-co-terephthalate) (PBAT) and poly(butylene succinate-co-adipate) (PBSA) into 80% PLA separately. Different Cloisite clays (Cloisite 30B and Cloisite 20A) were used as reinforcement for melt mixing with PLA and PLA-based blends. Rheology study was conducted in oscillatory mode at the processing condition of 190°C. It was observed that with the incorporation of the modified clays inside the blend systems, G' and G'' increased. It was also recorded that for 7 wt.% loading of Cloisite 30B clay, the increase was higher compared to others. An increasing trend was also noticed for complex viscosity with the loading of clay [38]. Similarly, the rheology investigation of PLA/PBAT blend was carried out and the effect of multi-walled carbon nanotube (MWNT) on the structure of the blend system was also investigated for composites. G' and G'' increased with an increase in the MWNT loading. Complex viscosity was observed to be dependent on the loading as well as the nature of polymer matrix [39]. In some of the investigations of solution processed polymeric film, melt rheology was investigated in order to check the impact of filler on the melt strength and flowability of bio-nanocomposites [40].

13.4.2.2 Dispersion of Reinforcement

The impact of reinforcement of filler sometimes alters the homopolymeric nature of matrix prepared through melt processing route. Through rheology investigation, it is possible to understand the nature of dispersion of filler within the polymer matrix. In some cases where different modified techniques are utilized prior to master batch dilution technique through melt processing, the effectivity of the process on the dispersion and flowability of matrix polymer chain can be studied. The dispersion of graphene in PLA matrix was studied using melt processing, and in this study, master batch of graphene was coated over the PLA pellets prior to extrusion [14]. Miscibility and dispersion were observed to be improved in case of master batch-coated nanocomposites from Han plot and Cole–Cole plot (Figure 13.4). The incorporation of master batch-coated graphene into PLA matrix was found to improve the flowability of the composites. The homopolymeric nature was not altered due to the nanocomposite fabrication. These effects

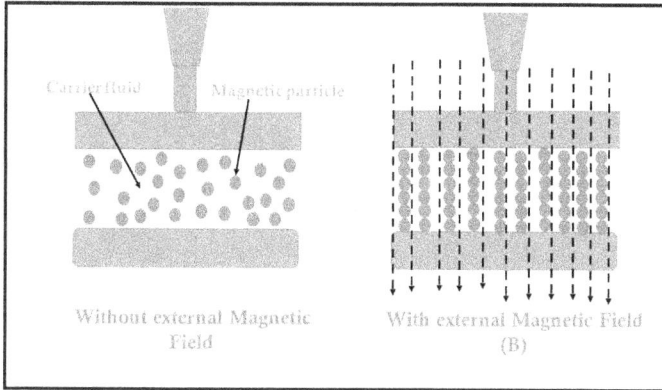

FIGURE 13.4 Alignment of magnetic nanomaterials inside a magnetorheological fluid [49].

are important issues during large-scale PLA film production. Rheology properties of PLA/chitosan-graft lactic acid oligomer (CH-g-OLLA) bio-nanocomposite films were studied in order to check the flow properties as well as the nature of dispersion of chitosan in the PLA matrix [41]. It was observed that G′ and G″ increased with ω when oscillatory rheology investigations were carried out for the composite melts at 180°C. However, the viscous nature of the composite melt was noticed from the higher value of G″ with respect to G′ possibly due to the inclusion of short-chain oligomers along with chitosan. Newtonian flow characteristics for PLA and PLA–CH-g-OLLA (1%) films were noticed up to angular frequency 0.3 rad s^{-1}. Shear-thinning flow behaviour was detected in the film samples.

13.4.2.3 Cross-linking, Plasticizing Effect and Compatibility Investigation

Application of melt rheology is further extending to processing to the composites where along with filler different chain-modifying substances are also added. The addition of polymer chain-modifying additives such as cross-linker, plasticizer or compatibilizer changes the properties of composites as well as alters the flow behaviour of melt. Most of the cases of cross-linking peroxides are used as an initiator for different biodegradable polymeric composite systems during melt processing. Cross-linking increases the modulus remarkably compared to the homopolymeric system. Biodegradable composites with polybutylene succinate (PBS) matrix and rice straw (RS) reinforcement were melt extruded into strip in the presence of cross-linking agent dicumyl peroxide (DCP) [42]. Rheological analysis was carried out in oscillatory mode using parallel palate arrangement. It was observed that G′ is less compared to G″ over the frequency range, which implies the viscous nature of the composite melts (Figure 13.5). The addition of RS (for both 5 wt.% and 10 wt.% RS) in the PBS matrix has led to the increment of both G′ and G″ with ω. The increment in the modulus was even higher with the incorporation of DCP along with RS into the PBS system. Shear-thinning

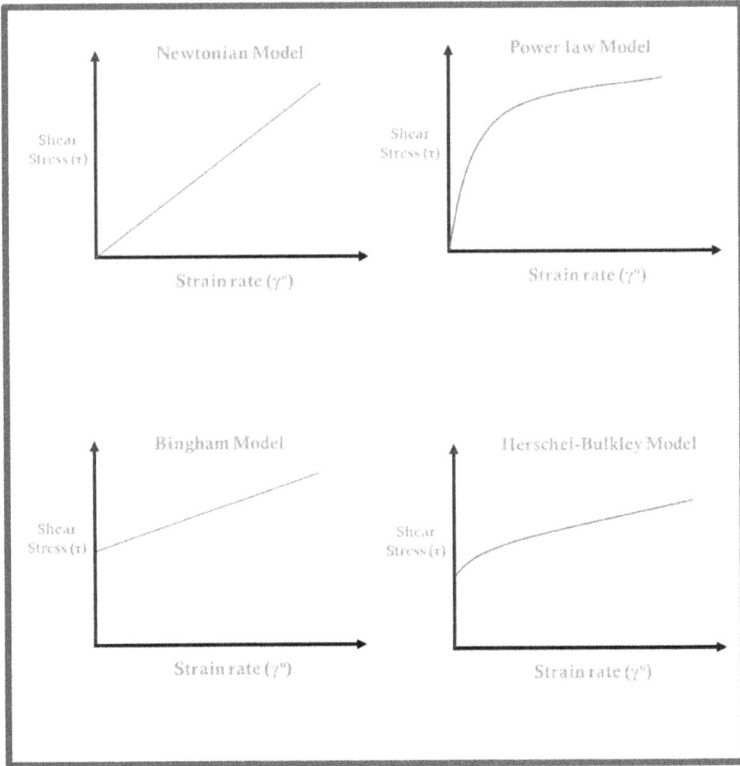

FIGURE 13.5 Different flow models of biodegradable polymeric composite fluids [53].

behaviour was noticed from the values of 'η' for the composite melts with different RS content. Crossover point was found to be shifted towards lower value for the PBS/RS composites, which is possibly due to the loss in molecular weight during melt processing. The presence of polydisperse system and cross-linking were observed from Han plot, Cole–Cole plot and phase angle analysis. The impact of cross-linking agent on the reactive extrusion of modified SNC-based PLA composite was studied through rheological parameters measurement by Melaku et al. [15]. It can be seen that rheological properties remarkably improved on the addition of grafted SNC and in the presence of cross-linking agent dicumyl peroxide (DCP); further, solid-like behaviour was noticed. Crossover point shifted towards low frequency possibly due to the presence of long-chain branching and cross-linking. Complex viscosity was found to be increasing with the incorporation of cross-linker agent into the PLA matrix. The Cole–Cole plot and Han plot indicated deviation from polymeric structure because of cross-linking, branching due to the addition of DCP into PLA/SNC nanocomposite during reactive extrusion. Plasticizers are also in some cases useful to improve the melt flow property for composites. The incorporation of plasticizer reduces flow activation energy and thus improves melt flowability. Additives such as plasticizers reduce the viscosity of melt and enhance the chain mobility. The incorporation of coconut oil (CO) into PLA matrix acted like plasticizer [43]. It is reflected from the lowering of complex viscosity at various loading values of coconut oil. Non-Newtonian behaviour of melts at higher frequencies suggested shear-thinning behaviour of the melts. This investigation implies the possibility of coconut oil as effective plasticizer in future composite fabrications. The implication of rheology investigation is very high in case of biodegradable polymeric blends. Similarly, ternary composite fabrication efficiency was also investigated using melt rheology parameters. Eslami et al. (2012) studied extensional rheology behaviour of clay-reinforced nanocomposite blend of PLA and PBSA. With the incorporation of Cloisite 30B clay into the PLA/PBSA blend complex viscosity and modulus were found to be increased. This implies increment of solid-like behaviour for the ternary nanocomposite in incorporation of clay [44].

Thus, different scientific investigations related to new composite fabrication need proper flow study for proper optimization of processing conditions, for obtaining information about dispersion nature and particularly for modified filler-based composites and indication of extent of impact on chains by chain modifiers.

13.4.3 Magnetorheological Fluids (MRF)

The flow behaviour of composite solution or melt under external magnetic field can be investigated using magnetic rheometer whereby altering current magnetic field strength can be applied during flow measurement. Magnetorheological materials are able to change their properties depending upon the strength of the magnetic field; particularly, shear stress–strain rate flow is going to alter based on the loading of magnetic nanomaterial within the fluid and field strength. MRF, elastomers and foams are different forms of magnetic rheological material [45].

MRF application covers various high-performance areas, namely shock absorbers, clutches, vibration dampers, control valves and artificial joints. Biomedical applications, accurate polishing, sound, magnetic advection in isothermal condition, and use as chemical sensing material are some of the other applications of such materials.

The elements constituting MRF are magnetic particles, carrier fluid and additives. Biodegradable polymers such as cellulosic materials are mostly used for the preparation of magnetic nanomaterial with high magnetism. The incorporation of these nanomaterials inside the carrier fluid such as silicon oil acts as ferrofluid. The addition of other additives generally stabilizes the nanoparticles inside the carrier fluid and prevents sedimentation, particularly in higher loading. Nowadays, utilization of biodegradable polymeric solution and melt as carrier fluid is the subject of extensive research. MRF was mostly prepared in two steps, namely solid phase and liquid phase. In solid phase, magnetic nanomaterials are synthesized, and in some cases, coating of these materials with biodegradable polymers such as guar gum is carried out. Then in the liquid phase, these solid magnetic particles are incorporated into carrier fluid by using sonication or homogenizer.

Sometimes, additives are used to stabilize the fluid, particularly in higher loading of filler. In some cases, during the casting of polymeric composites external magnetic field is used in order to prepare oriented magnetic material-reinforced nanocomposite.

It can be seen that for MRF the shear stress (τ) increases with the rise in shear strain rate ($\gamma°$). The yield stress which is a measure of deformation under zero strain rate of each MRF is obtained from 'τ–$\gamma°$' curve. This fluid exhibits shear-thinning behaviour, and with an increase in loading (φ), viscosity increases and in other way in increase in magnetic field (B) viscosity and dynamic field strength increases.

Utilization of nanocelluloses in the preparation of MRF is explored much. Different cellulosic materials such as CNC, cellulose nanofibre (CNF) and carboxymethyl cellulose (CMC) were used as thickener for the stabilization of carbonyl iron particles (CIPs) in MRF [46]. The presence of CNC and CNF because of their anisotropic property hinders the sedimentation of CIPs through the physical hindrance and electrostatic repulsion. Besides, those cellulosic nanoparticles based on their aspect ratio restrain the formation of structured CIP chains and thus endow the systems with remarkable thixotropic flowability. CNF at a comparative loading (0.3 wt.%) enhanced the magneto-responsive nature of MRF because of the formation of percolation network of CNF, which was observed to act as a stress amplifier. It was also noticed that CNC-based stabilization of MR that percolation network formation was not evident at 0.3 wt.% loading. Microwave-assisted synthesized magnetic nanoparticles were also used in the preparation of MRF, where CNFs were introduced as stabilization substance [47]. The rheological properties were found to be dependent on applied magnetic field strength and loading for MRF and CNF–MRF, and the values were greatly improved due to the formation of aligned network of magnetic particles under magnetic field. CNFs were found to improve the suspension of magnetic fluid; thus, the viscosity of CNF–MRF was observed to be higher compared to MRF under the application of external magnetic field. It was observed that after 100 cycles, the viscosity value was stable and reversible rate was 95.57%. These study results open up the possibility of use of bio-based degradable polymers such as CNFs as additive for the preparation of stable magnetorheological fluid. Rice husk-based microcrystalline cellulose (MCC) was also utilized as a stabilizer for MR suspensions, and similar flow property was recorded [48].

Utilization of biodegradable polymers can also be extended for the preparation of template-based magnetic material preparation, which itself can act as a magnetic material. The prospect of polymers as carrier liquids such as hydrogel or polymeric solution is also a subject of current research. As external magnetic field can create heating effect for magnetic packaging materials, the impact of magnetorheological melts is a subject of current research.

13.4.4 RHEOLOGICAL MODEL

Experimental investigations are explained using different theoretical models in order to get information about molecular-level changes under different flow conditions.

For theoretical investigation of using different models, the behaviour of shear stress (τ) in various strain rates (γ°) is examined and, from intercept, dynamic shear stress is calculated, which is a measure of resistance towards the deformation in no-flow condition. Thus, using different theoretically available models and corroborating with the experimental findings, the nature of non-Newtonian flow can be predicted. There are some models available where the viscosity can be expressed as a function of loading of filler (viscosity model). Some models used to predict polymeric system, particularly biodegradable nanocomposites, are as follows.

13.4.4.1   Power-Law Model

It is the most used model for the investigation of non-Newtonian fluid, which was proposed by Ostwald and de Waele [50] and has the following general form:

$$\eta = A\omega^n \tag{13.1}$$

where $A = A_o e\left[\dfrac{\Delta E}{R}\left(\dfrac{1}{T} - \dfrac{1}{T_0}\right)\right]$. A and η are represented as pre-exponential factor and flow index.

13.4.4.2   Herschel–Bulkley (HB) Model

This model is used for polymeric systems where for initial flow, yield stress is required.

$$\tau = \tau_y + k\gamma^n$$

And the viscosity can be calculated as

$$\eta = \tau_0/\gamma + k\gamma^n$$

Here τ is the shear stress and it is dependent of nanocomposite fluid composition and strain rate. τ_s is the yield stress, and it is a fluid internal property that depends on the polymer as well as loading of composite. k is a measure of resistance towards flow, and n is the flow index. $n > 1$ indicates shear-thickening behaviour, i.e. increase in viscosity with strain rate, which is normally not seen for biodegradable nanocomposite-based solutions and melts. It gets transformed into **Bingham plastic model** when $n = 1$. In most of the biodegradable nanocomposite-based systems, $n < 1$, which indicates the shear-thinning behaviour. When $\tau_0 = 0$, it reduces to power-law model [51].

13.4.4.3   Casson Model

It is also similar to that of HB model. Only the dependency of shear stress on strain rate is different; other terms have similar significance [51].

$$\tau^{1/2} = \tau_o^{1/2} + \eta_\infty^{1/2}\left(\gamma^o\right)^{1/2}$$

There are some viscosity models used in case of polymeric suspensions, such as [52]

Einstein model $\dfrac{\eta}{\eta_o} = 1 + 2.5\phi$

Brinkman model $\dfrac{\eta}{\eta_o} = \dfrac{1}{\left(1-\phi\right)^{2.5}}$

Here, η is the viscosity of composite, η_o is the viscosity of the composite suspension without filler, and φ is the wt.% of the filler within suspension.

There are some other models also available for the prediction of polymer flow property based on the nature of polymer and aim of investigation. These mechanical models and viscosity models can also be utilized for the prediction of magnetorheological behaviours of MRF.

13.5 SUMMARY AND OUTLOOK

This current chapter delineates the importance of biodegradable polymers and their composites where rheology plays an important role in gathering information on the extent of polymer–filler interactions and structure–property dependencies in polymer biocomposites. It is an important tool in understanding the viscoelastic properties of polymers. Next, we will be familiarized with the different instruments used for measuring different rheological properties. A detailed literature review of different biocomposites is summarized along with their rheological properties. Recently, a new class of materials are gaining importance, i.e. magnetorheological fluid, which is stated thoroughly in this chapter, and their detailed rheological characterization (magnetic rheology) is focused. Different rheological models are also stated in this chapter for obtaining the viscosity profile and the curve fitting parameters.

Future polymeric materials are approaching towards biodegradable polymers based on the view of environmental sustainability. In view of this, more precise processing and cost-effective production of such materials are also needed, which indicates the importance of measurement of rheology parameters, particularly melt rheology. Similarly, solution processing such as gel drawing, coating and electrospinning is dependent on solution flowability. In this respect, more investigation involving improvement in properties is highly required. More applications are need to be studied for biodegradable nanocomposite-based MRF for the development of sustainable ferrofluid.

REFERENCES

1. Rhim JW, Park HM, Ha CS. 2013. Bio-nanocomposites for food packaging applications. *Progress in Polymer Science* 38:1629–1652.
2. Reddy MM, Vivekanandhana S, Misra M, Bhatia SK, Mohanty AK. 2013. Biobased plastics and bionanocomposites: Current status and future opportunities. *Progress in Polymer Science* 38:1653–1689.

3. Chakraborty G, Katiyar V, Pugazhenthi G. 2021. Improvisation of polylactic acid (PLA)/exfoliated graphene (GR) nanocomposite for detection of metal ions (Cu²⁺). *Composites Science and Technology* 213:108877.

4. Ojijo V, Ray SS. 2013. Processing strategies in bionanocomposites. *Progress in Polymer Science* 38:1543–1589.

5. Auras R, Harte B, Selke S. 2004. An overview of polylactides as packaging materials. *Macromolecular Bioscience* 4:835–864.

6. Chakraborty G, Valapa RB, Pugazhenthi G, Katiyar V. 2018. Investigating the properties of poly (lactic acid)/exfoliated graphene based nanocomposites fabricated by versatile coating approach. *International Journal of Biological Macromolecules* 113:1080–1091.

7. Chakraborty G, Pugazhenthi G, Katiyar V. 2019. Exfoliated graphene-dispersed poly (lactic acid)-based nanocomposite sensors for ethanol detection. *Polymer Bulletin* 76:2367–2386.

8. Chakraborty G, Dhar P, Katiyar V, Pugazhenthi G. 2020. Applicability of Fe-CNC/GR/PLA composite as potential sensor for biomolecules. *Journal of Materials Science: Materials in Electronics* 31:5984–5999.

9. Lasprilla AJR, Martinez GAR, Lunelli BH, Jardini AL, Filho RM. 2012. Poly-lactic acid synthesis for application in biomedical devices — A review. *Biotechnology Advances* 30:321–328.

10. Ray SS, Okamoto M. 2003. New polylactide/layered silicate nanocomposites, 6a melt rheology and foam processing. *Macromolecular Materials and Engineering* 288:936–944.

11. Dai H, Huang H. 2017. Synthesis, characterization and properties of pineapple peel cellulose-g-acrylic acid hydrogel loaded with kaolin and sepia ink. *Cellulose* 24:69–84

12. Sabzi M, Jiang L, Liu F, Ghasemi I, Atai M. 2013. Graphene nanoplatelets as poly (lactic acid) modifier:linear rheological behavior and electrical conductivity. *Journal of Materials Chemistry A* 1:8253.

13. Jandas PJ, Mohanty S, Nayak SK. 2013. Rheological and mechanical characterization of renewable resource based high molecular weight PLA nanocomposites. *Journal of Polymers* 2013, Article ID 403467:1–11.

14. Chakraborty G, Gupta A, Pugazhenthi G, Katiyar V. 2018. Facile dispersion of exfoliated graphene/PLA nanocomposites via in situ polycondensation with a melt extrusion process and its rheological studies. *Journal of Applied Polymer Science* 46476:1–11.

15. Tesfaye M, Patwa R, Dhar P, Katiyar V. 2017. Nanosilk-grafted poly(lactic acid) films: Influence of cross-linking on rheology and thermal stability. *ACS Omega* 2:7071–7084.

16. Karakuş S. 2019. Preparation and rheological characterization of Chitosan-Gelatine@ZnOSi Nanoparticles. *International Journal of Biological Macromolecules* 137:821–828.

17. Yuankun Wang, Wenyuan Xie, Defeng Wu. 2020. Rheological properties of magnetorheological suspensions stabilized with nanocelluloses. *Carbohydrate Polymers* 231:115776.

18. Mezger. TG. 2019. *The Rheology Handbook*. Walter de Gruyter GmbH, Hanover, Germany.

19. Al-Itry R, Lamnawar K, Maazouz A. 2014. Reactive extrusion of PLA, PBAT with a multi-functional epoxide: Physico-chemical and rheological properties. *European Polymer Journal* 58:90–102.

20. Li M-C, Wu Q, Song K, Lee S, Qing Y, Wu Y. (2015a). Cellulose nanoparticles: Structure–morphology–rheology relationships. *ACS Sustainable Chemistry & Engineering* 3(5):821–832.

21. Wu Q, Meng Y, Wang S, Li Y, Fu S, Ma L, Harper D. 2014. Rheological behavior of cellulose nanocrystal suspension: Influence of concentration and aspect ratio. *Journal of Applied Polymer Science* 131:40525.
22. Lopez CG, Richtering W. 2021. Oscillatory rheology of carboxymethyl cellulose gels: Influence of concentration and pH. *Carbohydrate Polymers* 267:118117.
23. Baniasadi H, Ajdary R, Trifol J, Rojas OJ, Sepp¨ala J. 2021. Direct ink writing of aloe vera/cellulose nanofibrils bio-hydrogels. *Carbohydrate Polymers* 266:118114.
24. Markstedt K, Escalante A, Toriz G, Gatenholm P. 2017. Biomimetic inks based on cellulose nanofibrils and cross-linkable xylans for 3D printing. *ACS Applied Material Interfaces* 9:40878–40886.
25. Hodder E, Duin S, Kilian D, Ahlfeld T, Seidel J, Nachtigall C, Bush P, Covill D, Gelinsky M, Lode A. 2019. Investigating the effect of sterilisation methods on the physical properties and cytocompatibility of methyl cellulose used in combination with alginate for 3D-bioplotting of chondrocytes. *Journal of Materials Science: Materials in Medicine* 30:10.
26. Binder L, Mazál M, Petz R, Klang V, Valenta C. 2019.The role of viscosity on skin penetration from cellulose etherbased hydrogels. *Skin Research and Technology* 25:725–734.
27. Ju Y, Ha J, Song Y, Lee D. 2021. Revealing the enhanced structural recovery and gelation mechanisms of cation-induced cellulose nanofibrils composite hydrogels. *Carbohydrate Polymers* 272:118515.
28. Adhikari J, Perwez MS, Das A, Saha P. 2021. Development of hydroxyapatite reinforced alginate–chitosan based printable biomaterial-ink. *Nano-Structures & Nano-Objects* 25:100630.
29. Lakehal I, Montembault A, David L, Perrier A, Vibert R, Duclaux L, Reinert L. 2019. Prilling and characterization of hydrogels and derived porous spheres from chitosan solutions with various organic acids. *International Journal of Biological Macromolecules* 129:68–77.
30. Gholizadeh H, Messerotti E, Pozzoli M, Cheng S, Traini D, Young P, Kourmatzis A, Caramella C, Ong HX. 2019. Application of a thermosensitive in situ gel of chitosan-based nasal spray loaded with tranexamic acid for localised treatment of nasal wounds. *AAPS PharmSciTech* 20:299.
31. García MA, Pinotti A, Martino MN, Zaritzky NE, Huber KC, Embuscado ME. 2009. *Characterization of Starch and Composite Edible Films and Coatings Edible Films and Coatings for Food Applications.* Springer, New York.
32. Bertolo MRV, Martins VCA, Horn MM, Brenelli LB, Ana, Plepis AMG. 2020. Rheological and antioxidant properties of chitosan/gelatin-based materials functionalized by pomegranate peel extract. *Carbohydrate Polymers* 228:115836.
33. Mikesová J, Hasek J, Tishchenko G, Morganti P. 2014. Rheological study of chitosan acetate solutions containing chitin nanofibrils. *Carbohydrate Polymers* 112:753–757.
34. Pan G, Xu H, Mu B, Ma B, Yang Y. 2018. A clean approach for potential continuous mass production of high-molecular-weight polylactide fibers with fully stereocomplexed crystallites. *Journal of Cleaner Production* 176:151e158.
35. Liu C, Jia Y, He A. 2013. Preparation of higher molecular weight poly (l-lactic Acid) by chain extension. *International Journal of Polymer Science* 2013, Article ID 315917:1–6.
36. Tesfaye M, Patwa R, Gupta A, Kashyap MJ, Katiyar V. 2017. Recycling of poly (lactic acid)/silk based bionanocomposites films and its influence on thermal stability, crystallization kinetics, solution and melt rheology. *International Journal of Biological Macromolecules* 101:580–594.

37. Shojaeiarania J, Bajwa DS, Stark NM, Bajwa SG. 2019. Rheological properties of cellulose nanocrystals engineered polylactic acid nanocomposites. *Composites Part B* 161:483–489.
39. Ko SW, Gupta RK, Bhattacharya SN, Choi HJ. 2010. Rheology and physical characteristics of synthetic biodegradable aliphatic polymer blends dispersed with MWNTs. *Macromolecular Materials and Engineering* 295:320–328.
39. Lee S, Kim M, Song HY, Hyun K. 2019. Characterization of the effect of clay on morphological evaluations of PLA/biodegradable polymer blends by FT-rheology. *Macromolecules* 52:7904–7919.
40. Dhar P, Bhardwaj U, Kumar A, Katiyar V. 2015. Investigations on rheological and mechanical behavior of poly (3-Hydroxybutyrate)/cellulose nanocrystal based nanobiocomposites. *Polymer Composites* 1–10.
41. Pal AK, Bhattacharjee SK, Gaur SS, Pal A, Katiyar V. 2017. Chemomechanical, morphological, and rheological studies of chitosan-graft-lactic acid oligomer reinforced poly (lactic acid) bionanocomposite films. *Journal of Applied Polymer Science* 45546:1–10.
42. Bhattacharjee SK, Chakraborty G, Kashyap SP, Gupta R, Katiyar V. 2021. Study of the thermal, mechanical and melt rheological properties of rice straw filled poly (butylene succinate) bio-composites through reactive extrusion process. *Journal of Polymers and the Environment* 29:1477–1488.
43. Bhasney SM, Patwa R, Kumar A, Katiyar V. 2017. Plasticizing effect of coconut oil on morphological, mechanical, thermal, rheological, barrier, and optical properties of poly(lactic acid): A promising candidate for food packaging. *Journal of Applied Polymer Science* 45390:1–12.
44. Eslami H, Kamal MR. 2013. Elongational rheology of biodegradable poly(lactic acid)/poly[(butylene succinate)-co-adipate] binary blends and poly(lactic acid)/poly[(butylene succinate)-co-adipate]/clay ternary nanocomposites. *Journal of Applied Polymer Science* 127(3):2290–2306.
45. Carlson J, Jolly MR. 2000. Magnetorheological fluids, foam and elastomers devices. *Mechatronics* 10(4–5):555–569.
46. Wang Y, Xie W, Wu D. 2020. Rheological properties of magnetorheological suspensions stabilized with nanocelluloses. *Carbohydrate Polymers* 231:115776.
47. Liu C, Li MC, Mei C, Xu W, Wu Q. 2020. Rapid preparation of cellulose nanofibers from energy cane bagasse and their application as stabilizer and rheological modifiers in magnetorheological fluid. *ACS Sustainable Chemistry & Engineering* 8:10842–10851.
48. Baea DH, Choi HJ, Choi K, Nam JD, Islam MS, Kao N. 2017. Microcrystalline cellulose added carbonyl iron suspension and its magnetorheology. *Colloids and Surfaces A: Physicochemical and Engineering Aspects* 514:161–167.
49. Testa P, Chappuis B, Kistler S, Style RW, Heyderman LJ, Dufresne ER. 2020. Switchable adhesion of soft composites induced by a magnetic field. *Soft Matter* 16(25):5806–5811.
50. Power law (Ostwald-deWaele model). In Gooch JW (Ed.), 2007. *Encyclopedic Dictionary of Polymers.* Springer, New York, NY, 781e781. https://doi.org/10.1007/978-0-387-30160-0_9193.
51. Lv H, Chen R, Zhang S. 2018. Comparative experimental study on constitutive mechanical models of magnetorheological fluids. *Smart Materials Structures* (27)115037:1–9.
52. Genc S, Derin B. 2014. Synthesis and rheology of ferrofluids: A review. *Current Opinion in Chemical Engineering* 3:118–124.
53. Bingham plastic model, Explore the new Oilfield Glossary, Schlumberger.

14 Active Biodegradable Composites for Packaging Applications

Neha Singh and Meenakshi Garg
Bhaskaracharya College of Applied
Science, University of Delhi

Rajni Chopra
National Institute of Food Technology
Entrepreneurship and Management

CONTENTS

14.1 INTRODUCTION

The genesis of plastic or synthetic polymers has betided around the 1950s; since that period, they have ubiquitously been introduced in our day-to-day lives (viz. used as food packaging materials) because of their reasonable nature. A substantial amount of plastic produced globally is single-use plastic, which is discarded after its single use; hence, plastic packaging contributes nearly half of the plastic waste on the planet. About 9% of the 9 billion tonnes of synthetic plastic that has been produced worldwide were recycled, while the remaining were propelled in landfills, dumped, or littered in the environment (UNEP, 2018). The major drawback associated with the use of plastics is that they are not biodegradable; rather, they are steadily disintegrated into remnants, i.e., microplastics. Studies suggested that bottles made from high-density polyethylene (HDPE) take around 58 years to decompose in a marine environment. In contrast, bioplastics, particularly polylactic acid (PLA), degrade rapidly (about 20 times faster) than HDPE on land. However, the decomposition rate of both PLA and HDPE is almost the same in the marine conditions (Chamas et al., 2020). Thus, the disposition of plastic detritus has become the major subject of concern for our planet's environmental challenge and global health-threatening issues too. The plastic waste in the environment possesses a great threat to fauna subsisting on land and in the marine surroundings by choking the respiratory passages and stomachs of many different species. Besides, toxic chemicals used for the production of synthetic polymers find their way into animal tissues, which consequently enter the human food chain and may cause various health hazards. Thus, the governments of many countries have imposed a ban on single-use plastics. For instance, the government of Rwanda has banned the production, use, sale and importation of plastic bags in 2008. So, in 2008, Kigali, Rwanda's capital was nominated as cleanest city by UN Habitat (UNEP, 2018).

So, taking into consideration all of the aforementioned issues, a reliable, sustainable, and cost-effective outlook is imperative to produce eco-friendly, technically smart, and safe "active biodegradable composite packaging materials" that assure safe handling and distribution of processed food products from farm to fork. Even, packaging materials have received special recognition in the 12th sustainable development goal, which is related to sustainable consumption and production because around one-third of food produced has perished in the bins of consumers and retailers, or decay due to inadequate transportation and harvesting practices. So, reducing dependence on synthetic packaging and adopting smart and active bioplastics packaging would certainly help in accomplishing a few of the aims enclosed in the SDG 12 (such as climate change, plastic pollution in the ocean, sustainable transport, and food loss and waste). Therefore, packaging materials made from a renewable source or biodegradable composites receive much more attention to extend the shelf life of packaged foods and retain their nutritional aspects too. The expression biodegradable composites covers a broad spectrum of at least two-phase hybrid materials, where one of two fillers or matrices or both have to be selected from biodegradable origin. Accordingly, matrices made of biodegradable polymers are known as biodegradable matrices; in contrast, fillers being the minority phase exemplify as reinforcement in several biodegradable composites (Pegoretti et al., 2020). The inclusion of nanoparticles, pigments, enzymes, and other bioactive agents in the biopolymer matrix would improve the functional and mechanical aspects of the finished packaging material. Besides, the introduction of innovative technologies such as insertion of biosensors, bar code, and incorporation of bioactive constituents and pigments makes packaging smarter and intelligent, which in turn imparts various other functions, for instance antimicrobial, antioxidant, oxygen barrier, and light-blocking properties and monitors food quality. The composite film (C-CS/PVA) prepared by catechol-functionalized chitosan (C-CS)/polyvinyl alcohol (PVA) could be employed as an active packaging material as it has manifested remarkable mechanical, antibacterial, and UV barrier characteristics (Lei et al., 2021).

Active biodegradable packaging differs from intelligent packaging. The former one involves the integration of natural additives into biodegradable packaging materials, intending to increase the shelf life or maintain the quality and safety of foodstuff. On the contrary, the latter one aids in tracking product information, monitors product conditions, and promotes data access and information exchange by modifying the surroundings of the packaging and product (Lee et al., 2015). As defined in the

European Regulation (EC) No. 450/2009, AP systems are designed to *deliberately incorporate components that would release or absorb substances into or from the packaged food or the environment surrounding the food*. Active packaging materials are therefore *intended to extend the shelf-life or to maintain or improve the condition of packaged food*. Moreover, active packaging is categorized into three main categories based on the function it performs: antioxidants, antimicrobials, and scavengers/absorbers/blockers (for instance, oxygen scavengers, light blockers, moisture control compounds, and CO_2 and ethylene absorbers or emitters) (Vilela et al., 2018).

In consonance with EFSA, an intelligent packaging material is defined as *material and articles that monitors the condition of packaged food or the environment surrounding the food* (EFSA, 2009). The intelligent food packaging approach is a newly emerging field in the modern packaging era where packaging material perhaps sense, indicate, and document any variations in the product or the condition inside and outside of the package. This technological breakthrough would aid in monitoring the quality aspects of foodstuff and also provide principle facts about the product to the consumer, which consequently reduces food loss. Intelligent packaging is divided into three main categories, particularly sensors, indicators (freshness, time temperature, and gas indicators), and data carriers (bar codes and radio frequency identification (RIFD) technology) (Müller & Schmid, 2019). Figure 14.1 illustrates the benefits of bioactive components incorporated in active and intelligent biodegradable composites for packaging.

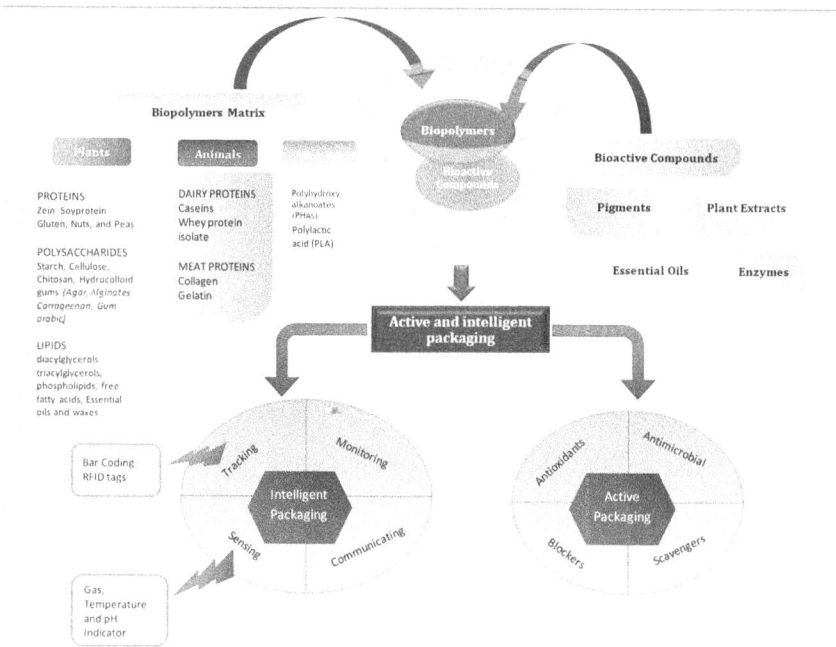

FIGURE 14.1 Benefits of bioactive components incorporated in active and intelligent biodegradable composites for packaging.

The dynamic nature of packaging industry is persuaded by regulations, legislation, and consumer trends. Packaging market is invariably driven by consumer needs and demands. In fact, the emergence of novel technology and consumer awareness about environment hiked the expectations of the end users from manufacturers to produce smart packaging materials that should be sustainable and lighter; extend the shelf life of food products; provide complete information about the packaged product, be cost-effective, and have an appealing design.

From the market point of view, bioplastics, viz. starch blends, polylactic acid, cellulose, and few others, have attained a markedly high growth rate in comparison with conventional standard plastics. It could be employed in various fields ranging from packaging to automotive and electronic areas. As per the recent Ceresana report, profits generated from biopolymer plastics may hike to about USD 4.4 billion by 2026. Further, biodegradable plastics, mainly polylactic acid and starch-based polymers, outreach a market share of 56% of the total market for biodegradable plastics in 2018, which may extend up to a volume of 7.1% per year till 2026, especially for the mentioned products. Surprisingly, the packaging industry was a predominant sector for the sale of bioplastics, where more than 60% of bioplastics were processed.

Although plastics of biodegradable origin have several advantages, manufacturers and processors of bioplastics are experiencing a broad spectrum of challenges. Few of them are a guarantee regarding the availability of high quality, reliable bioplastics, recycling and composting problem (infrastructure for recycling and composting, i.e., when plants' design is inadequate to segregate bioplastics or inefficacious to biodegrade them), and sustainability (proper knowledge about production, longevity, and disposal of bioplastics), and image and market prosperity whether the benefits of bioplastics are acknowledged or rewarded by consumers (Ceresana report, 2020).

However, the pervasive applications of biopolymers with "smart" or "intelligent" technology would aid in overcoming these challenges, reducing waste, monitoring the quality and safety of food products, and ensuring environmental sustainability. Subsequently, due to their biodegradable nature, they can enter the spontaneous rhythm of "birth from nature to end in nature". Hence, this chapter reviews the active biodegradable composites fabricated by incorporating various bio-based additives.

14.2   NATURAL PIGMENTS

Pigments or natural food colorants such as anthocyanin, chlorophyll, carotenoids, and tannins can be harnessed as a colorimetric indicator to determine the lifespan of different groups of polymeric materials. Naturally derived components integrated with polymeric material change color when exposed to various extrinsic factors, for instance UV radiation, humidity, and elevated temperature. Thus, a color change caused by natural dyes indicates the freshness of the product in the packaging material (Latos-Brozio & Masek, 2020).

Starch-based films formed by thermocompression technology with the addition of blueberry residues have the potential to be used as a pH change indicator, which works well in the pH range from 2 to 12 (Andretta et al., 2019). Biosensors or bio-based indicators also help in assessing the freshness of the fish and fishery products by detecting the volatile amine components (such as dimethylamine, ammonia, and trimethylamine), altogether recognized as total volatile basic nitrogen (TVBN). These products are formed by fish deterioration due to the growth of spoilage microbes and potent indicators of fish spoilage. The curcumin/bacterial cellulose membrane is an efficient TVBN biosensor as it is highly sensitive to pH change. The curcumin-based indicator changes color steadily from yellow to orange afterward, to reddish-orange in response to the production of TVBN in the headspace of the package; color change was visible to the eyes. Henceforth, curcumin/bacterial cellulose membrane can be effectively used as a sticker sensor to analyze shrimp spoilage in real time at ambient and chilled temperature (Kuswandi et al., 2012). Likewise, anthocyanin as a biosensor (extracted from rose and red cabbage) would assist in tracking the freshness of high-protein flesh foods (viz. fish, meat, shrimp, and other seafoods) as it is efficient to sense any alteration in pH value due to the generation of basic volatile amines, which consequently indicates a change in color from red to green as the pH increases.

Therefore, dye-based sensors could act as a simple and non-destructive shelf-life-indicating device to check the spoilage of flesh foods in storage conditions (Shukla et al., 2016). So, the use of biosensors will ultimately aid in minimizing the economic damage faced by manufacturers due to the spoilage of food during transport and storage conditions because they have a chance to remove their product from the market before it reaches the retailers and end users. Figure 14.2 outlines the various pigments that can be incorporated in active and intelligent packaging.

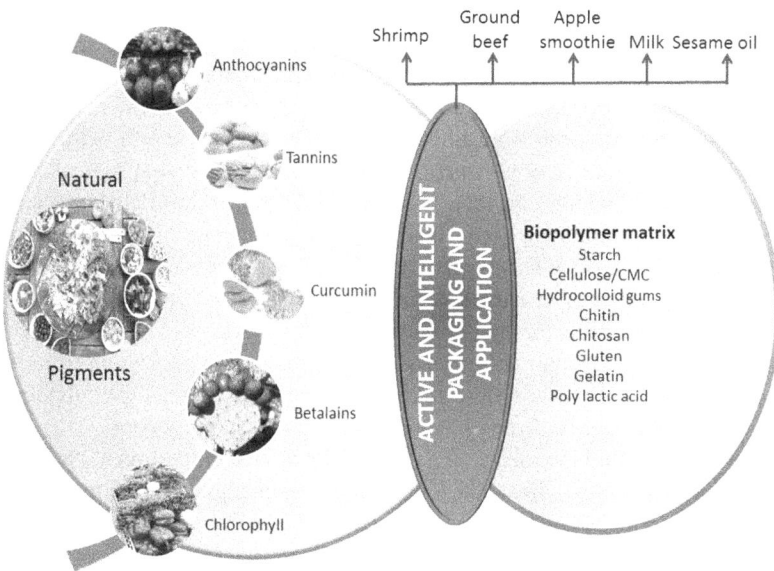

FIGURE 14.2   Pigments in active and intelligent packaging.

The evidence from the research study reveals that the inclusion of natural dyes (e.g., β-carotene 1% E160a; curcumin 8% E100; and lutein 0.9% E161b) in biopolymer matrices (i.e., polylactic acid and polyhydroxybutyrate) showed a significant color change under the influence of high temperature, UV radiation, and weathering. Alongside, natural food colorants improve the stability of packaging material by increasing the resistance of a material to oxidation. Even from a health standpoint, natural dyes are safe to use and do not cause any health hazards in case of migration to packaged foods (Latos-Brozio & Masek, 2020). Table 14.1 stipulates the functional properties offered by different pigments when integrated in biopolymer matrices.

14.3 PLANT EXTRACTS

Fruit and vegetable by-products (viz. peels, seeds, and leaves) are major sources of phytochemicals, pigments, and essential oils that can be efficiently integrated into the packaging material to generate active food package that can increase the lifespan of perishable food commodities. The polyphenolic compounds present in these by-products or food waste have the prospects to be utilized as antimicrobial and antioxidant vehicles. Liu et al. (2020) reported different types of polyphenolic compounds found in the pomegranate flesh extract (PFE). They are pelargonidin-3-O-(6″-ethylmalonyl)-glucoside (7.05%), peonidin-3-O-(6-malon)-glucoside (9.61%), cyanidin-3-O-glucoside (49.76%), isorhamnetin (9.18%), pelargonidin-3-O-(6-acetyl)-glucoside (9.51%), and malvidin-3-O-(6-acetyl)-glucoside (14.89%). In contrast, pomegranate peel extract (PPE) consists of gallic acid (15.73%), malvidin-3-O-(6-acetyl)-glucoside (4.07%), epicatechin gallate (3.66%), luteolin (5.69%), vanillic acid (7.82%), caffeic acid (4.37%), punicalagin (39.96%), and ellagic acid (18.70%). Also, the loading of active constituents of plant origin tends to show significant inhibition for a broad spectrum of spoilage and pathogenic microbes and reduces the dependence on chemical additives in the food industry that may impose a negative influence on human health. This can be attained by the extraction of chemical constituents from these by-products known as plant extracts, which can then be incorporated or encapsulated in the biopolymer matrix to fabricate active biodegradable composite films. For instance, mango peel extract, olive leaves extract, moringa leaf extract, and many more in a row could be utilized as natural additives to fabricate antimicrobial and antioxidant packaging films (Kanatt & Chawla, 2018; Martiny et al., 2020; Ju et al., 2019) and to enhance other characteristic aspects of active packaging films such as water vapor barrier properties, increased elongation at break (i.e., flexibility) with a decrease in tensile strength, and other mechanical properties.

Plant extract could also serve as a pH sensing device in intelligent packaging, which can successfully indicate the food spoilage by changing the color of the food packaging material and guide the end users about the safety and quality of the food product before consumption. For example, jamun extract has a dark violet color that turns cherry red in the presence of acidic pH, while showing brownish yellow color at alkaline pH and light green in a neutral medium (Jayakumar et al., 2019).

TABLE 14.1
Stipulated Functional Properties of Different Pigments Integrated in Biopolymer Matrix

Pigment/Bioactive Compounds	Polymer Matrix	Foods	Characteristic Properties	References
			Anthocyanins	
Anthocyanin from red cabbage and propolis extract (PE)	Polyvinyl alcohol and starch (PVA–starch(X)–anth-PE film)	Milk	The pH indicator or sensor: reddish to blue (pH 2–5); blue color (pH 7); blue to green to yellow (pH 8–14). Film tends to show good tensile strength, water-resistant properties, and moisture retention ability. All these properties significantly improved due to the increase in the amount of PE.	Mustafa et al. (2020)
Extract of red cabbage and sweet whey	Starch and glycerol	Ground beef	Adequate amount of antibacterial activity against *E. coli*. Films exhibited low solubility and water vapor permeability. Anthocyanins and phenolic compounds in the extract responsible for antioxidant properties.	Sanches et al. (2021)
Red cabbage	Polyvinyl alcohol (PVA) and chitosan (CS) with anthocyanin (ATH) and sodium tripolyphosphate (STPP) film	Partially wrapped slice of pork belly	The pH indicator: red color (pH 1); reddish to purple (pH 1–6); blue (pH 7–8); sea green (pH 9); yellow green (pH 12). Sequential change in color in partially wrapped slice of pork belly was observed under ambient temperature for 12 hours and 24 hours.	Vo et al. (2019)

(Continued)

TABLE 14.1 (*Continued*)
Stipulated Functional Properties of Different Pigments Integrated in Biopolymer Matrix

Pigment/Bioactive Compounds	Polymer Matrix	Foods	Characteristic Properties	References
Tannins				
Condensed tannins/pecan nut shells	Polylactic acid (PLA) films with pecan nut shell hydroalcoholic extract (PNSE)	Apple smoothie	Film manifested outstanding antioxidant activity and reduced the onset of browning in apple smoothies. PNSE was successful in inhibiting enzymatic browning and had anthocyanin-stabilizing activity; i.e., half-life of blackberry anthocyanin was increased up to 20%.	Moccia et al. (2020)
Curcumin				
Curcumin/zinc oxide (ZnO)	Carboxymethyl cellulose (CMC) (CMC/curcumin/ZnO composite films)		The composite films display satisfactory mechanical, UV-blocking, and water vapor barrier properties. They also exhibited high free radical scavenging and antimicrobial activities due to the addition of curcumin and ZnO, respectively.	Roy et al. (2020a)
Curcumin/sodium dodecyl sulfate (SDS) as emulsifier	Gelatin (gelatin/curcumin composite film)		The incorporation of 1.5% curcumin into films significantly improved the UV protection effect (>99%) and mechanical and water vapor barrier properties. Alongside, antioxidant activity and antimicrobial action against pathogenic microbes (i.e., *E. coli and L. monocytogenes*).	Roy and Rhim (2020b)
Chlorophyll				
Chlorophyll	Wheat gluten (WG) integrated with chlorophyll (Ch) (WG/Ch smart film)	Sesame oil	Addition of chlorophyll in active film increases the antioxidant capacity by 85%, which subsequently improves the expiration date of the oil and the pigment also changes color from green to yellow in response to oxidation of oil. So, this could act as smart sensing device that can indicate oil expiry time.	Chavoshizadeh et al. (2020)

(Continued)

TABLE 14.1 (*Continued*)
Stipulated Functional Properties of Different Pigments Integrated in Biopolymer Matrix

Pigment/Bioactive Compounds	Polymer Matrix	Foods	Characteristic Properties	References
Betalains				
Betalains (betacyanins; main components)	Extract of red pitaya peel blended with starch/polyvinyl alcohol	Shrimp	Active film containing 1% extract was ammonia sensitive and thus could be used as freshness indicator of shrimp as it indicates visible color change in response to TVBN accumulation. Films possessed UV barrier, water barrier, antimicrobial, and antioxidant properties.	Qin et al. (2020)
Betalains	Quaternary ammonium chitosan (QAC)/ polyvinyl alcohol (PVA) matrix blended with cactus pears extract (CPE)	Shrimp	The functional properties such as UV–Vis light barrier, flexibility, ammonia-sensitive, antimicrobial, and free radical scavenging activities of QAC/PVA active films were ameliorated due to the introduction of CPE at various levels (i.e., 1–3 wt.%). The film incorporated with CPE (at 2 wt.% and 3%) indicated color change to orange from purple in response to pH alteration suggesting shrimp spoilage.	Yao et al. (2020)
Betalains	Quaternary ammonium chitosan (QC)/fish gelatin (FG) matrix blended with amaranth extract (AE)	Shrimp	QC/FG films functionalized with AE possessed antioxidant and antimicrobial activities against foodborne pathogens. Films incorporated with AE (at 5 wt.% and 10 wt.% levels) were found to be successful in monitoring the shrimp freshness.	Hu et al. (2020)

So, pigments present in the plant extracts could be exploited as a pH indicator device for intelligent packaging. Constituents from nature are non-toxic in contrast to synthetic ones. Chemical dyes, mainly bromocresol compounds, cresol red, and chlorophenol red, have extensively been used in the food industry to indicate pH change, but color-changing agents of synthetic origin may defile the food and may expose human life to health hazards (Yadav et al., 2011). Functional properties of some of the plant extracts incorporated in the suitable biopolymer matrices are depicted in Table 14.2.

14.4   ESSENTIAL OILS AND ENZYMES

The utilization of essential oils (EOs) in active packaging system is an emerging field as they exhibit antimicrobial and antioxidant properties. Apart from that, they also impart other functional attributes to packaging films or coatings, viz. a barrier to UV light, increased flexibility, and reduced water wettability. The EOs' preponderance in the packaging industry has been isolated from cinnamon, mandarin, eucalyptus, thyme, cumin, oregano, clove, tea tree, lemongrass, and so on. As per Food and Drug Administration (2016), EOs from aromatic plants have received generally recognized as safe (GRAS) status (FDA, 2016). The volatile components in the essential oils contribute to antimicrobial and antioxidant capacity. Additionally, a biopolymer matrix blended with essential oil aggrandizes UV light-blocking ability of the active film, which prevents the oxidation of packaged foods by UV light. Further, as we increase the concentration of the essential oils, the tensile strength of the biocomposite film decreases causing expansion of pore size of the composite matrix. This consequently results in weakened structure and more rupture points, but enhances the flexibility of the film (Sharma et al., 2020). Pieces of evidence have also demonstrated that the inclusion of EOs such as *Zataria multiflora* Boiss essential oil, cinnamon oil, eucalyptus oil, and oregano oil in biodegradable films showed a significant reduction in tensile strength (Moradi et al., 2012; Hosseini et al., 2015). Nonetheless, the insertion of essential oil increases the hydrophobicity of the packaging material, which reduces the water wettability index and improves the quality and safety of the food product. It also maintains the integrity of the food wrapped in the film (Sharma et al., 2020).

In an active packaging system, EOs can be introduced via two ways. They are incorporated in the biodegradable matrix as either free EOs or encapsulated EOs. For example, 1,8-cineole-rich essential oils (i.e., extracted from bay and rosemary leaves) encapsulated in zein nanofiber films (ZNFs) have efficaciously reduced or inhibited the progression of gram-positive bacteria. The cheese slice coated with the active film was effective in reducing the growth of *L. monocytogenes and S. aureus* (up to ~2 log cycles) after storing it at 4°C for more than 28 days. Moreover, EOs–ZNF active films have also shown significant inhibition against aerobic mesophilic bacteria on cheese slices. Besides, an essential oil extracted from bay leaves (*Laurus nobilis*) was more efficacious in inhibiting or reducing the bacterial growth compared to rosemary leaves (*Rosmarinus officinalis*). The long-term antimicrobial effect of the active film was contributed due to the

TABLE 14.2
Depicted Functional Properties of Plant Extracts Incorporated in Biopolymer Matrix

Active Packaging Material	Biopolymer Matrix	Plant Extracts/ Polyphenols	Functional Properties	References
Starch–PVA-based composite film integrated with ZnO-NPs and phytochemicals (PSNZJ) film	Starch–PVA	Nutmeg and jamun	The nanocomposite film has improved thickness, tensile strength, and water barrier properties, which may be due to the presence of ZnO-NPs, nutmeg oil, and jamun extract.	Jayakumar et al. (2019)
			Outstanding UV-blocking property, which could prevent food spoilage caused by oxidation.	
			Good antimicrobial action; addition of ZnO-NPs and nutmeg found to decrease or inhibit the growth of foodborne pathogen (*S. typhimurium*).	
			pH indicator due to the presence of jamun extract.	
Mango peel extract incorporated in polyvinyl alcohol, gelatin, and cyclodextrin	Polyvinyl alcohol (PVA), cyclodextrin, and gelatin	Mango peel (MP) extract (Alphonso, Kesar, Langra, and Badami)	Composite film consists of mango peel of Langra variety and stipulates satisfactory UV light filtering property, better tensile strength, and good antioxidant and antimicrobial properties due to its high bioactive compounds.	Kanatt and Chawla (2018)
			Mango peel extract in active film was successful in improving the lifespan of the minced chicken during chilled storage. So, it can be exploited as a future smart packaging material for primary packaging.	
Carrageenan film containing olive leaves extract (CAR-OLE)	Carrageenan (CAR)	Olive leaves extract (OLE)	High amount of phenolic compounds in OLE shows antioxidant and antimicrobial activity against aerobic mesophiles and total coliforms (i.e., OLE at 100mg mL^{-1} displayed 100% inhibition of *E.coli* and *Salmonella enteritidis*).	Martiny et al. (2020)
			CAR-OLE films have high water vapor barrier property, increased film thickness, increased elongation at break, and flexibility properties.	

(*Continued*)

TABLE 14.2 (*Continued*)

Depicted Functional Properties of Plant Extracts Incorporated in Biopolymer Matrix

Active Packaging Material	Biopolymer Matrix	Plant Extracts/ Polyphenols	Functional Properties	References
Moringa leaf extract inserted in khorasan wheat starch (KWS)	Khorasan wheat starch	Extract of moringa leaf (MLE) at various concentrations (0, 0.4, 0.7, and 1% w/v)	Active film could probably be used as an antioxidant biodegradable packaging material due to its excellent free radical scavenging and UV light-blocking actions. Films added with 1% extract were biodegrade within 30 days.	Ju et al. (2019)
Extract of moringa leaves added in gelatin	Gelatin	Extract of moringa leaves	The active film was effective against *Listeria monocytogenes* and retarded lipid oxidation of gouda cheese. Therefore, it can be harnessed as an antimicrobial and antioxidant agent in active packaging system.	Lee et al. (2016)
Extract of cocoa nibs (CNE) incorporated in starch	Starch from adzuki bean	Extract from cocoa nibs (CNE) at different levels (0.3%, 0.7%, and 1%)	The film with 1% CNE manifested high antioxidant capacity, great flexibility, and reduction in tensile strength. Films were degraded after 28 days.	Kim et al. (2018)
Extract of grape seed, malic acid, EDTA, and nisin merged in whey protein isolate	Whey protein isolate (WPI)	Extract of grape seed (0.5%), malic acid (MA, 1%), nisin (6000IU g^{-1}), and EDTA	The study results delineated that nisin, organic acids, and natural extract in WPI coating could be used as antimicrobial agents as they were effective in reducing the growth of *E.coli* 0157:H7, *Salmonella typhimurium*, and *Listeria monocytogenes* in RTE poultry products.	Gadang et al. (2008)
Fish gelatin (FG) incorporated with haskap berries extract (HBE)	Gelatin from fish (FG)	Haskap berries extract (HBE)	FG-HBE active films exhibited high antioxidant activity that may be present due to the high amount of polyphenolic compounds. Anthocyanin-rich HBE extract incorporation in active film was also effective in indicating color change in response to pH. So, active film consisting of HBE (1 wt.%) could be employed as freshness indicator of shrimp.	Liu et al. (2019)

(Continued)

TABLE 14.2 (Continued)

Depicted Functional Properties of Plant Extracts Incorporated in Biopolymer Matrix

Active Packaging Material	Biopolymer Matrix	Plant Extracts/ Polyphenols	Functional Properties	References
Aqueous hibiscus extract incorporated in natural polymeric matrix	Gelatin, chitosan, and starch	Aqueous hibiscus extract (HAE)	Anthocyanins and other phenolic agents in HAE could be used as pH indicator device.	Peralta et al. (2019)
K-carrageenan matrix containing extract of pomegranate flesh (PFE) or extract of pomegranate peel (PPE)	K-carrageenan	Extract of pomegranate flesh (PFE) or extract of pomegranate peel (PPE)	The active films showed strong antioxidant, antimicrobial, and UV light-blocking abilities.	Liu et al. (2020)
Polyphenols				
Chitosan (CS) grafted with three hydrocinnamic acids (hydroxycinnamic acid-g-CS)	Chitosan (CS)	Three hydroxycinnamic acids are p-coumaric acid, caffeic acid, and ferulic acid	Caffeic acid-functionalized films possessed superior UV-blocking, water barrier, antioxidant, and antimicrobial attributes. They showed high thermal stability and improved mechanical strength compared to other hydroxycinnamic acids-grafted films. So, caffeic-g-CS film was effective in improving the lifespan of pork for 10 days at 4°C.	Yong et al. (2021)
Polyvinyl alcohol (PVA)/ oxidized maize starch (OMS) film blended with 7-hydroxy-4-methylcoumarin (7H₄MC)	Polyvinyl alcohol (PVA)/ oxidized maize starch (OMS) film	7-Hydroxy-4-methylcoumarin (7H₄MC)	Addition of 7H₄MC in PVA/OMS film was found to be more water resistant, reduce water absorption, and show significant free radical scavenging activities.	Hiremani et al. (2020)

concentration of EO (i.e., 10%) and the method adopted by a researcher to prepare the film. Therefore, the ZNF films nano-encapsulated with EOs offer a sustained delivery of active agents that can be a possible solution to extend the shelf life of perishable commodities (Göksen et al., 2020). Thus, not only natural additives, but also the technology adopted to transfer the active ingredients plays a crucial role in the effectiveness of the active packaging system. Table 14.3 shows some of the examples of EOs and enzymes used in the active food packaging industries.

14.5   NANOPARTICLES

Nanomaterials are a new thrust area in the research and development of novel packaging materials. The biopolymers-based packaging materials show unsatisfactory permeability for gases and vapors, and inadequate barrier and mechanical properties. Thus, consolidation of a nanomaterial as a filler or reinforcement material would provide an impetus for the development of nanocomposite packaging material with superior barrier and mechanical properties, because nanoparticles have a proportionally greater surface area and substantial aspect ratio compared to their micro-scale analogues (Majeed et al., 2013). Some of the most frequently employed nanoparticles in the food packaging industry are zinc oxide (ZnO-NPs), kaolinite, nanocellulose, nanostarch, silver nanoparticles (AgNPs), montmorillonite (MMT), and titanium dioxide (TiO_2-NPs). The addition of all of these nanoparticles in the biopolymer matrix would ultimately improve oxygen, CO_2, and UV blocking attributes and reinforce antimicrobial and antioxidant activities (Chaudhary et al., 2020). The experiment conducted by Sonkaew et al. (2012) reported that curcumin nanoparticles (Ccm-NPs) and ascorbyl dipalmitate nanoparticles (ADP-NPs) exhibited high antioxidant properties compared to curcumin and ascorbyl dipalmitate. Thus, the inclusion of curcumin and ascorbyl dipalmitate nanoparticles into cellulose-based films boosts up the antioxidant activities of both Ccm and ADP, which in turn decreases the amount of Ccm and ADP while improving their availability.

Likewise, soluble soybean polysaccharide (SSPS)-based nanocomposite films formed by the addition of SiO_2 nanoparticles at different concentrations (5%, 10%, and 15%) and curcumin indicated adequate antibacterial and antimicrobial properties. The results also showed that parameters such as water solubility, thickness, and water vapor permeability of the film decreases, whereas the mechanical performance enhances as the concentration of nanoparticle increases. Moreover, SSPS/curcumin/SiO_2 nanocomposite films demonstrated a color change in response to an increase in TVBN content in the shrimp, thus indicating shrimp spoilage. Therefore, the film containing nanoparticles can be employed as a smart packaging material (Salarbashi et al., 2021).

ZnO nanoparticles exhibit antimicrobial activity against a wide spectrum of bacteria and fungi, which is mainly due to their photocatalytic activity, where generation of reactive oxygen species target cell death caused by cytoplasmic oxidation of bacterial cell (Kim et al., 2020). Apart from that, ZnO nanoparticles are also known for their characteristic attributes; for instance, they are safe

TABLE 14.3

Showing EOs and Enzymes Employed in the Active Packaging

Active Packaging Material	Biopolymers	Essential Oils	Functional Properties	References
PLA/PBAT (polylactide/poly(butylene adipate-co-terephthalate) composite film containing essential oils	PLA/PBAT (polylactide/poly(butylene adipate-co-terephthalate)	Eucalyptus and Cinnamon oil	PLA/PBAT–cinnamon oil film was yellowish and more opaque in appearance, which could be due to its phenolic compounds (majorly eugenol). Thus, it manifested 80% UV-blocking capacity. Increased flexibility, lower tensile strength, increased thickness, and reduced water wettability of the film were observed in the film containing cinnamon oil. Cinnamon oil showed higher inhibitory action for *E.coli* and *S. aureus* compared to the eucalyptus oil in composite films.	Sharma et al. (2020)
Gelatin and chitosan consist of *Origanum vulgare* L. essential oil (OEO) at 0.4%, 0.8%, and 1.2% (w/v)	Fish gelatin and chitosan	*Origanum vulgare* L. essential oil (OEO) added at various concentrations of 0.4%, 0.8%, and 1.2% (w/v)	Active film was more effective against gram-positive bacteria (*Staphylococcus aureus* and *Listeria monocytogenes*) in contrast to gram-negative bacteria (*Salmonella enteritidis* and *Escherichia coli*). Improved UV barrier property.	Hosseini et al. (2015)
Gelatin nanofibers containing encapsulated angelical essential oil (AEO)	Gelatin nanofibers	Angelical essential oil (AEO)	The active film exhibited antioxidant and antimicrobial activities against both gram-positive and gram-negative bacteria.	Zhou et al. (2020)
Clove essential oil (CEO) or oregano essential oil (OEO) integrated into trays made from cassava bagasse and polyvinyl alcohol (PVA)	Cassava bagasse and polyvinyl alcohol (PVA)	Clove essential oil (CEO) or oregano essential oil (OEO)	Trays coated by OEO displayed inhibition against broad range of microbes such as mold, yeast, and gram-positive and gram-negative bacteria. OEO-incorporated trays improved the flexibility, but diminished water absorption and adsorption properties.	Debiagi et al. (2014)

(Continued)

TABLE 14.3 (Continued)
Showing EOs and Enzymes Employed in the Active Packaging

Active Packaging Material	Biopolymers	Essential Oils	Functional Properties	References
Ecklonia cava alginate (ECA) added with cinnamon leaf oil (CLO) or cinnamon bark oil (CBO)	Ecklonia cava alginate (ECA) and calcium chloride as cross-linking agent	Cinnamon leaf oil (CLO) or cinnamon bark oil (CBO) added in different amounts (0%, 0.4%, 0.7%, and 1%)	Addition of CBO to ECA film showed greater antioxidant capacity compared to CLO. ECA film consisting of CLO has shown antibacterial action against *E. coli O157:H7, Listeria monocytogenes, Staphylococcus aureus, and Salmonella typhimurium.*	Baek et al. (2018)
Burdock root inulin (INU) and chitosan (CHI) film containing Oregano and thyme essential oils (OT)	Composite film prepared by combining Burdock root inulin (INU) with chitosan (CHI)	Oregano and thyme essential oils (OT)	Composite film showed antioxidant and antimicrobial properties.	Cao et al. (2018)
Gelidium corneum (GC)–chitosan composite film inserted with java citronella essential oil (JCEO)	Composite film made from Gelidium corneum (GC)–chitosan	Java citronella essential oil (JCEO) incorporated at various levels (0.5%, 1.0%, and 1.5%)	GC–chitosan film incorporated with JCEO was efficient in retarding lipid oxidation of wrapped food as it has UV light-blocking and antioxidant capacity.	Go and Song (2020)

(Continued)

TABLE 14.3 (*Continued*)
Showing EOs and Enzymes Employed in the Active Packaging

Active Packaging Material	Biopolymers	Essential Oils	Functional Properties	References
			Enzymes	
Whey protein isolate (WPI) coatings incorporated with lactoperoxidase system (LPOS)	Whey protein isolate (WPI)	Lactoperoxidase system (LPOS)	Roasted turkey coated with LPOS–WPI coatings containing LPOS at concentrations 7% and 4% was successful in inhibiting *S. enterica and E.coli* O157:H7 up to 3- and 2-log cycles, respectively, in initial phase. Addition of LPOS at concentrations of 5% and 3% (w/wt) was effective in retarding the growth of *S. enterica and E.coli* O157:H7, respectively, in turkey for 42 days at 4 and 10°C. Moreover, LPOS–WPI coating also efficaciously reduced the growth of total aerobes during storage.	Min et al. (2006)
Zein film containing partially purified lysozyme and disodium ethylenediaminetetraacetic acid (Na₂EDTA)	Zein protein	Partially purified lysozyme (700 µg cm⁻²) and Na₂EDTA (300 µg cm⁻²)	Active film incorporated with lysozyme and Na₂EDTA efficaciously reduced the growth of foodborne pathogens (*E.coli* O157:H7, *S. typhimurium*, and *L. monocytogenes*) and significantly delayed oxidation in ground beef patties during storage conditions.	Ünalan et al. (2011)

and possess antimicrobial, barrier to UV light, and other mechanical properties. Therefore, they are extensively used in active packaging as a reinforcement or coating material to improve the efficiency of food packaging applications (Li et al., 2010). For example, chitosan and gelatin nanocomposite hybrid films reinforced with ZnO-NPs (at a rate of 2% and 4%) result in better flexibility and thermal stability and have smooth and compact surface in contrast to chitosan–gelatin hybrid films. Further, nanocomposite films had shown satisfactory antibacterial action against *E. coli*. Therefore, hybrid nanocomposite films can be used in active and intelligent packaging systems to preserve the freshness of fruits and vegetables (Kumar et al., 2020).

Alongside, the future prospects of green packaging embedded with biodegradable polymer matrices, bioactive components, and nanocomposite particles as fillers or reinforcement materials could provide an opportunity in narrowing down the negative influence of synthetic polymers on the environment.

Bacterial nanocellulose (BNC) is another rung on the ladder of active and intelligent packaging. It is a safe, biodegradable, and chemically pure biopolymer that is formulated by growing certain bacteria, namely *Pseudomonas, Dickeya, Achromobacter, Komagataeibacter (formerly Gluconacetobacter), Agrobacterium, and Sarcina*. But among them, gram-negative, aerobic bacteria belonging to the *Komagataeibacter* genus are widely used to develop BNC (Ryngajłło et al., 2019; Lin et al., 2020). The merits of employing BNC as an active and intelligent packaging material are biodegradability, non-toxicity, adequate barrier property, and high mechanical strength. Even, USFDA has recognized BNC as a generally recognized as safe (GRAS) material. Although BNC has some drawbacks such as the hydrophilic nature of the membrane and a dearth of antimicrobial and antioxidant action, these limitations can be conquered by modifying the BNC matrix with various fillers or reinforcing compounds (such as bioactive compounds, antimicrobial and antioxidant agents, oxygen scavengers, plasticizers, stabilizers, and nutrients). The composite synthesized from modifying the BNC matrix with a reinforcing material could be harnessed for smart packaging (Dıblan & Kaya, 2018; Ludwicka et al., 2020). The evidence from various studies reveals that BNC incorporated with epsilon-polylysine and lactoferrin tends to have bactericidal activity against *E. coli and Staphylococcus aureus*, while the addition of nisin (at a concentration of 2500 IU ml^{-1}) shows antibacterial activity against *L. monocytogenes* on processed meat (Wahid et al., 2019; Padrao et al., 2016; Nguyen et al., 2008). Further, the study stipulated the effect of bacterial cellulose membrane blended with extracts of pomegranate peel, green tea, and rosemary on the quality parameters of button mushroom stored for 15 days at 4°C. The results signify that extracts isolated from rosemary and pomegranate peel exhibited strong antioxidant and antimicrobial actions. Similarly, green tea extract was found to be effective in retaining the mushroom color (Moradian et al., 2018). Another one in the row of smart packaging is the composite film developed by using nanofibers of bacterial cellulose (BCNF) and chitin (CNF) with curcumin, as it holds high antioxidant capacity, good antibacterial action, and pH indicator (Yang et al., 2020).

14.6 CONCLUSIONS

Active biodegradable composites for green packaging are a great alternative to single-use plastics or conventional synthetic plastics, which pose a substantial burden on our plant and lives surviving on it. The biodegradable packaging materials developed from natural origin (viz. plant proteins, polysaccharides, animal proteins, and microbial products) have considerable potential to be employed in the packaging industry. The introduction of bioactive components in the biopolymers ultimately produces active packaging materials with superior functional properties to monitor the quality assurance and safety of the foodstuff, increase the lifespan of packed food, and curtail food waste. The natural pigments extracted from food by-products can be harnessed as biosensors to indicate any change that occurs due to temperature abuse, pH change, and alteration in gas levels. The color change due to alteration in pH, temperature, and gas levels can be easily visible to the eyes and help in real-time monitoring of the food product without opening the package. Moreover, other bioactive compounds, for instance plant extracts, polyphenols, essential oils, enzymes, and nanoparticles, can be exploited as powerful antimicrobial and antioxidant agents. Simultaneously, they have manifested great antioxidant capacity and show inhibition against a broad spectrum of pathogenic and spoilage microorganisms. The incorporation of these natural additives maintains the integrity and freshness of the food products. Alongside, they impart UV light barrier, gas barrier, decreased water solubility, water absorption, and antioxidant and antimicrobial characteristics to active film matrices. Henceforth, active and intelligent packaging materials synthesized from naturally derived compounds would provide an opportunity to replace the synthetic ones and go for sustainable and environment-friendly packaging.

REFERENCES

Andretta, R., Luchese, C. L., Tessaro, I. C., & Spada, J. C. (2019). Development and characterization of pH-indicator films based on cassava starch and blueberry residue by thermocompression. *Food Hydrocolloids*, *93*, 317–324.

Baek, S. K., Kim, S., & Song, K. B. (2018). Characterization of *Ecklonia cava* alginate films containing cinnamon essential oils. *International Journal of Molecular Sciences*, *19*(11), 3545. https://doi.org/10.3390/ijms19113545

Bioplastics Market Report (2021). Market study bioplastic. Available from https://www.ceresana.com/en/market-studies/plastics/bioplastics/market-study-bioplastics.html

Cao, T. L., Yang, S. Y., & Song, K. B. (2018). Development of burdock root inulin/chitosan blend films containing oregano and thyme essential oils. *International Journal of Molecular Sciences*, *19*(1), 131. https://doi.org/10.3390/ijms19010131

Chamas, A., Moon, H., Zheng, J., Qiu, Y., Tabassum, T., Jang, J. H.,... & Suh, S. (2020). Degradation rates of plastics in the environment. *ACS Sustainable Chemistry & Engineering*, *8*(9), 3494–3511.

Chaudhary, P., Fatima, F., & Kumar, A. (2020). Relevance of nanomaterials in food packaging and its advanced future prospects. *Journal of Inorganic and Organometallic Polymers and Materials*, *30*(12), 5180–5192.

Chavoshizadeh, S., Pirsa, S., & Mohtarami, F. (2020). Sesame oil oxidation control by active and smart packaging system using wheat gluten/chlorophyll film to increase shelf life and detecting expiration date. *European Journal of Lipid Science and Technology, 122*(3), 1900385.

Debiagi, F., Kobayashi, R. K., Nakazato, G., Panagio, L. A., & Mali, S. (2014). Biodegradable active packaging based on cassava bagasse, polyvinyl alcohol and essential oils. *Industrial Crops and Products, 52*, 664–670.

Diblan, S., & Kaya, S. (2018). Antimicrobials used in active packaging films. *Food and Health, 4*(1), 63–79.

European Food Safety Authority (EFSA). (2009). Guidelines on submission of a dossier for safety evaluation by the EFSA of active or intelligent substances present in active and intelligent materials and articles intended to come into contact with food. *EFSA Journal, 7*(8), 1208.

Gadang, V. P., Hettiarachchy, N. S., Johnson, M. G., & Owens, C. (2008). Evaluation of antibacterial activity of whey protein isolate coating incorporated with nisin, grape seed extract, malic acid, and EDTA on a Turkey frankfurter system. *Journal of Food Science, 73*(8), M389–M394. https://doi.org/10.1111/j.1750-3841.2008.00899.x

Go, E. J., & Song, K. B. (2020). Effect of java citronella essential oil addition on the physicochemical properties of *Gelidium corneum*-chitosan composite films. *Food Science and Biotechnology, 29*(7), 909–915. https://doi.org/10.1007/s10068-020-00740-8

Göksen, G., Fabra, M. J., Ekiz, H. I., & López-Rubio, A. (2020). Phytochemical-loaded electrospun nanofibers as novel active edible films: Characterization and antibacterial efficiency in cheese slices. *Food Control, 112*, 107133.

Hiremani, V. D., Sataraddi, S., Bayannavar, P. K., Gasti, T., Masti, S. P., Kamble, R. R., & Chougale, R. B. (2020). Mechanical, optical and antioxidant properties of 7-hydroxy-4-methyl coumarin doped polyvinyl alcohol/oxidized maize starch blend films. *SN Applied Sciences, 2*(11), 1–18.

Hosseini, S. F., Rezaei, M., Zandi, M., & Farahmandghavi, F. (2015). Bio-based composite edible films containing Origanum vulgare L. essential oil. *Industrial Crops and Products, 67*, 403–413.

http://www.fao.org/fileadmin/templates/est/COMM_MARKETS_MONITORING/Jute_Hard_Fibres/Documents/IGG_40/19-02_Plastic_Bags_Policy.pdf

Hu, H., Yao, X., Qin, Y., Yong, H., & Liu, J. (2020). Development of multifunctional food packaging by incorporating betalains from vegetable amaranth (Amaranthus tricolor L.) into quaternary ammonium chitosan/fish gelatin blend films. *International Journal of Biological Macromolecules, 159*, 675–684.

Jayakumar, A., Heera, K. V., Sumi, T. S., Joseph, M., Mathew, S., Praveen, G.,… & Radhakrishnan, E. K. (2019). Starch-PVA composite films with zinc-oxide nanoparticles and phytochemicals as intelligent pH sensing wraps for food packaging application. *International Journal of Biological Macromolecules, 136*, 395–403.

Ju, A., Baek, S. K., Kim, S., & Song, K. B. (2019). Development of an antioxidative packaging film based on khorasan wheat starch containing moringa leaf extract. *Food Science and Biotechnology, 28*(4), 1057–1063. https://doi.org/10.1007/s10068-018-00546-9

Kanatt, S. R., & Chawla, S. P. (2018). Shelf life extension of chicken packed in active film developed with mango peel extract. *Journal of Food Safety, 38*(1), e12385.

Kim, I., Viswanathan, K., Kasi, G., Thanakkasaranee, S., Sadeghi, K., & Seo, J. (2020). ZnO nanostructures in active antibacterial food packaging: Preparation methods, antimicrobial mechanisms, safety issues, future prospects, and challenges. *Food Reviews International*, 1–29.

Kim, S., Baek, S. K., Go, E., & Song, K. B. (2018). Application of adzuki bean starch in antioxidant films containing cocoa nibs extract. *Polymers*, *10*(11), 1210. https://doi. org/10.3390/polym10111210

Kumar, S., Mudai, A., Roy, B., Basumatary, I. B., Mukherjee, A., & Dutta, J. (2020). Biodegradable hybrid nanocomposite of chitosan/gelatin and green synthesized zinc oxide nanoparticles for food packaging. *Foods*, *9*(9), 1143.

Kuswandi, B., Larasati, T. S., Abdullah, A., & Heng, L. Y. (2012). Real-time monitoring of shrimp spoilage using on-package sticker sensor based on natural dye of curcumin. *Food Analytical Methods*, *5*(4), 881–889.

Latos-Brozio, M., & Masek, A. (2020). The application of natural food colorants as indicator substances in intelligent biodegradable packaging materials. *Food and Chemical Toxicology*, *135*, 110975.

Lee, K. Y., Yang, H. J., & Song, K. B. (2016). Application of a puffer fish skin gelatin film containing *Moringa oleifera* Lam. leaf extract to the packaging of Gouda cheese. *Journal of Food Science and Technology*, *53*(11), 3876–3883. https://doi. org/10.1007/s13197-016-2367-9

Lee, S. Y., Lee, S. J., Choi, D. S., & Hur, S. J. (2015). Current topics in active and intelligent food packaging for preservation of fresh foods. *Journal of the Science of Food and Agriculture*, *95*(14), 2799–2810.

Lei, Y., Mao, L., Yao, J., & Zhu, H. (2021). Improved mechanical, antibacterial and UV barrier properties of catechol-functionalized chitosan/polyvinyl alcohol biodegradable composites for active food packaging. *Carbohydrate Polymers*, *264*, 117997.

Li, X. H., Xing, Y. G., Li, W. L., Jiang, Y. H., & Ding, Y. L. (2010). Antibacterial and physical properties of poly (vinyl chloride)-based film coated with ZnO nanoparticles. *Food Science and Technology International*, *16*(3), 225–232.

Lin, D., Liu, Z., Shen, R., Chen, S., & Yang, X. (2020). Bacterial cellulose in food industry: Current research and future prospects. *International Journal of Biological Macromolecules*, *158*, 1007–1019.

Liu, J., Yong, H., Liu, Y., Qin, Y., Kan, J., & Liu, J. (2019). Preparation and characterization of active and intelligent films based on fish gelatin and haskap berries (Lonicera caerulea L.) extract. *Food Packaging and Shelf Life*, *22*, 100417.

Liu, Y., Zhang, X., Li, C., Qin, Y., Xiao, L., & Liu, J. (2020). Comparison of the structural, physical and functional properties of κ-carrageenan films incorporated with pomegranate flesh and peel extracts. *International Journal of Biological Macromolecules*, *147*, 1076–1088. https://doi.org/10.1016/j.ijbiomac.2019.10.075

Ludwicka, K., Kaczmarek, M., & Białkowska, A. (2020). Bacterial nanocellulose—A biobased polymer for active and intelligent food packaging applications: Recent advances and developments. *Polymers*, *12*(10), 2209.

Majeed, K., Jawaid, M., Hassan, A. A. B. A. A., Bakar, A. A., Khalil, H. A., Salema, A. A., & Inuwa, I. (2013). Potential materials for food packaging from nanoclay/natural fibres filled hybrid composites. *Materials & Design*, *46*, 391–410.

Martiny, T. R., Pacheco, B. S., Pereira, C. M., Mansilla, A., Astorga–España, M. S., Dotto, G. L.,... & Rosa, G. S. (2020). A novel biodegradable film based on κ-carrageenan activated with olive leaves extract. *Food Science & Nutrition*, *8*(7), 3147–3156.

Min, S., Harris, L. J., & Krochta, J. M. (2006). Inhibition of Salmonella enterica and Escherichia coli O157:H7 on roasted turkey by edible whey protein coatings incorporating the lactoperoxidase system. *Journal of Food Protection*, *69*(4), 784–793. https://doi.org/10.4315/0362-028x-69.4.784

Moccia, F., Agustin-Salazar, S., Berg, A. L., Setaro, B., Micillo, R., Pizzo, E.,... & Napolitano, A. (2020). Pecan (Carya illinoinensis (Wagenh.) K. Koch) nut shell as an accessible polyphenol source for active packaging and food colorant stabilization. *ACS Sustainable Chemistry & Engineering*, 8(17), 6700–6712.

Moradi, M., Tajik, H., Rohani, S. M. R., Oromiehie, A. R., Malekinejad, H., Aliakbarlu, J., & Hadian, M. (2012). Characterization of antioxidant chitosan film incorporated with Zataria multiflora Boiss essential oil and grape seed extract. *LWT-Food Science and Technology*, 46(2), 477–484.

Moradian, S., Almasi, H., & Moini, S. (2018). Development of bacterial cellulose-based active membranes containing herbal extracts for shelf life extension of button mushrooms (Agaricus bisporus). *Journal of Food Processing and Preservation*, 42(3), e13537.

Müller, P., & Schmid, M. (2019). Intelligent packaging in the food sector: A brief overview. *Foods*, 8(1), 16.

Mustafa, P., Niazi, M. B., Jahan, Z., Samin, G., Hussain, A., Ahmed, T., & Naqvi, S. R. (2020). PVA/starch/propolis/anthocyanins rosemary extract composite films as active and intelligent food packaging materials. *Journal of Food Safety*, 40(1), e12725.

Nguyen, V. T., Gidley, M. J., & Dykes, G. A. (2008). Potential of a nisin-containing bacterial cellulose film to inhibit Listeria monocytogenes on processed meats. *Food Microbiology*, 25(3), 471–478.

Padrao, J., Gonçalves, S., Silva, J. P., Sencadas, V., Lanceros-Méndez, S., Pinheiro, A. C.,... & Dourado, F. (2016). Bacterial cellulose-lactoferrin as an antimicrobial edible packaging. *Food Hydrocolloids*, 58, 126–140.

Pegoretti, A., Dong, Y., & Slouf, M. (2020). Biodegradable matrices and composites. *Frontiers in Materials*, 7, 265.

Peralta, J., Bitencourt-Cervi, C. M., Maciel, V. B., Yoshida, C. M., & Carvalho, R. A. (2019). Aqueous hibiscus extract as a potential natural pH indicator incorporated in natural polymeric films. *Food Packaging and Shelf Life*, 19, 47–55.

Qin, Y., Liu, Y., Zhang, X., & Liu, J. (2020). Development of active and intelligent packaging by incorporating betalains from red pitaya (Hylocereus polyrhizus) peel into starch/polyvinyl alcohol films. *Food Hydrocolloids*, 100, 105410.

Roy, S., & Rhim, J. W. (2020a). Carboxymethyl cellulose-based antioxidant and antimicrobial active packaging film incorporated with curcumin and zinc oxide. *International Journal of Biological Macromolecules*, 148, 666–676.

Roy, S., & Rhim, J. W. (2020b). Preparation of antimicrobial and antioxidant gelatin/curcumin composite films for active food packaging application. *Colloids and Surfaces B: Biointerfaces*, 188, 110761.

Ryngajłło, M., Kubiak, K., Jędrzejczak-Krzepkowska, M., Jacek, P., & Bielecki, S. (2019). Comparative genomics of the Komagataeibacter strains—Efficient bionanocellulose producers. *Microbiologyopen*, 8(5), e00731.

Salarbashi, D., Tafaghodi, M., Bazzaz, B. S. F., Mohammad Aboutorabzade, S., & Fathi, M. (2021). pH-sensitive soluble soybean polysaccharide/SiO_2 incorporated with curcumin for intelligent packaging applications. *Food Science & Nutrition*, 9(4), 2169–2179. https://doi.org/10.1002/fsn3.2187

Sanches, M. A. R., Camelo-Silva, C., da Silva Carvalho, C., de Mello, J. R., Barroso, N. G., da Silva Barros, E. L.,... & Pertuzatti, P. B. (2021). Active packaging with starch, red cabbage extract and sweet whey: Characterization and application in meat. *LWT*, 135, 110275.

Sharma, S., Barkauskaite, S., Jaiswal, S., Duffy, B., & Jaiswal, A. K. (2020). Development of essential oil incorporated active film based on biodegradable blends of poly (lactide)/poly (butylene adipate-co-terephthalate) for food packaging application. *Journal of Packaging Technology and Research, 4*(3), 235–245.

Shukla, V., Kandeepan, G., Vishnuraj, M. R., & Soni, A. (2016). Anthocyanins based indicator sensor for intelligent packaging application. *Agricultural Research, 5*(2), 205–209.

Sonkaew, P., Sane, A., & Suppakul, P. (2012). Antioxidant activities of curcumin and ascorbyl dipalmitate nanoparticles and their activities after incorporation into cellulose-based packaging films. *Journal of Agricultural and Food Chemistry, 60*(21), 5388–5399.

Ünalan, İ. U., Korel, F., & Yemenicioğlu, A. (2011). Active packaging of ground beef patties by edible zein films incorporated with partially purified lysozyme and Na_2EDTA. *International Journal of Food Science & Technology, 46*(6), 1289–1295.

UNEP (2018). Single-Use Plastics: A Roadmap for Sustainability, Rev. ed., pp. vi; 6.

Vilela, C., Kurek, M., Hayouka, Z., Röcker, B., Yildirim, S., Antunes, M. D. C.,... & Freire, C. S. (2018). A concise guide to active agents for active food packaging. *Trends in Food Science & Technology, 80*, 212–222.

Vo, T. V., Dang, T. H., & Chen, B. H. (2019). Synthesis of intelligent pH indicative films from chitosan/poly(vinyl alcohol)/anthocyanin extracted from red cabbage. *Polymers, 11*(7), 1088. https://doi.org/10.3390/polym11071088

Wahid, F., Wang, F. P., Xie, Y. Y., Chu, L. Q., Jia, S. R., Duan, Y. X.,... & Zhong, C. (2019). Reusable ternary PVA films containing bacterial cellulose fibers and ε-polylysine with improved mechanical and antibacterial properties. *Colloids and Surfaces B: Biointerfaces, 183*, 110486.

Yadav, S. S., Meshram, G. A., Shinde, D., Patil, R. C., Manohar, S. M., & Upadhye, M. V. (2011). Antibacterial and anticancer activity of bioactive fraction of Syzygium cumini L. seeds. *HAYATI Journal of Biosciences, 18*(3), 118–122.

Yang, Y. N., Lu, K. Y., Wang, P., Ho, Y. C., Tsai, M. L., & Mi, F. L. (2020). Development of bacterial cellulose/chitin multi-nanofibers based smart films containing natural active microspheres and nanoparticles formed in situ. *Carbohydrate Polymers, 228*, 115370.

Yao, X., Hu, H., Qin, Y., & Liu, J. (2020). Development of antioxidant, antimicrobial and ammonia-sensitive films based on quaternary ammonium chitosan, polyvinyl alcohol and betalains-rich cactus pears (Opuntia ficus-indica) extract. *Food Hydrocolloids, 106*, 105896.

Yong, H., Liu, Y., Yun, D., Zong, S., Jin, C., & Liu, J. (2021). Chitosan films functionalized with different hydroxycinnamic acids: Preparation, characterization and application for pork preservation. *Foods, 10*(3), 536.

Zhou, Y., Miao, X., Lan, X., Luo, J., Luo, T., Zhong, Z.,... & Tang, Y. (2020). Angelica essential oil loaded electrospun gelatin nanofibers for active food packaging application. *Polymers, 12*(2), 299.

15 Microplastic and Nanoplastic Pollution in Water Bodies from Conventional Packaging Materials
Need to Search for Biodegradable Polymers

Rahul Patwa
Bernal Institute, University of Limerick

Daisy Das
Indian Institute of Technology (Indian School of Mines), Dhanbad

CONTENTS

DOI: 10.1201/9781003227908-15

15.1 INTRODUCTION

In the year 1907, Leo H. Baekeland invented the first synthetic plastic, which marks the start of plastic age, and since then, the plastics have become an inevitable part of people's lives. Application of plastic as a packaging material and protection of cosmetics, medicine and food have gained immense popularity due to several enticing characteristics such as stability, extreme insulation, lighter weight and durability [1]. However, the concern that arises with the excessive utilization of plastic is its poor biodegradability due to environmental processes occurring in nature. Also, typical disposal of plastic waste by burying or burning them, or disposal in water bodies causes pollution of land, water and air, which poses a threat to biota. Plastic production relies heavily on non-renewable crude oil reserves; hence, large-scale production of plastic not only causes huge pollution, but also poses a serious threat to petroleum resources [1]. According to statistics, global plastics production was about 360 million tonnes (Mt) in 2018, which is expected to double in the coming two decades [2]. However, recycled plastics account for as low as 6% of total plastics, implying that 94% of plastics are likely to reach the environment through landfills or other pathways. In this way, the oceans may function as a vast plastic dumping ground. According to Plastics Europe's report, around 6–12 Mt of plastics are diverted to the oceans each year, and by the year

TABLE 15.1

Classification of Plastics, Characteristics and Their Uses

S. No.	Resin Code	Name	Characteristics	Uses
1	Resin code number 1	PET (polyethylene terephthalate)	Amorphous or semi-crystalline polymer	Bottles for water, juice and carbonated beverages; clothing
2	Resin code number 2	HDPE (high-density polyethylene)	Stronger intermolecular forces and tensile strength	Beauty products; containers for liquids and beverages
3	Resin code number 3	PVC (polyvinyl chloride)	Rigid and flexible	Pipes for plumbing, jackets and toys
4	Resin code number 4	LDPE (low-density polyethylene)	Flexible and tough	Cling films, bakery packaging, dairy cartons, coffee cups
5	Resin code number 5	PP (polypropylene)	Partially crystalline and non-polar	High-altitude clothing, autoparts, labware
6	Resin code number 6	PS (polystyrene)	Clear, hard and rather brittle	Packaging material, construction/building materials and fishing industry
7	Resin code number 7	Combinations of plastic resins	-	Large water containers and packaging

2025, more than 250 Mt of plastics will get accumulated in the oceans [3]. The major plastic compounds that are used in daily life for food storage, packaging, making large vessels and several other applications, their characteristics and classification by unique code, i.e., resin number, are presented in Table 15.1. It is worth mentioning that the natural deterioration of plastics is incredibly slow. With due course of time through physical, chemical, and/or biological activity, the plastic molecules will break down into microplastics (MPs) or nanoplastics (NPs). In recent decades, MPs and NPs have been identified as "emerging contaminants" and the most dangerous pollutants due to their increasing detection in the environment [4]. These are water-insoluble solid polymer particulates produced by a wide range of plastics for use in packaging, construction, cosmetics, personal care and pharmaceutical products (PPCPs), clothing industries, agriculture, detergents, transportation, commercial fishing, marine recreation, ink for 3D laser printing, the medical industries, among other applications [5].

MPs were defined by the majority of the researchers as plastic particles with a diameter of 5 mm that are continually broken down until they produce NPs with a diameter of 100–1000 nm [4,5]. Some researchers, on the other hand, further classified NPs into primary and secondary NPs. Primary NPs are plastic particles that are factory-made for various utilizations such as cosmetics, electronics, drug delivery, biomedical applications, whereas secondary NPs occur from unintended natural breakdown of MPs [5]. As a result, the definitions of MPs and NPs are now overlapping. Overall, we think that MPs are particles with a size range of 0.1 m–5 mm, whereas NPs have a size of less than 0.1 m.

15.2　OCCURRENCE OF MICROPLASTICS AND NANOPLASTICS IN THE ENVIRONMENT

Microplastics and nanoplastics have become commonplace in the marine environment, as well as in terrestrial water, freshwater and groundwater [6]. The primary sources of these developing contaminants include wastewater treatment plants (WTTPs), home waste disposal, cosmetics, food packaging, landfills, agricultural and irrigation runoff, industrial wastewater and domestic runoff [6]. They are also released into the environment due to various anthropogenic activities such as aquaculture, tourism, transportation, oil spillage from marine vessels and oil rigs [7]. With due course of time, these pollutant particles such as polycyclic aromatic hydrocarbons (PAHs), heavy metals and contaminants get collected in the deep oceans and pristine polar ice sheets and finally enter the human body through food chain [6]. The major sources, occurrence and transport to the MPs and NPs in aquatic and terrestrial environments is broadly illustrated in Figure 15.1. According to the literature, marine creatures are more

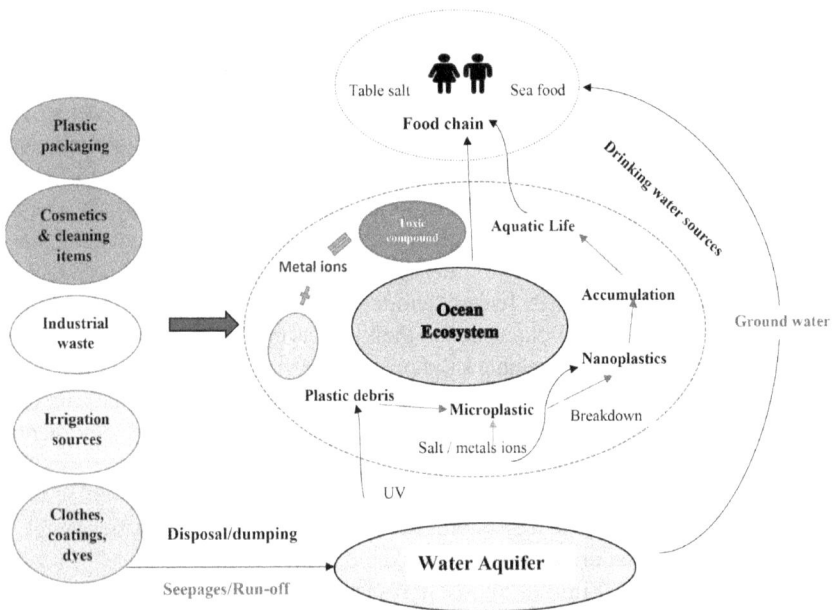

FIGURE 15.1　Sources, occurrence and pathway of microplastics and nanoplastics in the environment.

prone to swallow MPs and NPs due to their smaller dimensions, larger surface area and robust biological penetration capacity, resulting in the bioaccumulation of these pollutants within live species and a variety of detrimental ecological consequences [6]. Furthermore, because of the hydrophobic nature, large surface area and available charge species on the surface, MPs and NPs can function as a pathway for absorbing and transmitting coexisting hazardous contaminants such as harmful organic compounds, PPCPs, pathogens and heavy metals [8]. These newly generated toxic compounds tend to make more perplexing and complex cocktail compounds after undergoing various ionic interactions and reactions such as van der Waals electrostatic force, hydrogen bonding and other physico-chemical interactions [5]. However, the precise interaction mechanisms between pollutants and MPs/NPs are still unknown. The presence of MPs and NPs in WWTP effluent has recently gained attention. Nonetheless, the primary source of MPs and NPs in WWTPs is unknown because much of the work described has primarily focused on MP sampling, detection and purification, as well as knowing their physicochemical properties, which include size, color, shape and composition [9]. Additionally, it has been proved that the occurrence of MPs/NPs directly affects the treatment process used in WWTPs. Pesticides used in agricultural lands, biomedical/medical wastes, microfibers from polyester cloth materials, household items, cosmetics, food packaging, packaging materials and sewages, when discarded, are transported to water bodies and break down into microplastics and further nanoplastic particles due to various environmental conditions, photochemical reactions, and so on [10]. In addition, as a result of organism activity, chemical or biotic degradation of the plastic molecules may occur as a result of the hydrolytic process of enzymes produced by some micro-organisms, which reduce the molecular weight by breaking molecular chains. These contaminating particles (MPs and NPs) accumulate in aquatic life and eventually make their way into humans via the food chain. Furthermore, this hazardous trash is dumped and accumulates in landfills and in long term enters the freshwater streams and percolates and reaches groundwater [11]. In most of the places, due to growing population, people are dependent on surface water and groundwater as drinking water sources. Thus, MPs and NPs also affect humans and cause potential impacts. The long-term intake of these hazardous compounds, even at low levels, causes significant health problems and has been a source of worry.

15.3   SOURCES, PATHWAYS AND SINK OF MICROPLASTICS AND NANOPLASTICS IN WATER BODIES

15.3.1   Microplastics

The exact or precise size of microplastic (MP) is unknown since different researchers reported varying sizes of it. Microplastic is defined as plastic with a diameter of less than 5 mm. Furthermore, several authors reported MP sizes in the 1–5 mm [25], 2–6 mm [1], 10 mm [25] or 2 mm range [12]. Physical properties such as size, shape, density and color are important factors in deciding about the bioavailability and degradability of MPs [12]. MPs generally exist in irregular forms such as fragments, filaments, broken edges, pellets and granules. However, commonly found microplastics in the environment are microbeads and synthetic fibers originating from hygiene products, cleansing products, clothing or textiles, due to daily uses and laundering, respectively [2]. Once these microbeads and fibers come in contact with water or air as a result of drainage, they are most likely deposited in freshwater or surface water, as well as in the terrestrial environment. They've also been discovered in groundwater, perhaps due to surface runoff and delayed penetration through the surface. The primary factors that influence the morphologies of microplastics are the original form of the plastic, its degradability and exposure duration [12]. MP becomes highly dynamic because of its reactive nature, huge surface area and variety of shapes and forms. The various characteristics of micropollutant particles increase their bioavailability and make sampling and measurement extremely difficult. Plastics having a specific gravity <1 sink when they come into contact with water and float when their specific gravity >1; however, the hetero-aggregation and microbial colonization determine the density or specific gravity of plastic. As a result of activities of microorganisms, biofouling and biofilm evolution takes place, which intrigues the algae and invertebrates and further modifies the density or specific gravity of plastic. This, when comes in contact with biota and at influence of water current, permeates the water column in both freshwater and marine environmental conditions. The breakdown process of microplastics in the environment can be categorized as primary and secondary sources [6]. Microplastics derived from primary sources have specified shape, size and density and may be found as microbeads in exfoliants in hygiene and cosmetic goods, pellets of bigger plastic products, 3D printing emissions and industrial abrasives. Nonetheless, secondary sources of microplastics are derived from previously existing plastics in the environment as a result of breakdown and fragmentation by accidental littering. The fragmentation and degeneration of larger plastic molecules, i.e., macroplastics and mesoplastics, to microplastics leads to microplastics with irregular shape, size, density and particular chemical composition [6]. The identification of microplastic source in the environment whether it is from primary or secondary source is difficult task as the primary source is more likely to convert to secondary source in exposure to different elements, UV radiation and physical abrasion.

Microplastics enter the environment by either direct or indirect channels [13]. Microplastic contamination occurs directly in the environment owing to anthropogenic reasons such as excessive usage and exploitation of plastic in building, soil conditioning, and so on. Contamination in the water occurs as a result of indirect sources from land, powerful currents in the ocean and the accumulation of plastic waste in various zones, and the process is exacerbated by anthropogenic activities along the shore. This mechanism also promotes the migration of microplastics from the ocean to land or other regions as a result of oceanic movement.

15.3.2 Nanoplastics

Nanoplastics are particles that are less than 0.1 mm in size and cannot be seen with the naked eye. They are generated as a result of the breakdown and fragmentation of microplastics into nanometer-sized particles, which are generally the smallest sized particles. Commercial goods with exfoliating nature, cleaning supplies with abrasive qualities, air-blasting and plastic manufacturing are some of the most frequent sources of nanoplastics. NPs are typically difficult to define owing to their size and fluctuating physical characteristics, which has prompted several academics to conduct extensive research on their alterations under a variety of environmental conditions. Nanoplastics are generally amorphous or crystalline in nature and may occur as a single constituent material and sometimes embedded with other materials in the form of composite materials. Because of their larger surface area, NPs always tend to deposit or adsorb other organic materials or different ions present in water, leading to complicated cocktail compounds that enhance their harmful effects and make retrieval more difficult.

Fluvial networks such as river, its tributaries and lakes are the major sink for nanoplastics [2]. Contamination of nanoplastics in the environment occurs as a result of surface runoff from rural and urban landscapes, drainage of rainwater, rivers, lakes and streams containing microplastic and nanoplastic particles, which are then transferred to freshwater and the ocean. Movements of micro-/nanoplastics occur as a result of the effects of river currents, resulting in indirect deposition of these pollutants. In oceanic environment, gravity plays a major role in helping the nanoplastics to sink into deeper oceanic environment, but sometimes currents affect the process and complicate their transportation and circulation. Currents at the bottom of ocean generally contribute to the transport of the nanoplastic particles in surface sediments by shear force, drift deposits and current flows.

In addition, oceanic currents, internal tides and thermohaline stratification will also resuspend the plastics in the ocean, which may carry them to other areas. This process of resuspension is accelerated by the ingestion and defecation of the organism. Furthermore, Ekman and geostrophic currents, as well as Stokes drift, produce MPs/NPs accumulation regions, exacerbating the region's ecological danger. Furthermore, estuaries are greatly disturbed by bigger turbidity currents that include sufficient buoyant microplastic particles, producing micro-/nanoplastics hot spots in particular locations, as a result of the interaction of buoyancy, tides and wind under tidal currents.

15.4 EFFECTS OF MP AND NP ON AQUATIC ECOSYSTEM AND HUMAN LIFE

Microplastic or nanoplastic pollution in low concentrations can endanger the ecosystem and cause significant difficulties for aquatic life and human health. They are known to disrupt endocrine function in fish, resulting in reproductive impairment, reversal characteristics in the opposite sex and other abnormalities. Microplastics or microbeads derived from cosmetics also have an impact on the growth of common floating plants and other water plants. In comparison with leaf growth and photosynthesis, root growth is severely harmed by microplastics as the roots are immersed in water to collect nutrients. Microplastic also hampers the photosynthesis process and inhibits the development and fertility of algae and microalgae [14]. Also, the consumption of these polystyrene beads or generally microplastics by aquatic animals or different zooplanktons hampers their life stages and damages their intestine. Furthermore, with the excretion of these creatures or microbes, these microbeads enter the aquatic environment and produce additional hazardous compounds by sticking or absorbing with other sea algae or metal ions, promoting long-term accumulation and deposition of such poisonous compounds. Mussels' ingestion of high concentrations of high-density polyethylene (HDPE) plastic beads affects their oxygen consumption and survivability. Microplastic distress in baleen whales that rarely approach the near coast zone as it is found in their blubbers and skin is a matter of serious concern because this creature live at a depth of 200–300 m in the sea. In recent years, research in the field of harmful effects of these emerging contaminants in marine ecosystem as well as terrestrial ecosystem has been attaining importance due to their occurrence at various zones of sea and deep sea level. Microplastics have a degrading effect on animals, particularly filter and digestive processes, as evidenced by the behavior of numerous aquatic creatures or organisms. Reduced MS intake or increased defecation speed can help mitigate these negative effects. The presence of nanoplastics at various concentrations in both marine and freshwater environments and their toxicity is conflicting. Although appropriate quantifying methods are not available for the assessment of nanoplastic concentration in the environment, it is assumed that due to the degradation and fragmentation of microplastics and macroplastics and their release in waterbodies, it is possible that the concentration of nanoplastics increases up to 1014 times compared to the present microplastics concentration. Several studies revealed that polystyrene-based nanoplastics are more hazardous than polystyrene-based microplastic particles and mostly naturally occurring nanoplastics are found to be associated or entrapped with combination of PS, PE, PVC and PET in marine environment [11].

As per several studies reported, environmental effects of nanoplastic compounds are more harmful and obscure than those of microplastic compounds due to their nanometric dimensions and colloidal nature and can be viewed as a potentially more dangerous type of marine trash, allowing them to be absorbed by biological barrier as a result of its cell metabolism [15]. Nanoplastics are more inclined to entrap various components such as stabilizers, plasticizers, flame retardants, pigments, metal content and surface modifiers from the surrounding environment due to their greater affinity toward these compounds and form convulent cocktail compounds. The hazardous impact of nanoplastics may increase due to higher intake and uptake of nanoplastic particles by organism and due to their tendency to get adhered to other organic or biogenic matter due to their higher surface area and resulting in their increasing efficiency to transport and accumulate in the local environment. When nanoplastics come in association with common cyanobacteria, they produce microgels and polystyrene-based nanoplastic aggregates, which are deposited in aquatic environments and move to various estuaries and coastal regions under specific environmental circumstances [15]. Nanoplastic particles accumulate in brine shrimp's stomach, tail and appendages as a result of microalgae consumption, which is then transferred to humans via the food chain. The reproductive stages of oysters have also been discovered to be negatively impacted by polystyrene-based nanoplastic debris.

The accumulation of microplastics and nanoplastics in human body occurs due to the consumption of fish such as cod, mussel, oyster, mullet and shrimps, which are considered as animals of commercial importance [15]. Furthermore, animals from aquaculture in estuaries and coastal lagoons may swallow such micro- and nanoplastics, which then enter the human body through food. Additionally, aquaculture facilities feed their animals with microplastic-contaminated feed ingredients.

In European countries with high shellfish consumption, it is predicted that around 11,000 microplastic particles in the size range 5–1000 µm are consumed on average by a consumer per year, whereas in countries with low shellfish consumption, it is around 1800 microplastic particles per year, which is still frightening [2]. It is estimated that about 175 microplastic particles in size range 200–1000 µm are ingested through shrimp consumption only, per person per year [32].

In countries such as France, Italy, Denmark, Spain and the Netherlands, microplastics were found in mussels, which are consumed as food by humans [2]. Commercial mussels from Belgium were found to contain 3–5 microplastic fibers per 10 g of mussels. [14]. Similar studies have shown that microplastics were found in marine mussels as well. Microplastics (size > 500 m) have been found in popular fish species such as Atlantic cod, European hake, red mullet and European pilchard in amounts of 9% and 28% in fish gastrointestinal tracts in the USA and Indonesia, respectively. Likewise, plastic pieces of 0.5 and 1.4 per individual fish on average are found in the US and Indonesian samples, respectively. Also, liver of anchovies and sardines, which are generally consumed totally, is found to contain microplastics. The literature suggests a sufficient amount of records that microplastic is present in various synthetic forms and with organic matter in different sources of human foods, ingredients and drinking water [2,6]. Microplastic residues have been discovered in a variety of canned foods for human consumption, as well as salt, beer, sugar and honey. Furthermore, the use of plastic bottles and beverage cartoons adds to the presence of microplastics in drinking water and other liquids. As a result, there is an elevated danger associated with microplastic ingestion and long-term exposure. It is a well-known fact that microplastics are very persistent and are rapidly accumulating in many ecosystems. The buildup of microplastics in the environment has resulted in toxicological interactions between organisms and microplastics. The indirect impact of microplastic accumulation is causing global climate change, particularly temperature fluctuations, which is a major issue. Scientists think that microplastics larger than 150 m are not absorbed by the body, but microplastics less than 150 m are absorbed via the stomach into the lymph and blood flow, resulting in systemic exposure. It is also reported that microplastics of size $\leq 20\,\mu m$ penetrate into organs, while sizes in the range 0.1–10 µm are able to cross the cell membrane and blood–brain barrier. As a result, they are found to be present in organs such as the brain, liver and muscles, causing immunotoxicity. Recent research on microplastics and nanoplastics has revealed that they cause cytotoxicity and oxidative stress at the cellular level in humans [14]. The pollution of seas by microplastics has an ecological impact, but it also jeopardizes food safety since seafood is exposed to microplastics, which damage human health if ingested. Exposure to microplastics and nanoplastics floating in the air has a negative influence on human health.

15.5 DEVELOPMENT OF VARIOUS TECHNOLOGIES FOR THE REMOVAL OF MP AND NP FROM WATER

The presence of emerging pollutants in water, such as microplastics and nanoplastics, has become a major issue since it reduces the effectiveness of conventional wastewater treatment methods. The release of these chemicals into the aquatic environment has had an impact on all living species. The broad classification of the treatment processes used is presented in Figure 15.2. Among all the methods, membrane filtration has gained attraction due to its effective removal of 270–333 nm size NP and around 92% but limited to certain pollutant type. Electrocoagulation/coagulation has become a popular treatment method for the

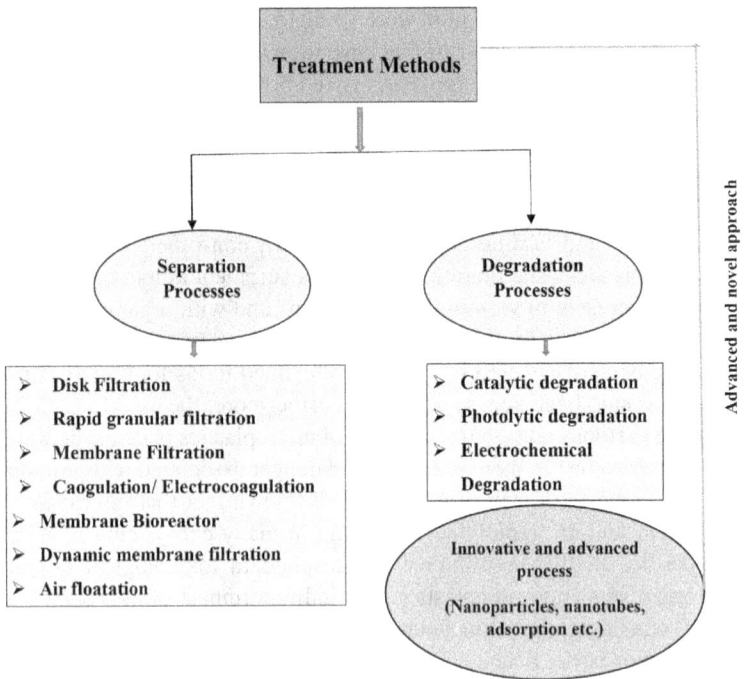

FIGURE 15.2 Developed treatment processes for the removal of MPs and NPs from water.

removal of MP and NP from water as it is proved to be an effective method to remove several types of MP and NP particles such as PE, PES and PS simultaneously with minimum production of sludge. Moreover, no addition of chemicals is required and it is easy to handle. Among the secondary treatment methods, electrochemical degradation is able to remove more than 90% of NP and MP contamination with any concentration along with organic pollutants. The other methods such as catalytic degradation and photocatalytic degradation have some advantages such as no side effects on marine life, 50% degradation under UV light and short reaction time. But recyclability and reusability of the catalyst are challenging, which implies the need for some advanced research in the area.

15.6 WORLDWIDE REGULATIONS AS A SUSTAINABLE DEVELOPMENT GOAL

The European Union (EU) has long been a leader in environmental sustainability and plastic pollution reduction. In 2018, the EU introduced a circular economy model, which states that all plastic goods would be developed for extended durability, reuse and effective recycling, and that all packaging materials in the EU market will be reusable and/or recyclable by 2030 [16,17].

The European Commission has mandated that companies involved in plastic manufacturing or recycling present ambitious and concrete set of voluntary commitments to support the strategy and its ambition for 2030 [18]. In other words, designers and producers must develop reusable and recyclable plastic items for consumers while discontinuing their present business models of single-use plastics [19]. In the last 5 years, European countries such as Belgium, France, Ireland, Italy, Sweden and the UK (UK), as well as Canada, New Zealand, South Korea and the USA have implemented restrictions on the intentional use of microplastics in cosmetics and personal hygiene products [19]. The European Chemicals Agency (ECHA) recommended significant restrictions and a ban on the deliberate use of microplastics in the EU on August 22, 2019. The proposal is aimed at items manufactured of or containing microplastics, such as fertilizers and their additives, plant protection products and coated seeds, cosmetic products, detergents and cleaning supplies and waxes. Furthermore, numerous international organizations, including the G7, G20, the UN and the Arctic Council, are working to promote policy to minimize or avoid plastic pollution in the environment. The G7 Ocean Plastics Charter, the Arctic Council Desktop Study on Marine Litter including Microplastics in the Arctic, and the G20 Implementation Framework for Actions are just a few of these projects.

15.7 SOLUTION TO PLASTIC PACKAGING MATERIALS

Over the past decade, due to daily lifestyle, the utilization of synthetic packaging materials which are conventional synthetic polymer has increased rapidly due to several enticing advantages such as easy process, low cost and low density.

But excessive utilization of such synthetic polymers in various forms is causing a serious concern in the environment due to their non-degrading behavior. In order to meet such environmental crises arising from non-biodegradable packaging polymers, demands have been raised for alternative sustainable packaging. Biodegradable polymers from natural resources can be an effective raw material to develop packaging materials that will be eco-friendly biodegradable packaging and will have an important impact on the market of packaging industry and applications in the coming years [20]. Hence, replacing the synthetic polymer by biodegradable packaging polymers may be an ideal solution to reduce the environmental impact and petro-dependence. Biopolymers, as an alternative bio-packaging material, enable packaging products to be entirely biodegradable or compostable. The hydrolytic or enzymatic breakage of polymer bonds occurs during the biodegradation of biopolymers. Biodegradation is frequently described as an occurrence that occurs as a result of enzyme action and/or chemical breakdown linked with living organisms (bacteria, fungi, etc.). It is worth noting that the other processes such as photodegradation, oxidation and hydrolysis may also have an impact on the structure and chains of polymers prior to or during biodegradation [21].

Generally, these biodegradable polymers are classified as (i) natural polymer and (ii) synthetic polymer. The natural polymers are derived from various natural or plant-based resources, for example proteins, chitosan, cellulose, starch and lipids, which are used as raw materials for the synthesis of biodegradable materials [20]. Also, some renewable sources such as microorganisms, biotechnology and biomass also contribute to this category. The synthetic polymers are conventional polymers produced from petroleum-based monomers such as aliphatic or aromatic polyesters, polyesteramides and polycaprolactones prepared from synthetic monomers, which are non-degradable in nature. Some of the nanocomposites of biodegradable polymers from synthetic sources widely used as packaging material basically food packaging are polylactic acid (PLA), polyvinyl alcohol (PVA), polycaprolactone (PCL), polybutyl succinic acid-butyl adipate (PBSA), poly(butylene adipate-co-terephthalate) (PBAT), furcellaram (FUR), poly(3-hydroxybutyrate-co-3-hydroxyvalerate) (PHBV) and polyethylene glycol (PEG) [22]. The decomposable nature of biodegradable polymers is advantageous for packaging materials, which is why they have been widely utilized in packaging. There are several natural and synthetic biodegradable polymers available today that are utilized in packaging applications due to their ease of breakdown and recyclable nature.

Although the newly developed biodegradable films derived from various natural sources are not able to replace the synthetic polymers in the market, their partial replacement acts as a major stress factor. These biocomposite films also provide an additional factor in food packaging application for preservation by controlling several conditions such as external or external–internal exchange of gases, humidity, undesirable aromas and/or compounds migration and also act as natural antimicrobial or antioxidant agents. Some of the widely and commonly used biodegradable polymer materials are discussed below.

15.7.1 STARCH

Starch are polysaccharides obtained from plants and are renewable in nature and most abundantly found. It consists of repeating glucose moieties of amylose and amylopectin. Amylose d-glucose unit consists of α-1,4 bonds, whereas amylopectin consists of α-1,4-linked d-glucose units that are branched by α-1,6 bonds [23]. Starch is abundant, of low cost and easy to handle, and it has low oxygen permeability, which makes it a suitable biodegradable packaging film candidate. However, hydrophilicity and brittleness limit its usage for use in plastic bags and food packaging. To overcome these limitations, thermoplastic starch is prepared by mixing starch with plasticizers such as glycerol, polyethylene glycol and sorbitol under the application of shear and heat during extrusion stage.

15.7.2 POLYLACTIC ACID (PLA)

PLA is prepared from lactic acid obtained upon bacterial fermentation of renewable crops such as corn and sugar beet. This polyester has attained popularity and gained attention as a packaging material because of its biodegradability, transparency, processability, abundance and cheaper cost. Low molecular weight PLA is obtained using lactic acid by condensation polymerization, whereas high molecular weight PLA with better mechanical properties is made using lactide monomer through ring-opening polymerization or lactic acid azeotropic condensation polymerization [24]. PLA has a melting temperature of 150°C, which is higher than PHA, PEG and PCL, making it suitable for a variety of processing techniques such as injection molding, blow molding, compression molding and thermoforming. However, PLA suffers from poor barrier properties as compared to PET and has a low elongation at break of ~10%, making it brittle [22]. One approach discusses the use of exfoliated clay and TPS to prepare films with improved mechanical properties [22]. The addition of PHB to PLA can significantly improve material properties of PLA such as barrier properties, hydrophobicity and mechanical strength [24]. Such approaches can become a valuable alternative to biopolymer-based food packaging applications.

15.7.3 POLYCAPROLACTONE (PCL)

PCL obtained from polymerization of ε-caprolactone has a low melting point and low viscosity. It has the advantages of biodegradability and high thermal processability, but has poor stress fracture resistance, barrier properties, dyeability and adhesion, which limits its usage in packaging sector. This limitation can be overcome by blending it with other polymers such as PLA and cellulose-based polymers [25].

15.7.4 POLYHYDROXYBUTYRATE/POLYHYDROXYALKANOATE (PHB/PHA)

PHA is another product of microbial fermentation, which is a polyester of different hydroxyalkanoates and has shown great promise in the field of packaging, biomedical and agricultural sectors. PHAs are biocompatible, crystalline thermo-plastic elastomers with excellent UV resistance and material properties. However, PHAs suffer from low thermal stability, weak mechanical strength and low pro-cessability [26]. PHB has the highest crystallinity (~70%) of all PHAs, which is responsible for high magnetic properties on a par with PET. It breaks up into (R)- and (S)-hydroxybutyrates and other non-harmful chemicals under aerobic/anaerobic conditions [26]. PHB has a lamellar microstructure, which is beneficial for better barrier properties. In addition, the processing temperature is within PLA processing temperature, which makes it an ideal candidate for blending. One such example is PLA/PHB blend containing catechin, which is further melt processed to create plasticized PLA/PHB mixes that are ideal as active packaging materials for fatty foods [27].

15.7.5 ADDITIVES USED TO IMPROVE OVERALL PROPERTIES OF BIODEGRADABLE MATERIALS

15.7.5.1 Glycerol

To enhance the plasticizing effect of biodegradable materials, glycerol ($C_3H_8O_3$) is commonly used. Generally, it is commonly used during the fabrication of starch-based biocomposite films as a plasticizer. The addition of glycerol improves the elongation of biodegradable materials by weakening the intermo-lecular hydrogen bonding and eventually reduces the viscosity of polymer chain. A biocomposite film, especially starch-based one, shows inferior barrier proper-ties, better elastic modulus and tensile strength with glycerol as an additive and has comparatively a less compact structure than the film without plasticizers or additives [1,28].

15.7.5.2 Cellulose

Cellulose and materials obtained from cellulose derivatives such as carboxy-methyl cellulose (CMC), cellulose nanocrystals (CNCs) and cellulose nanofibers (CNFs) are frequently used as additives for biocomposites as they are biocom-patible, biodegradable, abundant and cheap. Cellulose derivatives are suitable as organic additives as they are rich in surface hydroxyl groups, which interact with polymer molecules and form a dense network structure and enhance several properties such as thermal properties, hydrophobicity and tensile strength of the film.

15.7.5.3 Gelatin

Gelatin is a protein-based polymer that is derived from collagen of pigs and cows. It has a random coil conformation at elevated temperatures, whereas at lower tem-peratures, it forms helix. Its sensitiveness to temperature allows it to form viscous sols and thermo-reversible gels when heated and cooled, respectively. The incor-poration of gelatin improves the barrier properties, transparency and mechanical properties of starch-based films and other biocomposite films [1,29].

15.7.5.4 Chitosan

Chitosan is obtained by deacetylation of chitin and is known to have good bio-compatibility and antibacterial properties. In one such study where chitosan is blended with starch, the composite films have a robust structure, smoother texture and improved oxygen/water vapor barrier. This enhancement in properties is due to inter-/intramolecular hydrogen bonding among starch and chitosan molecules. This may be related to the formation of the intermolecular and intramolecular hydrogen bonds between the chitosan and starch molecules. Likewise, the -OH groups available on starch can bind with $-NH_2$ groups of chitosan [30].

15.7.5.5 Citric Acid

Citric acid has been around as a commonly used cross-linking agent due to the affinity of the carboxyl (-COOH) groups to attach with hydroxyl (-OH) groups on polysaccharides and forms diester linkages. Also, it is non-toxic and cheap, which attracts its use in food packaging materials. The addition of citric acid to biocomposite film tends to enhance humidity barrier and impedes the degeneration of packaging during storage. Moreover, some studies also report that the addition of citric acid in starch/polyvinyl alcohol/films shows strong antibacterial properties, making them suitable for food packaging application [31].

15.7.6 TECHNIQUES USED TO MANUFACTURE BIODEGRADABLE COMPOSITE

To fabricate biodegradable polymer films several methods are used, such as solution casting method for the formation of single-layer films, by continuous casting or coating in laboratory scale. For industrial-scale manufacturing, methods such as extrusion, blown casting, foaming process and calendaring are generally used. Also some new upcoming techniques such as electrospinning, 3D printing, nanotechnology and reactive extrusion are gaining interest in packaging industries due to their several attractive advantages over conventional methods of manufacturing [1]. This section will discuss briefly the widely used processes for polymer film manufacturing. The advantages, disadvantages and application of various widely used manufacturing methods of polymer film are presented in Table 15.2.

15.7.6.1 Casting Evaporation Approach

It is a facile, low-cost technique to prepare composite films. A mixture of solvent and polymer is stirred at elevated temperatures; upon dissolution, the solution is degassed to get rid of air bubbles and then poured onto a smooth surface such as glass or Teflon. Then the solvent is allowed to evaporate at room temperature or at higher temperatures in an oven. Upon complete evaporation, the film is peeled off from the plate [1,32].

15.7.6.2 Foaming Processing

Foaming essentially is of four main types, viz. baking, molding, supercritical fluid foaming and extrusion. In baking, the polymer and foaming agent are put in a mold and heated. Once the water evaporates, the resultant material is a foam.

TABLE 15.2

Various Methods Developed to Manufacture Biodegradable Composite Films, Their Applications, Advantages and Disadvantages [1]

Methods	Advantages	Disadvantages	Applications
Film solution casting	• Simple and inexpensive to implement • Easy to operate	• High solvent content • High production costs due to energy consumption for drying • Use in industry is restricted except starch-based films • Difficulty to control film thickness and uniformity	Biodegradable films from corn starch, cassava and potato starch
Foaming processing	• Can be manufactured on-site using basic extruder • Flexible technique	• Operating conditions can alter mechanical properties	Food packaging material
Extrusion processing	• Useful for industrial application • Making complex plastic products at low cost	• Reduced crystallinity due to thermal stresses during extrusion	Polymer processing, plastic product making, biodegradable starch-based films
Electrospinning	• Starch is good candidate because of its good biocompatibility • Electrospinning enables biopolymers to have some unique characteristics such as antioxidant and antibacterial properties	• Uses harmful solvents such as DMSO and formic acid Additives such as other polymers, plasticizers or cross-linking agents to improve overall properties of nanofibers	Food and biomedical applications • Food packaging, drug delivery, wound dressings
3D printing	Can be used as an efficient method for food manufacturing technology, especially important for environmental reasons	• Inferior supporting and molding characteristics	• Applied in many sectors such as medical, aeronautics, mobile and construction • Complex shapes and textures possible especially useful in food sector

(Continued)

TABLE 15.2 (*Continued*)
Various Methods Developed to Manufacture Biodegradable Composite Films, Their Applications, Advantages and Disadvantages [1]

Methods	Advantages	Disadvantages	Applications
Reactive extrusion	• Improves production efficiency due to the combination of enzymatic action and mechano-thermal effects during extrusion	• Long reaction times due to milder conditions • Short-lived reactants	Used to produce high-quality carbon sources from starch matrices for alcoholic beverages
Nanotechnology	Nanomaterials change the crystallization kinetics of the biopolymer and improve mechanical and barrier properties	Unappropriated addition of nanofibers sometimes produce weaker composites due to agglomeration	Used in various sectors such as food, agriculture, biomedical and automobile

Molding is similar to baking, just it uses a temperature-controlled hydraulic press, and foaming is caused by heating and pressure changes. Supercritical fluid foaming uses supercritical carbon dioxide and foaming material. Once the contents are depressurized, they form bubbles and lead to the formation of foam. Extrusion foaming involves mixing the contents under heat and pressure and then extruding them out to produce a foam [1].

15.7.6.3 Extrusion Processing

Extrusion processing does not use any solvent; hence, it is called the dry processing method. It is mainly of three kinds, viz. blowing, compression and injection molding. Extrusion is essentially used to prepare master batches where a mixture of materials under the influence of shearing and heating are pushed to the die head using a rotation motion of screw. The extruded material is cut in the form of pellets and used as raw material called master batches. Blowing process uses pressurized air to blow films out of molten polymer and cools down to form films with required thickness. Compression molding creates film by hot pressing the molten polymer. Injection molding is a combination of extrusion and molding processes [33].

15.7.6.4 Electrospinning

Electrospinning uses electrostatic interaction to produce nanofibers mainly in food and biomedical sectors. The fiber morphology can be modified based on requirements, such as solid, hollow, cylindrical and flat. The process essentially requires formation of a Taylor cone, and then the fibers are drawn out from the polymer jet toward the grounded electrode once the electrostatic force exceeds the surface tension. During the flow, the solvent dries up and the collected fibers are rolled up on a collection drum, which forms a nano-mat made from nanofibers [34].

15.7.6.5   3D Printing

In 3D printing, the material to be fabricated is converted into a 3D model and broken into multiple 2D layers, which are then printed layer by layer and ultimately
form the material. 3D printing has seen a rapid demand in the food industry where
it can be used to prepare food products with specific nutrient profiles and as per
special needs of people. The gelling profile, rheology, mechanical properties and
heat stability are key factors that decide the suitability of a material as an inking
material. One such combination is starch/acrylonitrile/butadiene/styrene copoly-
mer-based printing ink, which has good extrudability and gelling properties [23].

15.7.6.6   Reactive Extrusion

In reactive extrusion, extrusion occurs in combination with a chemical or enzymatic reaction, which leads to modifications in polymer structure by grafting,
cross-linking, polycondensation, etc. Extruder not only does the polymer processing, but also acts as a reactor where reaction occurs under the influence of physical fields such as shear, pressure and temperature. The screw is the main area
where the reaction occurs at various areas such as conveying element, mixing,
kneading, pressurization and melting [35].

15.7.6.7   Nanotechnology

The science of materials within size range 1–100 nm is known as nanotechnology.
The characteristics of nanoparticles are very different from larger-sized materials, and they can provide unique functionalities. Major applications include food
industry and pharmaceutics, where nanocarriers are used to deliver bioactive
agents and drugs to the target site, respectively [1].

15.8   CHALLENGES FACED IN BIODEGRADABLE PACKAGING

Biopolymers such as PLA, PVA and chitosan have undergone a remarkable
growth and gained extensive attention in recent years in food packaging as
well as other packaging applications due to their biodegradability, environmental sustainability and lower production cost by several new technologies. A
detailed flow diagram of various types of eco-friendly plastics from different
sources is shown in Figure 15.3. However, due to their brittleness, low thermal
stability and low barrier properties for gas and solvent, they are insufficient to
use in their pure form. Cellulose in the form of nanofibers and nanocrystals has
gained attention in packaging applications due to its ability to carry antioxidant
and antibacterial agents and also has a high thermal resistance, ultraviolet rays
shield, decomposability, renewable nature and non-toxicity. Apart from having
such attractive advantages and essential properties, it can't be used in pure form
due to its hydrophilic nature. It can be used in preparation of different polymer
composites. Few studies have reported that cellulose nanofibers and nanocrystals, some nanoparticles, carbon nanotubes, etc., can be encapsulated with PLA/
PVA/CS composite to overcome such drawbacks. Also with due course of time,

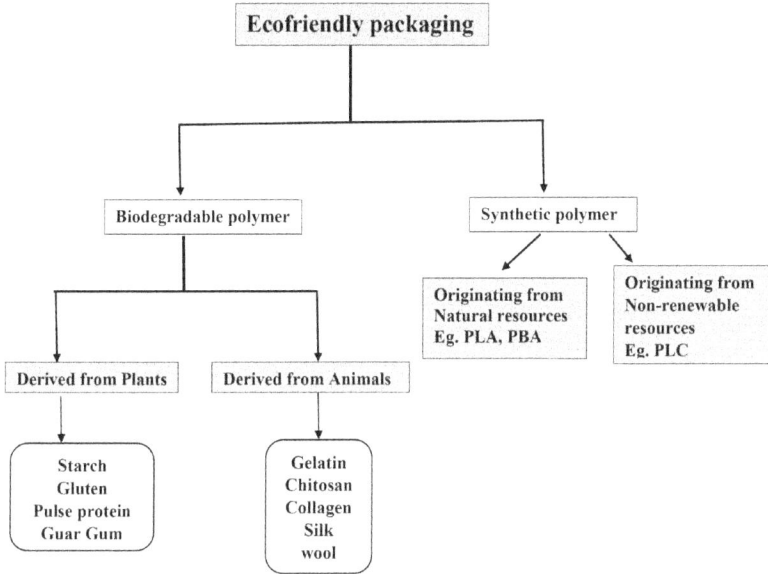

FIGURE 15.3 A detailed flow diagram of various types of eco-friendly packaging plastics from different sources is shown.

the quality of food may degrade due to microbial delay, which will increase the pH and also release several gases due to oxidation, and changes in temperature may occur due to some reaction within it. Such reinforcement of additives or nanomaterials in conventional biopolymer not only improves the mechanical properties of the composite film and enhances their barrier properties, but also acts as a thermo-responsive, pH-responsive and antioxidant material. The additives used for imposing the responsive behavior of conventional film to minimize the food waste and active packaging can be derived from natural sources or from various food wastes such as pulp, oil cakes, peels and husks and constitute about 30%–50% of the total food weight. The utilization of a particular food waste may be decided by the composition and cost of extraction of useful materials from it, such as flavanols, oils and tannins [20]. Recently, smart biodegradable packaging materials have been reported, which produce metabolites (amines/acids) upon microbial degradation of food material inside packaging which can display real-time visual changes in the form of color upon sudden changes in pH, atmospheric gases and temperature [36]. Some of the technologies and innovations in enhancing the performance and responsive properties of biodegradable polymer in food packaging are presented in Table 15.3. In this section, some of the key parameters such as mechanical properties, thermal properties, barrier properties and biodegradability need to be improved in conventional biopolymers for real-time applications.

TABLE 15.3

Technologies and Innovations in Enhancing the Performance and Responsive Properties of Biodegradable Polymers in Food Packaging [39]

Conventional Biodegradable Polymer	Modification with Additives/Nanomaterials	Properties Improved of Conventional Film
PLA	Carbon nanofibers, graphene oxide, MWCNT, graphene, nanoplatelets, silver nanoparticles	Antimicrobial properties, thermal stability and tensile strength
PVA	MWCNT, graphene oxide–zinc oxide nanoparticles	Antimicrobial, antioxidant, heavy metal, fungi entrapment, UV barrier, color effect, recyclability
PHBV	Cellulose nanofibrils/nanocrystals, graphene oxide, zinc oxide nanoparticles	Antimicrobial, antioxidant, antibacterial, wettability, barrier properties, compression, thermal stability, plasticizer effect
PLA/CS	MWCNT	Antimicrobial, swelling ratio, mechanical stability
CS	Graphite naturals, TiO_2 nanoparticles	Non-cytotoxicity, antifungal and mechanical stability
PVA/starch	Graphene oxide, silver nanoparticles	Antimicrobial, better tensile strength, thermal stability and barrier properties
PCL	Graphite oxide, silver nanoparticles	Improves electrical properties; biocompatible
PEG	Silver/zinc oxide nanoparticles, graphene oxide	Inhibits biofilm formation, mechanical and antimicrobial
FUR	Carbon quantum dots, graphene oxide, iron nanoparticles (γ-Fe_2O_3)	Antimicrobial, antioxidant, mechanical, UV-blocking, hydrophobicity
CS/PVP	Graphene nanolayers	Antimicrobial, flexibility, optical transmittance, thermal and water vapor permeability improvement

#PLA: polylactic acid, PVA: polyvinyl alcohol, PHBV: poly(3-hydroxybutyrate-co-3-hydroxy valerate), CS: chitosan, PCL: poly(ε-caprolactone), PEG: polyethylene glycol, FUR: furcellaram, PVP: polyvinylpyrrolidone, MWCNT: multi-walled carbon nanotubes.

15.8.1 MECHANICAL PROPERTIES

By mechanical properties, we mean the resistance of packaging material toward deformation. Common instruments for mechanical testing are universal testing machines, dynamic mechanical analyzers (DMAs) and flexural testing machine. Few of the significant mechanical property indicators are tensile strength, modulus, elongation at break, stress, strain, storage modulus (E') and loss factor (tan δ). Optimum mechanical properties are much needed for a material to be suitable for food packaging [37].

15.8.2 THERMAL PROPERTIES

Thermal characteristics of packaging are critical for their manufacture, use and disposal. Packaging materials are critical for their manufacture, use and disposal. Two most common thermal analytical techniques are thermogravimetric analysis (TGA) and differential scanning calorimetry (DSC). DSC essentially helps in determining the phase transition of the materials with respect to temperature change such as glass transition segment and melting point. The glass transition temperature (T_g) of a packaging plastic is governed by the compatibility/miscibility of its individual components present in it. The melting temperature (T_m) decides the thermal stability and thus the processing temperature of the material [1]. The TGA helps in determining the degradation temperature of material and the composition of material as well.

15.8.3 BARRIER PROPERTIES

To decide on the shelf life and quality of packaged food material, it needs to provide a good barrier against water vapor and oxygen molecule movement. Hence, it is highly essential to measure the water vapor permeability (WVP) and oxygen permeability (OP), which are collectively called barrier properties of the packaging films [1,38]. Few of the ASTM accepted test methods are ASTM E-96-95, ASTM E398-03, ASTM D3985 and ASTM F2622. The instruments used to analyze the barrier properties give results in the form of water vapor transmission rate (WVTR) and oxygen transmission rate (OTR). Incorporation of cellulose and starch nanoparticles to starch-based films has lowered the WVP and OP by increasing the tortuosity within the matrix for the movement of water and oxygen molecules. The improved barrier properties mean an extended shelf life as it prevents moisture uptake, oxidation and microbial buildup. In addition, it reduces the consumption of conventional plastics which contribute toward plastic pollution in the form of microplastics and nanoplastics.

15.8.4 BIODEGRADABILITY

A material's biodegradability is a measurement of how readily it can be digested by microbial enzymes secreted by microbes such as fungi and bacteria. Biodegradation occurs either aerobically, which results in carbon dioxide and water, or anaerobically resulting in hydrogen and methane. The extent of biodegradation can be measured by the amount of CO_2 released by the microbes present in the soil and also by the reduction in molecular weight and actual weight upon the soil burial tests. The above two tests are the benchmark tests to decide upon the biodegradability of a biopolymer-based packaging material [1].

15.9   CONCLUSIONS AND PERSPECTIVES

Plastics have by far been the most common environmental pollutants found in nearly all aquatic and terrestrial environments. They occur in so many distinct forms, with so many additions, diverse sources, routes and sinks that quantifying is almost impossible. The micron- and nanometer-scaled dimensions of microplastics (MPs) and nanoplastics (NPs) pose a serious challenge during sampling and analysis, thus limiting the ability to tackle with them. Based on the scant studies on plastic contamination, it has become clear that the marine ecosystem has suffered the most. It has become inevitable to determine the impact of micro- and nanoplastics in marine and terrestrial biomes in terms of transport, quantities and fate. Upon intake by organisms, their presence affects reproductive functions, overall growth and development and affects the bodily functions to deal with the foreign materials. Hence, it is imperative to come up with an alternative solution such as biodegradable plastics along with minimizing the usage of conventional plastics. Biodegradable plastics have shown great potential for use toward eco-friendly and sustainable packaging materials. The major drawback with the biodegradable plastics as a packaging material is their weak mechanical and thermal stability and barrier properties, which can be improved by reinforcing with various fillers of biological and inorganic origin such as cellulose, lignin and carbon. Such approaches have improved the film properties drastically and highly comparable to those obtained from synthetic polymers. To enhance the characteristics and functionalities, further additives such as antibacterial agents and antioxidants have been incorporated. As a result, in the future packaging business, active or smart packaging materials will be conceivable. This will not only assist in extending the use of these materials in the packaging sector, but will also help to reduce emerging contaminants in water bodies as a result of plastic use and improve environmental sustainability.

REFERENCES

1. Cheng, H., Chen, L., McClements, D.J., Yang, T., Zhang, Z., Ren, F., Miao, M., Tian, Y., Jin, Z. 2021.Starch-based biodegradable packaging materials: a review of their preparation, characterization and diverse applications in the food industry. *Trends in Food Science & Technology* 114, 70–82.
2. Huang, D., Tao, J., Cheng, M., Deng, R., Chen, S., Yin, L., Li, R. 2021. Microplastics and nanoplastics in the environment: macroscopic transport and effects on creatures. *Journal of Hazardous Materials* 407, 107329.
3. Wright, S.L., Kelly, F.J. 2017. Plastic and human health: a micro issue? *Environmental Science and Technology* 51, 6634–6647.
4. Kumar, M.N., Sudha, M.C., Damodharam, T., Varjani, S. 2020. Chapter 3- Micropollutants in surface water: impacts on the aquatic environment and treatment technologies. *Current Development of Biotechnology and Bioengineering*. Emerging Organic Micro-pollutants, 41–62.
5. Ali, I., Ding, T., Peng, C., Naz, I., Sun, H., Li, J., Liu, J. 2021. Micro- and nanoplastics in wastewater treatment plants: occurrence, removal, fate, impacts and remediation technologies – A critical review. *Chemical Engineering Journal* 403, 130205.
6. Tang, Y., Liu, Y., Chen, Y., Zhang. W., Zhao, J., He, S., Yang. C., Zhang, T., Tang, C., Zhang, T., Zhang, C., Wang, Z. 2021. A review: research progress in microplastic pollutant in aquatic environment. *Science of the Total Environment* 766, 142572.

7. Rahman, A., Sarkar, A., Yadav, O.P., Achari, G., Slobodnik, J. 2021. Potential human health risks due to environmental exposure to nano- and microplastics and knowledge gaps: a scoping review. *Science of Total Environment* 757, 143872.

8. Rist, S., Hartmann, N.B. 2018. Aquatic ecotoxicity of microplastics and nanoplastics: lessons learned from engineered nanomaterials. *Freshwater Microplastics*, 25–49. doi: 10.1007/978-3-319-61615-5_2.

9. Schmiedgruber, M., Hufenus, R., Mitrano, D.M. 2019. Mechanistic understanding of microplastic fiber fate and sampling strategies: synthesis and utility of metal doped polyester fibers. *Water Research* 155, 423–430.

10. Alimi, O.S., Farner Budarz, J., Hernandez, L.M., Tufenkji, N. 2018. Microplastics and nanoplastics in aquatic environments: aggregation, deposition, and enhanced contaminant transport. *Environmental Science and Technology* 52, 1704–1724.

11. Gaylarde, C.C., Neto, J.A.P., Fonseca, E.M. 2021. Nanoplastics in aquatic systems - are they more hazardous than microplastics? *Environmental Pollution* 272, 115950.

12. Jiang, J.Q., Zhou, J., Sharma, V.K. 2013. Occurrence, transportation, monitoring and treatment of emerging micro-pollutants in waste water — A review from global views. *Microchemical Journal* 110, 292–300.

13. Wang, J., Liu, X., Liu, G., Zhang, Z., Wu, H., Cui, B., Bai, J., Zhang, W. 2019. Size effect of polystyrene microplastics on sorption of phenanthrene and nitrobenzene. *Ecotoxicology and Environmnetal Safety* 173, 331–338.

14. Barboza, L.G.A., Vethaak, A., Lavorante, B.R.B.O., Lundebye, A.K., Guilhermino, L. 2018. Marine microplastic debris: an emerging issue for food security, food safety and human health. *Marine Pollution Bulletin* 133, 336–348.

15. Davranche, M., Veclin, C., Pierson-Wickmann, A.C., Hadri, H.E, Grassl, B., Rowenczyk, L., Dia, A., Halle, T.A., Blancho, F., Reynaud, S., Gigault, J., 2019. Are nanoplastics able to bind significant amount of metals? The lead example. *Environmental Pollution* 249, 940–948.

16. Estabanati, M.R.K., Keindreibenego, M., Mostafazadeh, M.K., Drgoi, P., Tyagi, R.D. 2021. Treatment processes for microplastics and nanoplastics in waters: state-of-the-art review. *Marine Pollution Bulletin* 168, 112374.

17. Scientific Advice Mechanism (SAM). Microplastic Pollution: The Policy Context— Background Paper. 2018. https://ec.europa.eu/research/sam/pdf/topics/microplastic_pollution_policy-context.pdf

18. European Commission. Communication from the Commission to the European Parliament, the Council, the European Economic and Social Committee and the Committee of the Regions: A European Strategy for Plastics in a Circular Economy; European Commission: Brussels, Belgium, 2018; Volume SWD.

19. [ECHA] European Chemicals Agency. Annex to the Annex XV Restriction Report; European Chemicals Agency: Helsinki, Finland, 2019.

20. Bhargava, N., Sharanagat, V.S., Mor, R.S., Kumar, K. 2020.Active and intelligent biodegradable packaging films using food and food waste-derived bioactive compounds: a review. *Trends in Food Science and Technology* 105, 385–401.

21. Amin, U., Khan. M.U., Majeed, Y., Ravezov, M., Khayrullin, M., Bobkova, E., Shariati, M.A., Chung, M., Thiruvangadam, M. 2021. Potentials of polysaccharides, lipids and proteins in biodegradable food packaging applications. *International Journal of Biological Macromolecules* 183, 2184–2198.

22. Zhong, Y., Godwin, P., Jin, Y., Xiao, H. 2020. Biodegradable polymers and green-based antimicrobial packaging materials: a mini-review. *Advanced Industrial and Engineering Polymer Research* 3, 27–35.

23. Ramírez, M.G.L., Satyanarayana, K.G., Iwakiri, S., Muniz, G.B., Tanobe, V., Flores-Sahagun, T.S. 2011. Study of the properties of biocomposites. Part I. Cassava starch-green coir fibers from Brazil. *Carbohydrate Polymers* 86, 1712–1722.

24. Justine, M., Chelo, G.M., Amparo, C. 2017. Combination of poly(lactic) acid and starch for biodegradable food packaging. *Materials* 10, 952.

25. Navarro-Baena, I., Sessini, V., Dominici, F., Torre, L., Kenny, J.M., Peponi, L. 2016. Design of biodegradable blends based on PLA and PCL: from morphological, thermal and mechanical studies to shape memory behavior. *Polymer Degradation and Stability* 132, 97–108.

26. Li, Z., Yang, J., Loh, X.J. 2016. Polyhydroxyalkanoates: opening doors for a sustainable future. *NPG Asia Material* 8, e265.

27. Seoane, I.T., Manfredi, L.B., Cyras, V.P., Torre, L., Fortunati, E., Puglia, D. 2017. Effect of cellulose nanocrystals and bacterial cellulose on disintegrability in composting conditions of plasticized PHB nanocomposites. *Polymers (Basel)* 9, 561.

28. Aghazadeh, M., Karim, R., Abdul Rahman, R., Sultan, M.T., Paykary, M., Johnson, S. 2018. Effect of glycerol on the physicochemical properties of cereal starch films. *Czech Journal of Food Sciences* 36, 403–409.

29. Fakhouri, F.M., Martelli, S.M., Bertan, L.C., Yamashita, F., Innocentini Mei, L.H., Collares Queiroz, F.P. 2012. Edible films made from blends of manioc starch and gelatin - influence of different types of plasticizer and different levels of macromolecules on their properties. *Lwt-Food Science and Technology* 49, 149–154.

30. Wu, H., Lei, Y., Lu, J., Zhu, R., Xiao, D., Jiao, C., Rui, X., Qing, Z.Z., Hui, S.G., Tao, L.Y., Shan, L.S., Liang, L.M. 2019. Effect of citric acid induced crosslinking on the structure and properties of potato starch/chitosan composite films. *Food Hydrocolloids* 97, 105208.

31. Ortega-Toro, R., Collazo-Bigliardi, S., Talens, P., Chiralt, A. 2016. Influence of citric acid on the properties and stability of starch-polycaprolactone based films. *Journal of Applied Polymer Science* 133, 42220.

32. Gao, W., Wu, W., Liu, P., Hou, H., Li, X., Cui, B. 2020. Preparation and evaluation of hydrophobic biodegradable films made from corn/octenylsuccinated starch incorporated with different concentrations of soybean oil. *International Journal of Biological Macromolecules* 142, 376–383.

33. Chen, W., Ma, S., Wang, Q., McClements, D.J., Liu, X., Ngai, T., Liu, F. 2021. Fortification of edible films with bioactive agents: a review of their formation, properties, and application in food preservation. *Critical Reviews in Food Science and Nutrition*, Article 1181435.

34. Kouhi, M., Prabhakaran, M.P., Ramakrishna, S. 2020. Edible polymers: an insight into its application in food, biomedicine and cosmetics. *Trends in Food Science & Technology* 103, 248–263.

35. Xu, E., Campanella, O.H., Ye, X., Jin, Z., Liu, D., BeMiller, J.N. 2020. Advances in conversion of natural biopolymers: a reactive extrusion (REX)–enzyme-combined strategy for starch/protein-based food processing. *Trends in Food Science & Technology* 99, 167–180.

36. Dobrucka, R., Cierpiszewski, R. 2014. Active and intelligent packaging food–research and development–a review. *Polish Journal of Food and Nutrition Sciences* 64, 7–15.

37. Bastos, M.d.S.R., Laurentino, L.d.S., Canuto, K.M., Mendes, L.G., Martins, C.M., Silva, S.M.F., et al. 2016. Physical and mechanical testing of essential oil-embedded cellulose ester films. *Polymer Testing* 49, 156–161.

38. Evangelho, J.A. do., Silva Dannenberg, G. da., Biduski, B., El Halal, S.L.M., Kringel, D.H., Gularte, M.A., Fiorentini, A.M., Zavareze, E.R. 2019. Antibacterial activity, optical, mechanical, and barrier properties of corn starch films containing orange essential oil. *Carbohydrate Polymers* 222, 114981.

39. Carvalho, A.P.A., Junior, C.A.C. 2020. Green strategies for active food packagings: a systematic review on active properties of graphene-based nanomaterials and biodegradable polymers. *Trends in Food Science and Technology* 103, 130–143.

16 Developments in Food Packaging for Enhancing Food Quality and Safety

Priyanka Prajapati and Meenakshi Garg
Bhaskaracharya College of Applied
sciences, University of Delhi

Rajni Chopra
National Institute of Food Technology
Entrepreneurship and Management

CONTENTS

16.1   INTRODUCTION

Traditionally, packaging performs four basic functions, i.e. storage, preservation, containment, and protection, which are still indispensable for the food sector. Packaging of food creates a barrier against physical damage, chemical damage, and microbiological spoilage (Altieri et al. 2014). In addition to these basic functions, food package performs several other functions such as providing nutritional composition and calorie content, how to use the food product, and giving manufacturer information to the consumer (Galić et al. 2011). In recent decades, shifts in consumer demands, modernization of social and scientific society, and globalization have led to the development of novel techniques to manufacture and preserve the food product. Moreover, modifications in retail practices such as the distribution of food product around the globe resulting in longer travelling time and centralization of activities (packaging of cheese slice and diced vegetables on same process line) create hurdles for the food packaging industry to develop new packaging techniques that extend food's shelf life without compromising with food safety and quality. Moreover, further augmentation of shelf life of food product with application of traditional packaging techniques could seem impossible (Kruijf et al. 2002, Han et al. 2018).

Modern food packaging technologies focus on the extension of shelf life with improved food safety and quality. Another focus of packaging industry is to create packaging material that has elevated barrier properties, is cost-effective, and is biodegradable, thus helping in waste reduction (Kruijf et al. 2002, .Galić et al. 2011). In that field, evolution of novel packaging techniques (active, intelligent packaging and bioactive packaging) is required, which protect the food and give real-time information about the condition of food. It reduces the risk of foodborne illness and early food recalls as consumer/retailer is aware about the condition inside the food package. All this has create an impulse to develop systems that can be added inside the food package during storage period (Kruijf et al. 2002, Galić et al. 2011, Yousefi et al. 2019).

New technologies such as intelligent packaging, active packaging, bioactive packaging, and modified atmosphere packaging are the promising packaging techniques that help in improving food safety and quality using minimum artificial preservatives (Yousefi et al. 2019). The smart packaging industry is anticipated to witness significant growth in the coming years. In 2020, active and intelligent packaging market was valued at USD 17.5 billion in 2020. The expected growth of this market is estimated at USD 25.16 billion by 2026, registering a CAGR (compound annual growth rate) of 6.78% during the forecast period 2021–2026 (Mordor Intelligence) (Figure 16.1).

Strict guidelines are formulated and regulated for packaging material, labelling, and testing as for other aspects in food sector. The materials and articles including active and intelligent materials should be effective for their intended use and should be manufactured in accordance with good manufacturing practices (GMP). These materials or articles should carry information about permitted use or the amount of substance released by the active component. Smart packaging

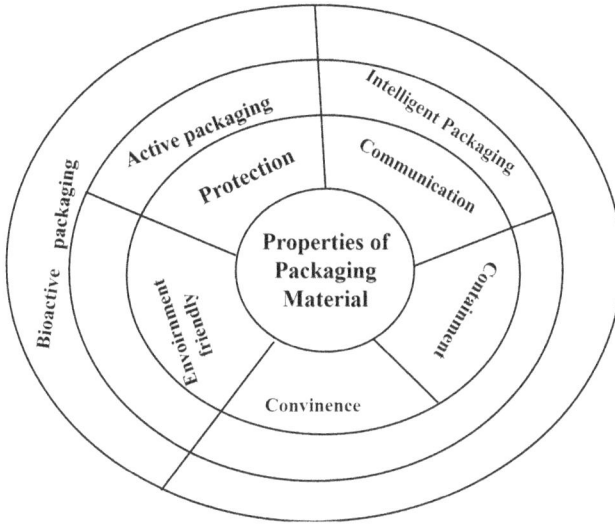

FIGURE 16.1   Novel technologies of food packaging.

devices should carry mandatory "DO NOT EAT" label to guide the consumer to distinguish between edible and non-edible part of food package. This information must be clearly visible, indelible, and legible (Majid et al. 2016).

16.2   NOVEL PACKAGING TECHNOLOGIES

16.2.1   Active Packaging

In spite of the advancements in manufacturing processes, preservation methods, and packaging films with elevated barrier properties, food products are prone to microbial attack and biochemical changes. This acts as a driving force to develop new packaging technologies that help to protect the food from external environment, growth of microorganism, and biochemical changes that occur in food (Wyrwa & Barska 2017). The paradigm shift in role of packaging from passive functions to active functions makes active packaging more popular in food sector. Previously, the main passive function of packaging is to preserve the food product, but recently, new packaging materials have been developed to provide "active" protection to the food throughout the supply chain.

Active packaging is an indispensable vehicle in food preservation as it interacts with the food package in desirable way and protects it from external environment and microbial contamination (Otoni et al. 2016, Bhardwaj et al. 2019). The new packaging techniques account for the packaging system to hold 35% volume (Wyrwa & Barska 2017). It is definitely an innovative technology that can be explained as a type of packaging system that modifies the packaging condition, thus helping in extending the product shelf life as well as improving its quality while maintaining food safety (Singh et al. 2011). Active packaging is defined by European Union under regulation (EC) No 450/2009 as a packaging system that interacts with the food in such a way as to "deliberately incorporate components that would release or absorb substances into or from the packaged food or the environment surrounding the food." Active packaging generally accompanies with intelligent packaging as major a function of active packaging is to protect the food product, while the role of intelligent packaging role is to communicate with the consumer.

16.2.2 MECHANISM OF ACTIVE PACKAGING

In active packaging system, the synthesis of active agents can be achieved by incorporating the active agent (moisture absorber, oxygen scavenger, antimicrobial within the food package or at the primary packaging where food product is directly in contact with the packaging material). The most common technologies used to manufacture active packaging system are **surface immobilization**, where active substances can be immobilized on the surface of the whole packaging material used for packaging food. **Controlled release coatings**, in this type of packaging system, the active agents are incorporated within the polymer of packaging film, from which they are expected to transfer within the package food and perform their specific function. **Photografting** is a technique which works in UV light exposure in the range of 315–400 nm, where polymerization of monomers takes place with the help of photointiators. **Layer-by-layer assembly** is an adaptive method that utilizes the attraction forces to deposit alternating polyelectrolytes onto a solid support (Bastarrachea et al. 2015).

16.2.3 TYPES OF ACTIVE SUBSTANCES

Active substances are classified in two categories based on their mechanism of action, i.e. releasing systems (emitters) and scavenging systems (absorbers). Scavenging systems (absorbers) are believed to remove unwanted compounds (carbon dioxide, moisture, oxygen, and ethylene) from the food or its surroundings. On the other hand, releasing system works on the principle of releasing active compounds (antioxidant, antimicrobial ethanol) to the food or from the headspace of the food package (Singh et al. 2011, Wyrwa & Barska 2017).

Active packaging systems are grouped as:

1. Oxygen scavengers
2. Carbon dioxide emitters and scavengers
3. Ethylene scavengers
4. Moisture absorbers
5. Antioxidant agents
6. Taint scavengers and flavour absorbers.

A. Oxygen Scavengers

Commercially, oxygen scavengers systems are one of the crucial active packaging technologies that help in the preservation of food by extracting residual oxygen present in the food package. The removal of residual oxygen is necessary, as high level of oxygen starts deteriorating the quality characteristics of the food product (Majid et al. 2016). The remaining parts of oxygen in headspace react with sensitive foods (cuts of meat, spices, poultry product, dairy product, etc.) within the package and accelerate the food spoilage, which ultimately decreases food's shelf life and its nutritional quality. It oxidizes the ascorbic acid, foils, and fats, which ultimately decreases the nutritional quality of food and encourages microbial growth. Oxygen molecules can be converted to a variety of transitional species (hydrogen peroxide, hydroxyl radical, water, and superoxide) by the insertion of one or four electrons into their outer shell.

Initially, oxygen scavengers are added in the food package as an independent system. They are placed besides the food product in the package in the form of bags, strips, or labels. They are the most widely used system. However, these independent oxygen scavenging systems have certain drawbacks. These bags, sachets, and strips cannot be incorporated in liquid foods because these kinds of active packaging system will lose their activity when comes in contact with liquid food product and a slurry of oxygen scavenger is formed, which might leach into the food product and ultimately spoil the food appearance. They also need separate packaging to place the oxygen scavenger system.

Nowadays, to overcome this problem, a new system is developed, in which oxygen scavengers (iron, ascorbic acid, and other low-molecular-weight ingredients) are integrated within the matrix of packaging material; thus, oxygen scavenger sachets or pads will not visually perceive as distinct element to customer (Majid et al. 2016).

Iron salts, ascorbic acid, unsaturated fatty acids, photosensitive dyes, and glucose oxidase are the major active elements that are used as oxygen scavenger systems. These methods are efficient and decrease the oxygen levels up to 0.01% within the food package, which is much below the typical oxygen level achieved with modified atmosphere packaging, i.e. 0.3%–3.0%. Depending upon the gaseous resistance of packaging film, the volume of oxygen can be maintained for prolonged period. Oxy-Guard™, FreshPax®, ATCO®, Ageless®, and OMAC® are examples of some commercially available oxygen scavenging systems used to protect different food products (frozen muscles food) (Table 16.1).

B. Moisture scavengers

The control of excess moisture in food package prevents the food spoilage by hindering the growth of spoilage microorganisms. Water activity and moisture content are the critical parameters that govern deterioration of food products (Haghighi-Manesh & Azizi 2017). Choosing the correct type of packaging material for different types of food products is also one of the major factors that inhibit the accumulation of moisture in the package. If the food package is permeable to water vapour, accumulation of water inside the package takes place, which

TABLE 16.1

Oxygen Scavenging Systems and Their Application in Food

S. No.	Oxygen Scavenger System	Food Product	Reference
1.	Activated carbon and sodium L-ascorbate	Raw meat loaves	Lee et al. (2018a)
2.	Gallic acid (GA) and sodium carbonate	High water activity food ($a_w > 0.86$)	Pant et al. (2017)
3.	Sodium metabisulphite-based oxygen scavenger	Kimchi	Lee et al. (2018b)
4.	Palladium-based oxygen scavenger	Ham	Hutter et al. (2016)
5.	Aerobic microorganism	–	Altieri et al. (2004)

further starts the deterioration process. Relative humidity needs to be carefully controlled inside the food package to prevent drip loss from meat cuts and fish poultry products and to prevent drying in case of fresh fruits and vegetables (Ozdemir & Floros 2004). Traditional methods that are employed to reduce moisture content and to maintain correct relative humidity in food are vacuum packaging and modified atmosphere packaging (MAP). Active packaging systems that are used to control moisture are desiccant and adsorbent (Haghighi-Manesh & Azizi 2017). Moisture scavenging system binds with water either in vapour phase or in liquid phase and thus inhibits the accumulation of water in food package (Bhardwaj et al. 2019).

Moisture scavenger systems are classified into two different groups: RH controllers and moisture removers. Desiccants are the type of RH controllers that help in scavenging humidity in the headspace of food package. Silica gel, clays, calcium oxide, sorbitol, magnesium chloride, sodium chloride, and calcium sulphate are some of the examples of desiccant. Silica gel is the most preferred desiccant at commercial level because it remains free flowing and dry even when get saturated (Haghighi-Manesh & Azizi 2017, Bhardwaj et al. 2019).

Moisture removers are available in the form of blankets, pads, and sheets that help in absorbing liquid or drip loss from the food product. Moisture remover pads are combined with antimicrobial compounds or pH controllers or odour removers to prevent the spoilage of food product and maintain its quality characteristics (Prasad & Kochhar 2014). Absorbent pads coated with pinosylvin to prevent the growth of pathogen microorganism (*Campylobacter* spp.) in chicken meat (Silva et al. 2018).

C. Carbon dioxide emitters and scavengers

Carbon dioxide plays a major role when added to the packaging environment. It helps in suppressing the growth of microorganism in certain products such as bakery products, fish, fresh meat, cheese, and poultry. High level of carbon dioxide (60%–80%) helps in maintaining the quality of food by reducing the respiration rate of fresh produce.

Holck et al. (2014) conducted a study to compare the traditional modified atmosphere packaging and CO_2-emitting sachets. They found that CO_2-emitting sachets are more efficient in controlling the growth of spoilage organisms and in expanding the shelf life up to 7 days as compared to control. The storage of food product in pure carbon dioxide results in the collapse of package due to the dissolution of CO_2 in meat. Carbon dioxide emitters decrease the gas to product volume ratio as compared to traditional modified atmosphere packaging (MAP) and thus improve the transport efficiency. CO_2-generating systems help to increase the shelf life of food product by inhibiting the growth of spoilage microorganisms, particularly Listeria, and maintain the food quality by controlling oxidation of the food product (Yildirim et al. 2017).

Carbon dioxide scavengers are mainly used in roasted coffees to prevent the sealed package from bursting. Packaging of coffee beans directly after roasting leads to bursting of package and loss of volatile aroma. Utilization of CO_2 scavengers replaces the "ageing" process and thereby maintains the coffee quality. Mitsubishi Gas Chemical Company manufactures sachets specifically for CO_2 scavenging (Prasad & Kochhar 2014).

D. Taint scavengers and flavour emitters

During the processing of food products, loss of certain aromatic compounds is evident (Saffarionpour & Ottens 2018). To improve the quality of food, artificial flavour and aromatic compounds are added in the food product to compensate for the loss occurred during processing. However, in some food products undesirable flavours or off-odours are produced during storage due to the biochemical changes. Taint scavengers and flavour emitters are the type of active packaging system used to absorb off-odours in the food package. The active material used as a taint scavenger or flavour emitter should not bring any change in composition or organoleptic properties of the food product. Only few active compounds are commercially used to remove desirable flavour or aroma compounds from food products (Yildirim et al. 2017).

Naringin, a bitter compound found in citrus fruit juice, can be removed by naringinase enzyme. This enzyme is immobilized in cellulose acetate film, which is used a packaging material to pack citrus fruit juice (Soares & Hotchkiss 1998).

The ANICO bags (Japan) manufactures bags that are made from film containing ferrous salt and an organic acid (ascorbic acid or citric acid), which has the potential to oxidize amines and other aroma-producing compounds (Vermeiren et al. 1999). Moris developed an odour-impermeable package made up of gas-impermeable plastic such as polyethylene terephthalate (PET) or polyethylene. The package contains a sachet that has a mixture of sachet charcoal and nickel to absorb odour and a hole to allow the passage of respiratory gases (Haghighi-Manesh & Azizi 2017). To absorb the off-odours of mercaptans and hydrogen sulphide formed during the storage inside the food package, certain odour-absorbing sachets (MINIPAX1 and STRIPPAX1, Multisorb Technologies, the USA) are available in the market (Vermeiren et al. 1999).

E. Ethylene scavengers

Ethylene is a phytohormone of plant that is mainly responsible for the post-harvest loss of fruits and vegetables. Softening of fruit, accelerated respiration of fruits and vegetables, and early fruit ripening are some major effects of ethylene. Zeolite, titanium dioxide, activated carbon base with transition metals, and potassium permanganate are some of the widely used ethylene scavenging compounds (Wei et al. 2020). SedoMate® (Japan), Neupalon™ (Japan), Hatofresh® (Japan), and KIF Scrubber ES 657 (India) are some of the commercial sachets used for scavenging ethylene from the food package (Haghighi-Manesh & Azizi 2017).

F. Antioxidant agents

Oxidation of food product is one of the main factor in deteriorating food quality. To prevent oxidation, various natural and artificial antioxidants are used in the food product. Utilization of artificial antioxidants (butylated hydroxytoluene and butylated hydroxyanisole) in food is still being questionable due to the harmful effects of these compounds (Kahl & Kappus 1993). Antioxidants are encapsulated in packaging films to minimize the oxidation of food and thus to extend the shelf life of food and maintain its quality (Haghighi-Manesh & Azizi 2017). Natural antioxidant-rich packaging film was developed to reduce oxidative changes in beef. Brewery waste containing polyvinylpolypyrrolidone resin is used as a natural antioxidant and is found to reduce lipid oxidation up to 80% as compared to control in cold storage (Barbosa-Pereira et al. 2014).

16.3   INTELLIGENT PACKAGING

Packaging of food has a crucial role in ensuring food safety by avoiding contamination through external environment. However, the basic functions of food package remain the same in spite of so many developments in the packaging field. Smart packaging technologies witness high growth throughout the world due to the increase in demand for ready-to-eat food, stringent food safety regulations, and demand for reducing packaging waste. Among other food packaging techniques, active packaging and intelligent packaging systems haves the ability to reduce the waste generated by food sector (Bastarrachea et al. 2015). Consumer awareness about nutritious food, reduction in food wastage, manufacturer concern for increased shelf life, and loop holes in supply chain are some other factors that fuel the growth of smart packaging systems. Time temperature indicators and smart labels are becoming more popular among the food manufacturers. The increase in demand for intelligent packaging systems had almost doubled its share in market with a value of $1.5 billion in 2019 (Poyatos-Racionero et al. 2018).

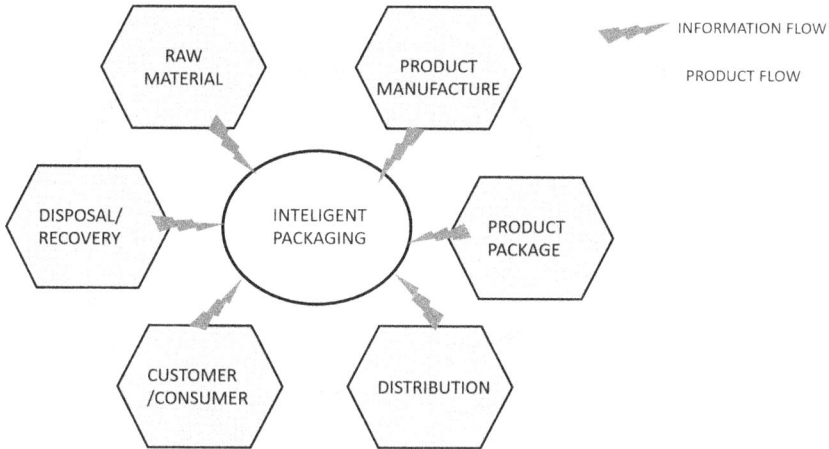

FIGURE 16.2 Product flow and information flow in intelligent packaging.

The function carried out by intelligent packaging systems is similar to its name as it works intelligently to provide accurate estimation about the food product's shelf life. The basic functions carried out by intelligent packaging systems are tracking, sensing, detecting, and recording the food product throughout the supply chain. It is a type of packaging system that gives information about the real-time changes that occur in food. This leads to a major paradigm shift in food packaging systems from an ordinary communicator to a smart communicator (Müller & Schmid 2019). Intelligent packaging systems should be positioned on the tertiary layer of food package that can be easily visible to consumers, and its constituent should not migrate into the food substances. European Union defines intelligent packaging as follows: "Intelligent packaging systems provide the user with information on the conditions of the food and should not release their constituents into the food." Apart from providing food product information (origin, theoretical expiration date, and composition), intelligent packaging systems also provide the history of food (temperature and pH during transportation, headspace composition within the package, microbial growth, etc.) to the customer/retailer. Thus, intelligent packaging might be able to fill many loop holes in food supply chain as it improves food logistics and traceability. Indicators, sensors, and smart tags are devices that are used in intelligent packaging systems to communicate information about the packaging system (Yam et al. 2005) (Figure 16.2).

16.3.1 SMART PACKAGE DEVICES

Smart package devices are the main tools of intelligent packaging that act like a communicator between the food product and consumer/retailer. They are inexpensive small labels or tags that are attached either onto the primary packaging that is directly in contact with food products such as bottles and trays, or usually on the tertiary packaging (container or cardboard). These tags convey information regarding the ongoing changes in the food product so that appropriate measures should be taken to enhance food safety and quality.

Smart packaging devices are basically of two types:

1. Data carriers are small devices that used to store and transmit data about the expiry date, manufacturing process, and storage condition of the food product. Barcode labels and radio frequency identification [RFID] tags are the example of data carriers that are often placed on tertiary packaging to store and transmit data. Data carriers not only provide information about the food quality, but also help in tracking and protection from theft in the food supply chain.
2. Indicators are the type of devices that provide information about whether a substance is present or not and the amount of that substance through various chemical reactions. A change in the chemistry of that particular substance within the food package can be easily identified with the help of smart indicators. Time temperature indicators, freshness indicators, and integrity indicators are the common examples of indicators used in intelligence packaging systems (Fuertes et al. 2016).

16.3.1.1 Barcodes

Barcodes are the machine-readable storage database that is used to trace the information about the food product. These are inexpensive and most common form of data carriers used in grocery stores. It was introduced in the 1970s to store product information such as manufacture identification and item number. The UPC (Universal Product Code) barcode is a unique pattern containing parallel bars and spaces that represent data of 12 digits. Barcode scanning is very popular in food stores and supermarkets as it helps in stock reordering, checkout, and inventory control (Smits et al. 2017) (Figure 16.3).

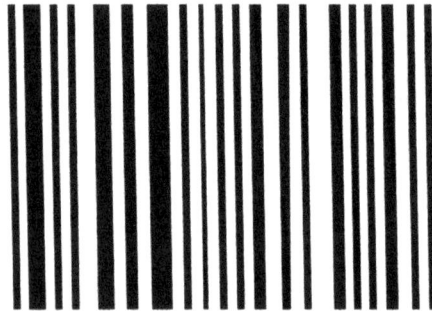

FIGURE 16.3 Barcode.

16.3.1.2   Radio Frequency Identification Tags

Radio frequency identification tag is an advanced form of data carrier used in intelligent packaging systems. It provides large storage capacity to store more information about food product, such as storage temperature, relative humidity data, and nutritional and calorific information. Current RFID tags have chip provided with memory that is wirelessly read out to identify the product (Smits et al. 2017). They work on radio waves, the reader emits radio waves to capture data from the RFID tag, and the data are then passed onto a host computer to get the food product information (Want 2004). Some advanced RFID tags also provide read and write system, which is useful in obtaining real-time information about the product as it moves thorough the supply chain.

16.3.1.3   Time–Temperature Indicators

Maintaining right temperature of the food product throughout the supply chain is the most important task as it influences the kinetics of chemical and physical deterioration and also triggers the microbial spoilage in the food. Time–temperature indicators (TTIs) are self-adhesive food quality recording tools that can be attached to food or food package. TTIs are relatively small, inexpensive, consumer-friendly quality indicators, which irreversibly indicate the change in time and temperature of the food. These labels provide visual indications of temperature accumulation effect on food and the remaining shelf life of the food product (Taoukis & Labuza 1989, Gao et al. 2020).

The commercially available TTIs have different working mechanisms. The operating principle of TTIs are chemical, physical, enzymatic, and biological in nature to show irreversible colour change (Realini, & Marcos 2014). Lifelines Freshness Monitor® and Fresh-Check (Lifelines Technologies, the USA) are chemical TTIs that are based on polymerization of diacetylene crystal into a highly coloured polymer. This reaction of polymerization is temperature dependent, and colour change can be measured by using laser pen (Fu & Labuza 1992). 3M MonitorMark® 10 TTI is an example of physical mechanism TTI. Temperature change is measured through the distance travelled by the blue dye from origin, much like reading a thermometer through diffusion process. Fatty acid ester moves through a porous wick (filaments) made of high-quality blotting paper. The working temperature of MonitorMark indicator ranges from −17°C to 48°C, and the maximum running time is up to 1 year (Fu & Labuza 1992, Roya & Elham 2016).

VITSAB® 13 is an enzyme-based TTI that hydrolyses lipid substrate. This enzymatic TTI has two different zones: One of the zone contains a fluid containing lipolytic enzymes and the other zone contains fat molecule having triglyceride structure and a pH indicator. The hydrolysis of lipolytic enzyme from triglyceride substrate reduces the pH of the medium, which brings the colour change from dark green to bright yellow (Fu & Labuza 1992, Realini & Marcos 2014).

With the help of TTI-based systems, realistic control over cold chain can be achieved as temperature is a crucial parameter in maintaining food quality and safety of chilled or frozen food during the transport and distribution of chilled or frozen food products (Roya & Elham 2016, Gao et al. 2020).

16.3.1.4   Gas Indicators

Gas indicators are useful devices that give information about the composition of gases present in the headspace of food package. This type of indicator helps in sensing the gases (carbon dioxide, oxygen, hydrogen sulphide, and water vapour) that are produced through decomposition of food, the presence of microbes, and biochemical reactions that occur in the food product. The tag is composed of two compartments: One is sensor that detects and reacts to changes in the atmosphere inside the food package and the other is the indicator that reflects the quality changes onto the food package. Most of the gas indicators monitor carbon dioxide and oxygen concentration within the package. To bring the colour change, indicator should be placed in the vicinity of targeted gas, which is present in the headspace surrounding the food product. Redox dyes, an alkaline compound, and a reducing compound are the main components of gas indicators. Nowadays, gas indicators find their application in modified atmosphere packaging (MAP) to detect CO_2 in food package. One of the major disadvantage of gas indicators is that they leach dye when come in contact with the moisture from the package. However, studies are going to utilize encapsulation or coating techniques and to develop colourimetric indicators that work in the presence of UV rays to minimize dye leaching in the food product (Roya & Elham 2016, Fuertes et al. 2016, Müller & Schmid 2019).

16.3.1.5   Freshness Indicators

A freshness indicator provides direct information about food product quality. Freshness indicators detect quality changes in the food product and transform these changes into a colour response. Basically, these indicators work on the principle of detecting first kind of changes (pH, gas composition, lactic acid, amines, etc.). Changes in the concentration of metabolites such as nitrogen compounds, amines, volatile sulphuric compounds, glucose, organic acids, carbon dioxide, and ethanol can be predicted as the initiation of microbial growth, which ultimately correlates with the freshness of food (Fu & Labuza 1992, Roya & Elham 2016). Intelligent packaging systems that monitor the freshness of food work on two principles, either estimation of metabolites using biosensors or indirect identification of metabolites using colour markers (Müller & Schmid 2019).

SensorQ™ sticker from FQSI (Food Quality Sensor International Inc., Lexington, MA, the USA) is an example of freshness indicator used to detect biogenic amines in meat and poultry packaging. The freshness indicator is applied inside the food packaging and shows colour change when microorganism starts spoiling the food product. This change is produced by anthocyanins dyes, which sense the change in pH because of the change in the composition of biogenic amines (Fu & Labuza 1992, Fuertes et al. 2016).

16.3.2   BIOACTIVE PACKAGING

With the growing demand to provide quality food product, packaging industry tries to fabricate packaging materials loaded with active molecules that are also highly responsive towards any changes in the surrounding condition. Bioactive packaging is an innovative technique that has the potential to provide health bene-fits to consumer by modifying food package materials. The active compounds that are encapsulated to the packaging material are phytochemicals, antimicrobials, vitamins, nanofibres, and prebiotics. Apart from fulfilling all passive functions, bioactive packaging contains functional compounds that may provide additional health benefits to the consumer. Gluconal Cal (Glucona America Inc., Janesville, Wis, U.S.A.) is a mixture of calcium lactate, gluconate, and α-tocopheryl ace-tate and is encapsulated in edible coatings made up of protein and carbohydrate. These edible coatings have shown increased nutritional value, good bioavailabil-ity of nutrients, and better barrier properties (Mei & Zhao 2002).

Bioactive packaging technology has made functionalities in the food sector through the following process:

i. Bioactive or functional compound should be delivered through poly-meric matrix composed of biomolecules that is easily biodegradable.
ii. Incorporation of active compound into food or packaging material that helps in delivering health benefits to the consumer.
iii. Encapsulation of enzymes in packaging material, which help in modi-fying some of the food components in order to increase the nutritional value of food.

Microencapsulation, nanoencapsulation, and electrospinning are some of the novel techniques used for the incorporation of active compounds in polymeric films. These technologies help in protection and sustained or fast release of func-tional compounds encapsulated in food matrices.

16.4   EFFECT OF NOVEL TECHNOLOGIES ON PACKAGING OF FOOD

Advancements in packaging technologies arose consumer's desire for wholesome, fresh, ready-to-eat, minimally processed food with elevated shelf life. These new packaging technologies help in maintaining the quality of food without compro-mising the food safety. These innovative technologies also focused on the uti-lization of biomolecules for packaging material, thus promoting environmental sustainability. These innovative technologies play a major role in the development of food sector to minimize food loss by using tracking devices, indicators which assess the real-time quality of food, and emitters and absorbers that help in the maintenance of suitable environment within the food package (Table 16.2).

16.5   CONCLUSIONS

Food is a perishable commodity that needs to be handled carefully to avoid soci-etal damage, which include food loss, food waste generation, and foodborne dis-ease outbreaks. Packaging of food plays a major role in the preservation of food from microbial, chemical, and physical changes. Packaging sector is looking for

TABLE 16.2

Novel Packaging Techniques and Their Application to Food Products

S. No.	Food Product	Type of Packaging	Remarks	References
	Minced beef meat	Time temperature indicator – intelligent packaging	TTI indicator was developed using *Janthinobacterium* sp., which produce violet pigment during early growth, depending on temperature	Mataragas et al. (2018)
2.	Frozen food products	Time temperature indicator – intelligent packaging	A self-healing nanofibre mat was developed, which becomes opaque under refrigeration temperature and scatter light at room temperature	Choi et al. (2020)
3.	Chicken breast	Freshness indicator – intelligent packaging	Nano-sized colourimetric freshness indicator was developed using sugarcane bagasse to monitor spoilage of chicken breast	Oudjedi et al. (2020)
4.	Kimchi	Freshness indicator – intelligent packaging	A freshness indicator system was developed, which was sensitive to change in pH by sensing the change in volatile acids and CO_2	Baek et al. (2020)
5.	Chicken	Freshness indicator – intelligent packaging	Two-faced Janus (emoji) colour indicator label was developed using cellulose acetate and black carrot anthocyanin to monitor real-time freshness of chicken	Franco et al. (2021)
6.	Fried potatoes	Bioactive packaging	Multilayer packaging film was developed using Algerian Sage and bay leaf extracts	Lu et al. (2018)
	–	Bioactive packaging	A multifunctional food packaging film was developed based on chitosan (CS), nano-sized TiO_2, and black plum peel extract (BPPE). It has ethylene scavenging, antimicrobial, antioxidant, and pH-sensitive properties	Zhang et al. (2019)

new solutions that help in fulfilling the consumer demands for nutritious, wholesome, ready-to-eat packaged food with improved quality and food safety. Novel food technologies, viz. active packaging, intelligent packaging, and bioactive packaging, provide opportunities to fulfil all the consumer demands and yet provide a good barrier to the food. These technologies help in monitoring the freshness of food and in indicating the condition of the food product throughout the food supply chain. The devices used in these technologies interact with the food product or its surroundings and perform desired function to persevere the food product. Researches are also carried out to utilize more and more biomolecules from sustainable resources to produce either the packaging material or devices used in the packaging material.

16.6   FUTURE TRENDS

The evolution in the packaging sector has led to the development of stimuli-responsive packaging systems. Factors such as traceability, sustainability, food safety, and food wastage are also taken care of by the novel food packaging technology. More researches need to be carried out to combine two or more packaging devices for the same food product for better food quality. Potential challenges faced by this industry are higher production cost and compliance with the food safety regulations. Application of these innovative technologies to beverages and ready-to-eat segment is the potential future target for the packaging industry and food industry.

REFERENCES

Active and Intelligent Packaging Market - Growth, Trends, Covid-19 Impact, And Forecasts (2022 - 2027). https://www.mordorintelligence.com/industry-reports/active-and-intelligent-packaging-market-industry

Altieri, C., Sinigaglia, M., Corbo, M., Buonocore, G., Falcone, P., & Del Nobile, M. (2004). Use of entrapped microorganisms as biological oxygen scavengers in food packaging applications. LWT - *Food Science and Technology*, 37(1), 9–15. doi:10.1016/s0023-6438(03)00115-4

Baek, S., Maruthupandy, M., Lee, K., Kim, D., & Seo, J. (2020). Freshness indicator for monitoring changes in quality of packaged kimchi during storage. *Food Packaging and Shelf Life*, 25, 100528. doi:10.1016/j.fpsl.2020.100528

Barbosa-Pereira, L., Aurrekoetxea, G. P., Angulo, I., Paseiro-Losada, P., & Cruz, J. M. (2014). Development of new active packaging films coated with natural phenolic compounds to improve the oxidative stability of beef. *Meat Science*, 97(2), 249–254. doi.org/10.1016/j.meatsci.2014.02.006

Bastarrachea, L., Wong, D., Roman, M., Lin, Z., & Goddard, J. (2015). Active packaging coatings. *Coatings*, 5(4), 771–791. doi:10.3390/coatings5040771

Bhardwaj, A., Alam, T., & Talwar, N. (2019). Recent advances in active packaging of agrifood products---a review. *Journal of Postharvest Technology*, 07(1), 33–62.

Choi, S., Eom, Y., Kim, S., Jeong, D., Han, J., Koo, J. M., … Oh, D. X. (2020). A self-healing nanofiber-based self-responsive time-temperature indicator for securing a cold-supply chain. *Advanced Materials*, 1907064. doi:10.1002/adma.201907064

Commission Regulation (EC) No 450/2009on active and intelligent materials and articles intended to come into contact with food. (2009). https://eur-lex.europa.eu/LexUriServ/LexUriServ.do?uri=OJ:L:2009:135:0003:0011:EN:PDF

Franco, M. R., da Cunha, L. R., & Bianchi, R. F. (2021). Janus principle applied to food safety---An active two-faced indicator label for tracking meat freshness. *Sensors and Actuators B---Chemical*, 333. doi:10.1016/j.snb.2021.129466

Fu, B., & Labuza, T. (1992). Considerations for the application of time–temperature integrators in food distribution. *Journal of Food Distribution Research*, 23, 1–10.

Fuertes, G., Soto, I., Carrasco, R., Vargas, M., Sabattin, J., & Lagos, C. (2016). Intelligent packaging systems---sensors and nanosensors to monitor food quality and safety. *Journal of Sensors*, 2016, 1–8. doi:10.1155/2016/4046061

Galić, K., Ščetar, M., & Kurek, M. (2011). The benefits of processing and packaging. *Trends in Food Science & Technology*, 22(2–3), 127–137. doi:10.1016/j.tifs.2010.04.001

Gao, T., Tian, Y., Zhu, Z., & Sun, D.-W. (2020). Modelling, responses and applications of time-temperature indicators (TTIs) in monitoring fresh food quality. *Trends in Food Science & Technology*. doi:10.1016/j.tifs.2020.02.019

Haghighi-Manesh, S., & Azizi, M. H. (2017). Active packaging systems with emphasis on its applications in dairy products. *Journal of Food Process Engineering*, 40(5), e12542. doi:10.1111/jfpe.12542

Han, J.-W., Ruiz-Garcia, L., Qian, J.-P., & Yang, X.-T. (2018). Food packaging---A comprehensive review and future trends. *Comprehensive Reviews in Food Science and Food Safety*, 17(4), 860–877. doi:10.1111/1541-4337.12343

Holck, A., Pettersen, M., Moen, M., & Sørheim, O. (2014). Prolonged shelf life and reduced drip loss of chicken filets by the use of carbon dioxide emitters and modified atmosphere packaging. *Journal of Food Protection*, 77, 1133–1141. doi:10.4315/0362-028X.JFP-13-428

Hutter, S., Rüegg, N., & Yildirim, S. (2016). Use of palladium based oxygen scavenger to prevent discoloration of ham. *Food Packaging and Shelf Life*, 8, 56–62. doi:10.1016/j.fpsl.2016.02.004

Kahl, R., & Kappus, H. (1993). Toxikologie der synthetischen Antioxidantien BHA und BHT im Vergleich mit dem natürlichen Antioxidans Vitamin E [Toxicology of the synthetic antioxidants BHA and BHT in comparison with the natural antioxidant vitamin E]. *Zeitschrift für Lebensmittel-Untersuchung und -Forschung*, 196(4), 329–338. doi:10.1007/BF01197931

Keep it fresh, KIF Scrubber ES 657, https://keep-it-fresh.com/kif-scrubber-es-657/

Kruijf, N. D., Beest, M. V., Rijk, R., Sipiläinen-Malm, T., Losada, P. P., & Meulenaer, B. D. (2002). Active and intelligent packaging---applications and regulatory aspects. *Food Additives & Contaminants*, 19(sup1), 144–162. doi:10.1080/02652030110072722

Lee, J.-S., Chang, Y., Lee, E.-S., Song, H.-G., Chang, P.-S., & Han, J. (2018a). Ascorbic acid-based oxygen scavenger in active food packaging system for raw meatloaf. *Journal of Food Science*, 83(3), 682–688. doi:10.1111/1750-3841.14061

Lee, J.-S., Jeong, S., Lee, H.-G., Cho, C. H., & Yoo, S. (2018b). Development of a sulfite-based oxygen scavenger and its application in kimchi packaging to prevent oxygen-mediated deterioration of kimchi quality. *Journal of Food Science*. doi:10.1111/1750-3841.14374

Lu, P., Yang, Y., Liu, R., Liu, X., Ma, J., Wu, M., & Wang, S. (2020). Preparation of sugarcane bagasse nanocellulose hydrogel as a colourimetric freshness indicator for intelligent food packaging. *Carbohydrate Polymers*, 116831. doi:10.1016/j.carbpol.2020.11683

Majid, I., Ahmad Nayik, G., Mohammad Dar, S., & Nanda, V. (2016). Novel food packaging technologies---Innovations and future prospective. *Journal of the Saudi Society of Agricultural Sciences*. doi:10.1016/j.jssas.2016.11.003

Mataragas, M., Bikouli, V. C., Korre, M., Sterioti, A., & Skandamis, P. N. (2018). Development of a microbial time temperature indicator for monitoring the shelf life of meat. *Innovative Food Science & Emerging Technologies*. doi:10.1016/j.ifset.2018.11.003

Mei, Y., & Zhao, Y. (2002). Using edible coating to enhance nutritional and sensory qualities of baby carrots. *Journal of Food Science*, 65, 1964–1968.

Müller, P., & Schmid, M. (2019). Intelligent packaging in the food sector---A brief overview. *Foods*, 8. doi:10.3390/foods8010016

Otoni, C. G., Espitia, P. J. P., Avena-Bustillos, R. J., & McHugh, T. H. (2016). Trends in antimicrobial food packaging systems---Emitting sachets and absorbent pads. *Food Research International*, 83, 60–73. doi:10.1016/j.foodres.2016.02.018

Oudjedi, K., Manso, S., Nerin, C., Hassissen, N., & Zaidi, F. (2018). New active antioxidant multilayer food packaging films containing Algerian Sage and Bay leaves extracts and their application for oxidative stability of fried potatoes. *Food Control*. doi:10.1016/j.foodcont.2018.11.018

Ozdemir, M., & Floros, J. D. (2004). Active food packaging technologies. *Critical Reviews in Food Science and Nutrition*, 44(3), 185–193. doi:10.1080/10408690490441578

Pant, A., Sängerlaub, S., & Müller, K. (2017). Gallic acid as an oxygen scavenger in bio-based multilayer packaging films. *Materials*, 10(5), 489. doi:10.3390/ma10050489

Poyatos-Racionero, E., Ros-Lis, J. V., Vivancos, J.-L., & Martínez-Máñez, R. (2018). Recent advances on intelligent packaging as tools to reduce food waste. *Journal of Cleaner Production*, 172, 3398–3409. doi:10.1016/j.jclepro.2017.11.075

Prasad, P., & Kochhar, A. (2014). Active packaging in food industry---A review. *IOSR Journal of Environmental Science, Toxicology and Food Technology*, 8, 01–07. doi:10.9790/2402-08530107.

Realini, C. E., & Marcos, B. (2014). Active and intelligent packaging systems for a modern society. *Meat Science*, 98(3), 404–419. doi:10.1016/j.meatsci.2014.06.031

Roya, A. Q., & Elham, M. (2016). Intelligent food packaging---Concepts and innovations. *International Journal of ChemTech Research*, 9(6), 669–676.

Saffarionpour, S., & Ottens, M. (2018). Recent advances in techniques for flavor recovery in liquid food processing. *Food Engineering Reviews*, 10, 1–14. doi:10.1007/s12393-017-9172-8.

Silva, F., Domingues, F., & Nerín, C. (2017). Control microbial growth on fresh chicken meat using pinosylvin inclusion complexes based packaging absorbent pads. *LWT - Food Science and Technology*, 89. doi:10.1016/j.lwt.2017.10.043

Singh, P., Abas Wani, A., & Saengerlaub, S. (2011). Active packaging of food products---recent trends. *Nutrition & Food Science*, 41(4), 249–260. doi:10.1108/00346651111151384

Smits, E., Schram, J., Nagelkerke, M., Kusters, R. H. L., Heck, G., Acht, V., … Schoo, H. (2012). Development of printed RFID sensor tags for smart food packaging. *14th International Meeting on Chemical Sensors*, 403–406. doi:10.5162/IMCS2012/4.5.2

Soares, N. F. F., & Hotchkiss, J. H. (1998). Naringinase immobilization in packaging films for reducing naringin concentration in grapefruit juice. *Journal of Food Science*, 63, 61–65. doi:10.1111/j.1365-2621.1998.tb15676.x

Taoukis, P. S., & Labuza, T. P. (1989). Applicability of time-temperature indicators as shelf life monitors of food products. *Journal of Food Science*, 54(4), 783–788. doi:10.1111/j.1365-2621.1989.tb07882.x

Vermeiren, L., Devlieghere, F., van Beest, M., de Kruijf, N., & Debevere, J. (1999). Developments in the active packaging of foods. *Trends in Food Science & Technology*, 10(3), 77–86. doi:10.1016/s0924-2244(99)00032-1

Want, R. (2004). RFID---A key to automating everything. *Scientific American*, 290, 56–65. doi:10.1038/scientificamerican0104-56

Wei, H., Seidi, F., Zhang, T., Jin, Y., & Xiao, H. (2020). Ethylene scavengers for the preservation of fruits and vegetables---A review. *Food Chemistry*, 337. 127750. doi:10.1016/j.foodchem.2020.127750

Wyrwa, J., & Barska, A. (2017). Innovations in the food packaging market---active packaging. *European Food Research and Technology*, 243, 1681–1692. doi:10.1007/s00217-017-2878-2

Yam, K. L., Takhistov, P. T., & Miltz, J. (2005). Intelligent packaging---concepts and applications. *Journal of Food Science*, 70(1), R1–R10. doi:10.1111/j.1365-2621.2005.tb09052.x

Yildirim, S., Röcker, B., Pettersen, M., Nilsen-Nygaard, J., Ayhan, Z., Rutkaite, R., … Coma, V. (2017). Active packaging applications for food. *Comprehensive Reviews in Food Science and Food Safety*, 17. doi:10.1111/1541-4337.12322

Yousefi, H., Su, H.-M., Sara., M. I., Alkhaldi, K., Filipe, C., & Didar, T. (2019). Intelligent food packaging---A review of smart sensing technologies for monitoring food quality. *ACS Sensors*, 4. doi:10.1021/acssensors.9b00440

Zhang, X., Liu, Y., Yong, H., Qin, Y., Liu, J., & Liu, J. (2019). Development of multifunctional food packaging films based on chitosan, TiO_2 nanoparticles and anthocyanin-rich black plum peel extract. *Food Hydrocolloids*. doi:10.1016/j.foodhyd.2019.03.009

Index

For Product Safety Concerns and Information please contact our EU
representative GPSR@taylorandfrancis.com
Taylor & Francis Verlag GmbH, Kaufingerstraße 24, 80331 München, Germany